Plant-Based Diets: Working towards a Sustainable Future

Plant-Based Diets: Working towards a Sustainable Future

Editors

Winston Craig
Ujué Fresán

MDPI • Basel • Beijing • Wuhan • Barcelona • Belgrade • Manchester • Tokyo • Cluj • Tianjin

Editors
Winston Craig
Loma Linda University
Loma Linda
USA

Ujué Fresán
Instituto de Salud Global
Barcelona (ISGlobal),
Barcelona
Spain

Editorial Office
MDPI
St. Alban-Anlage 66
4052 Basel, Switzerland

This is a reprint of articles from the Special Issue published online in the open access journal *Nutrients* (ISSN 2072-6643) (available at: https://www.mdpi.com/journal/nutrients/special_issues/plant_diets).

For citation purposes, cite each article independently as indicated on the article page online and as indicated below:

LastName, A.A.; LastName, B.B.; LastName, C.C. Article Title. *Journal Name* **Year**, *Volume Number*, Page Range.

ISBN 978-3-0365-7450-9 (Hbk)
ISBN 978-3-0365-7451-6 (PDF)

Contents

Preface to "Plant-Based Diets: Working towards a Sustainable Future"

There is a steadily growing interest in plant-based diets, both for personal health reasons and for the health of the planet. People, especially flexitarians, are increasingly looking for alternatives to meat and dairy that look and taste like the original. They also want their food choices to be kind to the environment and have a nutritional value similar to animal foods. This book of eleven chapters was written for health professionals who need answers about the safety and nutritional adequacy of plant-based diets and the motivations and barriers for consuming such a diet. The book will also be invaluable for research scientists and graduate nutrition students who are investigating these diets both from a nutritional and sustainability standpoint. Questions arise as to whether these plant-based meat and dairy alternatives help to lessen environmental degradation.

In the lead chapter, the book discusses the advantages of a plant-based diet, both for reducing chronic diseases and for reducing the carbon footprint of the food supply all the way from the farm to the plate. Other chapters deal with the factors that motivate people to or inhibit them from switching from meat to plant-based alternatives; the nutritional consequences of changing from an animal-based to a plant-based diet; the nutritional value of non-dairy beverages, yogurts, and cheeses; the value of plant-based diets during pregnancy and for people with chronic kidney disease; the effect of using pea protein on protein status and metabolism; and the design of nutritionally adequate and climate-friendly diets for adolescents of all dietary persuasions. The editors hope that the book will contribute to the field and answer some of the questions surrounding the current shift from animal foods to plant foods.

Winston Craig and Ujué Fresán
Editors

nutrients

MDPI

Review

The Safe and Effective Use of Plant-Based Diets with Guidelines for Health Professionals

Winston J. Craig [1,*], Ann Reed Mangels [2], Ujué Fresán [3], Kate Marsh [4], Fayth L. Miles [1,5], Angela V. Saunders [6], Ella H. Haddad [1], Celine E. Heskey [1], Patricia Johnston [1], Enette Larson-Meyer [7] and Michael Orlich [5]

1 Center for Nutrition, Healthy Lifestyles, and Disease Prevention, School of Public Health, Loma Linda University, Loma Linda, CA 92354, USA; fmiles@llu.edu (F.L.M.); ehaddad@llu.edu (E.H.H.); cheskey@llu.edu (C.E.H.); pjohnston@llu.edu (P.J.)
2 Vegetarian Resource Group, Baltimore, MD 21203, USA; reedmangela@gmail.com
3 eHealth Group, Instituto de Salud Global Barcelona (ISGlobal), 08036 Barcelona, Spain; ujue.fresan@isglobal.org
4 Private Practice, Chatswood, NSW 2067, Australia; drkatemarsh@gmail.com
5 School of Medicine, Loma Linda University, Loma Linda, CA 92354, USA; morlich@llu.edu
6 Nutrition Insights, Sanitarium Health Food Company, Berkeley Vale, NSW 2261, Australia; angela.saunders@sanitarium.com.au
7 Human Nutrition, Foods, and Exercise Virginia Tech, Blacksburg, VA 24061, USA; enette@vt.edu
* Correspondence: wcraig@llu.edu

Abstract: Plant-based diets, defined here as including both vegan and lacto-ovo-vegetarian diets, are growing in popularity throughout the Western world for various reasons, including concerns for human health and the health of the planet. Plant-based diets are more environmentally sustainable than meat-based diets and have a reduced environmental impact, including producing lower levels of greenhouse gas emissions. Dietary guidelines are normally formulated to enhance the health of society, reduce the risk of chronic diseases, and prevent nutritional deficiencies. We reviewed the scientific data on plant-based diets to summarize their preventative and therapeutic role in cardiovascular disease, cancer, diabetes, obesity, and osteoporosis. Consuming plant-based diets is safe and effective for all stages of the life cycle, from pregnancy and lactation, to childhood, to old age. Plant-based diets, which are high in fiber and polyphenolics, are also associated with a diverse gut microbiota, producing metabolites that have anti-inflammatory functions that may help manage disease processes. Concerns about the adequate intake of a number of nutrients, including vitamin B12, calcium, vitamin D, iron, zinc, and omega-3 fats, are discussed. The use of fortified foods and/or supplements as well as appropriate food choices are outlined for each nutrient. Finally, guidelines are suggested for health professionals working with clients consuming plant-based diets.

Keywords: plant-based diets; vegetarian; vegan; sustainability; microbiome; vitamin B12; CV disease; diabetes; bone health; life cycle

Citation: Craig, W.J.; Mangels, A.R.; Fresán, U.; Marsh, K.; Miles, F.L.; Saunders, A.V.; Haddad, E.H.; Heskey, C.E.; Johnston, P.; Larson-Meyer, E.; et al. The Safe and Effective Use of Plant-Based Diets with Guidelines for Health Professionals. *Nutrients* **2021**, *13*, 4144. https://doi.org/10.3390/nu13114144

Academic Editor: Stephen Ives

Received: 27 October 2021
Accepted: 17 November 2021
Published: 19 November 2021

1. Introduction

Interest in plant-based diets has soared in the past decade for a myriad of reasons [1]. People are concerned about issues such as their health, climate change, the sustainability of the food production system, and the welfare of animals. A plant-based diet is defined in various ways. For some it means eating foods mostly, but not entirely, of plant origin, while for others it means eating only plant-based foods. In this manuscript, we chose to restrict the term plant-based to include both vegan diets (total plant-based nutrition) and lacto-ovo-vegetarian diets (this allows for the consumption of dairy products and eggs). We do not include a discussion of flexitarian, semi-vegetarian, or pesco-vegetarian eating patterns, as they do not fit into our definition of plant-based diets.

This paper will discuss the environmental issues and the benefits for the planet of significantly reducing or eliminating meat and dairy foods from our diet. In addition, it outlines the therapeutic advantages of a plant-based diet for managing the chronic diseases of Western society, such as obesity, cardiovascular disease (CVD), cancer, and diabetes. A plant-based diet is also shown to have a substantial impact upon the composition and function of the gut microbiome, which in turn influences our overall health. Furthermore, the safety of following a plant-based diet during all stages of the life-cycle is addressed. Finally, questions were raised about the adequacy of a plant-based diet with respect to eight key nutrients. These are discussed in detail, with solutions suggested as to how one can meet the dietary requirements through food choices and/or supplementation. Some simple guidelines are given for health professionals to effectively serve the growing population of those consuming a plant-based diet.

2. Current Trends

Internationally, the prevalence of following a vegetarian diet varies by country, but it is generally estimated to be less than 10% of the population. The exception is India, where 20% or more of adults are vegetarian [2,3]. In the United States, a nationwide poll in 2020 found that approximately 6% of adults followed a vegetarian diet, with half of them being vegans [4]. A similar U.S. poll found that approximately 2% of 8- to 17-year-old children followed a vegan diet, and 3% followed a non-vegan vegetarian diet [5].

Globally, the market for alternatives to dairy products is expected to reach $US 25 billion by 2026 [6]. U.S. retail sales of plant-based foods (plant-based dairy alternatives and plant-based meats) increased 27% between 2019 and 2020, with a total plant-based market value estimated at $7 billion [7], suggesting a growing consumer interest in non-animal products.

The 2020–2025 Dietary Guidelines for Americans endorses a "Healthy Vegetarian Dietary Pattern" as one of three dietary patterns that can "be tailored to meet cultural and personal preferences" [8]. There are versions of this plan for ages one year and older. The Guidelines also encourage all Americans to eat more plant foods, including dried beans, whole grains, fruits, vegetables, and nuts. Many other countries promote plant-based diets in their dietary guidelines [9].

With a growing interest in vegetarian eating, establishments such as colleges and universities, school food services, airlines, restaurants, prisons, employee food services, nursing homes, and hospitals are increasingly providing vegetarian options [10–12].

3. Environmental Sustainability of Vegetarian Diets

The production of different foods can have very diverse environmental impacts. There is a large variation in the ecological footprint of animal-based products, with ruminant meat being especially detrimental for the environment as compared with other products such as pork, white meat, or eggs [13–15]. An increasing body of data provides evidence that environmental degradation, through the emissions of greenhouse gas (GHG) and other pollutants, and the use of earth's resources, such as water and land, in the production of plant-based foods are significantly lower than that from animal-based foods [13–15]. Certainly, the effects of the lowest-impact animal products are typically greater than those of plant-based alternatives, even in the case of highly processed plant-based meat analogs [14,16]. The production of plant-based products is more efficient regardless of whether the comparison is made by weight of product, per serving, per calories, or even protein content [14,15,17,18]. Producing the same amount of protein from tofu (soybeans) in comparison to beef protein requires 74 times less land and eight times less water, while the GHG emissions are 25 times lower and the eutrophication (a process driven by the enrichment of water by nutrients, especially compounds of nitrogen and/or phosphorus, leading to an increased growth, primary production and biomass of algae; changes in the balance of organisms; and water quality degradation [19]) potential is reduced by 39 times [14,18]. Even if compared to egg protein, tofu protein requires almost three times less land and six times less water, while the GHG emissions are only half of that from egg protein, and the eutrophication potential is five times lower [14,18].

Likewise, a reduction of animal-based foods in the diet goes hand in hand with a decrease in the dietary environmental impact [20,21]. Vegetarian diets, both lacto-ovo-vegetarian and vegan, have been described as more environmentally sustainable than those diets including meat. A review study concluded that the adoption of lacto-ovo-vegetarian diets could reduce the dietary GHG emissions by 35%, land use by 42%, and freshwater use by 28% [20]. Adopting a vegan diet would lead to around one-half of both GHG emissions and land use of that of current dietary patterns. One should note that there is substantial variability in the dietary environmental impact of those consuming vegetarian diets. In the final analysis, any environmental benefits would depend on the quantity and the specific foods consumed. Overconsumption of calories, a high intake of fruits transported by plane, or the consumption of large quantities of fatty dairy products, such as cheese or butter, in lacto-ovo-vegetarian diets could jeopardize any potential benefit from the avoidance of meat.

Studies suggest that the adoption of nutritionally balanced vegetarian diets, both in developed and developing countries, could be an effective strategy for reducing GHG emissions worldwide [22,23]. This dietary transition would be moderately effective in reducing fertilizer application and would decrease, although to a lesser extent, cropland and fresh water use [23]. Altogether, embracing a balanced vegetarian diet, especially in developed countries, could be an effective strategy for reducing the food system's environmental degradation and reducing our use of the earth's resources.

4. Plant-Based Diets and Chronic Diseases

4.1. CVD, including Hyperlipidemia, Ischemic Heart Disease, Hypertension, and Stroke

CVD continues as the most common cause of death and disability in the U.S. and globally [24,25]. The leading risk factors for CVD include dyslipidemia, excess weight, hypertension, glucometabolic disorders, and diabetes and are attributed to poor diets [26,27]. Compared to omnivorous diets, vegetarian and plant-based diets rich in whole grains, legumes, vegetables, fruits, nuts, and seeds have been associated with substantial reductions in several modifiable risk factors, including body mass index (BMI) and waist circumference [28,29], atherogenic lipoprotein concentrations [29,30], blood glucose [28], inflammation [31], and blood pressure [32].

The results of randomized controlled trials (RCT) of vegan and vegetarian interventions along with systematic reviews and meta-analyses of such studies show improvements in several intermediate cardiometabolic risk markers, including body weight and blood lipids [33–37], and cardiometabolic risk profiles [38]. Data from relatively long-term (years) clinical intervention studies with intensive low-fat vegetarian [39] and vegan diets [40] show reversals in coronary artery disease in individuals with CVD. Due to lower saturated fat and cholesterol levels and more optimal plant sterol and fiber content, greater favorable effects of vegan diets on heart disease risk factors are expected. A vegan diet, compared to the American Heart Association (Dallas, TX, USA) diet for coronary heart disease (CHD), resulted in similar reductions in BMI, waist circumference, markers of glycemic control, blood lipids, and a 32% lower high-sensitivity C-reactive protein (a pro-inflammatory marker) [31].

In the Adventist Health Study-2 (AHS-2), vegetarians had a 13% and 19% lower risk of CVD and ischemic heart disease (IHD) mortality, respectively, compared with non-vegetarians. This difference occurred in spite of the fact that the non-vegetarians in the cohort consumed less meat than the general population. Blood pressure levels in vegans and vegetarians were also lower than those of the omnivores. Metabolic syndrome and type 2 diabetes (T2D) are prime risk conditions for CVD and stroke. A reduced prevalence of these conditions was observed in vegan and vegetarian participants of AHS-2 [28,41].

The EPIC-Oxford study of vegetarians, vegans, and health-conscious individuals reported that the risk of incident IHD hospitalizations and deaths caused by circulatory disease was 32% lower in vegetarians than in non-vegetarians [42]. The 18-year follow-up showed lower rates of IHD in vegetarians but higher rates of hemorrhagic and total stroke [43]. Red meat intake, both processed and unprocessed, was associated with CHD risk in male health professionals [44]. In the large prospective cohort of men and women of the US National Institutes of Health—AARP Diet and Health Study, higher plant protein intake was associated with reduced CVD mortality [45]. In the Atherosclerosis Risk in Communities study, higher intakes of plant-based diets scored as healthy were associated with a lower risk of incident CVD and CVD mortality [46].

Pooled data from seven prospective cohort studies showed a reduced CHD incidence and mortality of 28% and 22%, respectively, associated with vegetarian diets. No association with CVD or stroke mortality was seen [47]. Similarly, a comprehensive review and meta-analysis of 10 prospective cohort studies showed a 25% reduced risk of incidence or mortality from IHD in vegetarian and vegan diets but not of total CVD and stroke mortality [48]. CVD and stroke mortality outcomes may be influenced by lifestyle factors other than diet and by access to cardiovascular healthcare.

4.2. Type 2 Diabetes

Observational studies in a variety of populations have consistently shown that compared to non-vegetarians, those following a vegetarian or vegan diet have a significantly lower risk of T2D [41,49–53]. A 2017 systematic review and meta-analysis of 14 studies found a pooled odds ratio for diabetes in vegetarians vs. non-vegetarians of 0.73 [54]. A 2020 systematic review similarly found that a vegan diet was associated with lower prevalence or incidence of T2D, although in some studies it was not possible to determine if the benefits were due to the vegan diet alone or combined with other healthy lifestyle habits [55].

A 2018 systematic review of nine RCTs found that, compared to control diets (including those of several diabetes associations), plant-based diets were associated with significant improvement in emotional well-being, physical well-being, depression, quality of life, general health, HbA1c levels (a measure of long-term blood glucose levels), weight, and total and LDL cholesterol levels [56]. An earlier systematic review and meta-analysis of six studies found that the consumption of vegetarian diets was associated with a significant reduction in HbA1c compared to control diets [57]. Similarly, a reduction in HbA1c has been observed with plant-based diets, including vegetarian, vegan, Mediterranean, and Dietary Approaches to Stop Hypertension (DASH) diets, compared to control or conventional diets [58].

There are several possible explanations for the benefits of plant-based diets for diabetes prevention and management. Compared to most western diets, vegetarian and vegan diets are generally higher in dietary fiber and are likely to include more whole grains, legumes and nuts, all of which have been associated with a reduced risk of T2D [59]. There is also evidence for an inverse association between higher intakes of green leafy vegetables and fruit and the risk of T2D [60–63].

The absence or limited intake of animal protein and red meat also likely plays a role. At least 25 studies have been published assessing the relationship between meat intake and T2D risk, with the majority showing a positive association between red meat and/or processed meat intakes. A 2013 meta-analysis found an association between higher intakes of total meat, unprocessed red meat, and processed meat and T2D risk [64]. There is also consistent evidence for an association between total dietary heme iron intake and heme iron intake from red meat and risk of T2D, and high serum ferritin levels are associated with insulin resistance and T2D risk [65]. A 2019 meta-analysis of prospective cohort studies looking at dietary protein intake and subsequent risk of T2D found high total protein and animal protein intakes to be associated with an increased risk of T2D, while a moderate plant protein intake was associated with a decreased risk [66]. An earlier systematic review and meta-analysis of 13 RCTs in people with diabetes found that replacing animal protein with plant protein (around 35% of total protein/day) resulted in significant reductions in HbA1c, fasting glucose, and fasting insulin levels compared to control groups [67].

Excess weight is a significant contributor to insulin resistance and T2D risk, and weight loss is a key component of the management of T2D [68]. Following a vegetarian or vegan diet, one is less likely to be overweight [69].

4.3. Cancer

Each of the plant food-groups has shown that they possess chemo-protective properties. Systematic reviews and meta-analyses have shown that an increased nut consumption was associated with both a decreased risk of all cancers combined [70] and decreased cancer mortality [71]. In the same manner, an increased intake of fruits and vegetables and of whole grains was shown to decrease the risk of total cancer incidence [72] and total cancer mortality [73], respectively. Furthermore, a higher intake of legumes (beans and lentils) was associated with a decrease in the risk of gastro-intestinal cancers and all cancer sites combined [74]. Many plant foods are rich in health-promoting phytochemicals, some of which have been shown to be useful in the treatment of human cancer [75,76].

On the other hand, the consumption of 100–120 g/day of red meat significantly increased the risk of many cancers (compared to eating no meat): 11% for breast cancer, 17% for colorectal cancer, and 19% for advanced prostate cancer [77]. For the consumption of 50 g/day of processed meat, the risk was increased 4% for total prostate cancer, 9% for breast cancer, 18% for colorectal cancer, 19% for pancreatic cancer, and 8% for cancer mortality [77]. In the French NutriNet-Santé cohort study, red meat intake was associated with increased risk of overall cancers (HR 1.31) and breast cancer (HR 1.83), but not prostate cancer [78]. In the National Institutes of Health (Rockville, MD, USA)—AARP Diet and Health Study cohort of half a million people, aged 50 to 71 years at baseline, higher red and processed meat intakes were associated with modest increases in total and cancer mortality [79].

With the elimination of meat and a greater use of protective plant foods, vegetarians may have a reduced risk of cancer. Epidemiologic cohort studies in the U.S. and UK have provided high-quality evidence regarding the association of vegetarian dietary patterns with cancer risk. In the US-based AHS–2, vegans had lower overall cancer risk compared to non-vegetarians (HR 0.84); overall cancer risk for lacto-ovo vegetarians was not significantly different from non-vegetarians [80]. Vegans showed a lower risk of prostate cancer (HR 0.65) [81] and a lower (but not statistically significantly lower) risk of breast cancer [82]. Neither lacto-ovo-vegetarians or vegans had a significantly lower risk of colorectal cancer [83].

In the UK-based EPIC-Oxford study, compared with meat eaters, vegans (HR 0.82) and lacto-ovo vegetarians (HR 0.90) had lower risk of all cancers combined [84]. For prostate cancer, while vegans (HR 0.61) and vegetarians (HR 0.86) had lower risk, they were not significantly different from meat eaters [84]. For colorectal cancer and female breast cancer, risk for the vegetarian groups again did not significantly differ from meat eaters [84]. In the UK Women's Cohort Study, compared with red meat eaters, the risk of breast cancer for vegetarians was not significantly lower (HR 0.85) [85].

Taken as a whole, such results seem to support the idea that vegetarians (including vegans and lacto-ovo vegetarians) have a modest but potentially important reduced overall cancer risk compared to their non-vegetarian counterparts. Findings for common individual cancers (colorectal, prostate, breast) are less consistent and warrant further study.

4.4. Overweight and Obesity

Over 70% of adults in the U.S. are overweight or obese [86], and trends show that overweight and obesity are increasing worldwide [87]. Observational studies show that vegans and vegetarians typically have a lower BMI than omnivores [88,89], and vegetarian diets or plant-based type dietary patterns are protective against adult weight gain and/or the risk of overweight or obesity [90,91]. Vegans typically have the lowest BMI or lowest prevalence of overweight or obesity in studies that compare multiple dietary patterns, including vegetarians and omnivores [88,92]. Gogga et al. noted differences in percent body fat between vegans, lacto-ovo-vegetarians, and omnivores, even though all group BMI values were within the normal range [93]. Interventions using vegan [94–99], vegetarian [97,100], or whole-food plant-based dietary [101] treatments have been found to lower BMI, weight, or fat-mass compared to subjects on a meat-based diet. A 4.8% weight loss was reported for overweight and obese subjects randomized to a vegan or vegetarian diet for 2 months, compared to a 2.2% loss seen in those consuming an omnivorous diet [97]. Weight loss of 3 to 5% is clinically meaningful and may contribute to chronic disease risk factor reduction [102].

The quality of the plant-based diet is also an important consideration. Subjects who adhere to a healthy plant-based diet are reported to have a lower BMI, waist circumference, and visceral fat than those who adhere to 'unhealthy' plant-based diets [103,104]. Researchers have noted that diet quality may be more important than dietary patterns when comparing vegans, vegetarians, and omnivores, as the adiposity values did not differ significantly between these groups [105]. The weight loss experienced on a hypocaloric lacto-ovo-vegetarian diet was similar to that observed with a hypocaloric Mediterranean diet [106].

Mechanisms that explain the weight management benefits of plant-based diets include relatively higher fiber, fruit, and vegetable consumption compared to omnivorous diets [88,107]. This food pattern may lead to beneficial alterations to appetite hormones [93,108] and the gut microbiota [109], both of which may have an impact on body weight.

4.5. Bone Health

Healthy bones require a variety of essential nutrients and healthy lifestyle practices to maximize peak bone mass during growth and minimize bone loss later in life [110]. While calcium and vitamin D are well recognized as important contributors to bone health, other nutrients, including magnesium, potassium, vitamin K, vitamin C, and zinc, as well as bioactive compounds found in fruits and vegetables, have been suggested as contributing to bone health and/or reduced risk of fracture [111–114]. Some have reported greater benefit from vegetables, especially cruciferous and allium vegetables, than from fruit [115,116].

The relationship of protein intake to bone status is complex. Earlier studies showed high intakes cause a loss of calcium, while a recent review found "no adverse effects of higher protein intakes" and some positive trends at most bone sites [117]. A recent review and meta-analysis found no difference between soy and animal protein on bone mineral density (BMD) and certain markers of bone turnover [118]. Others suggest the low acid load of vegetarian diets, partly due to the potassium and magnesium content from an increased fruit and vegetable intake, is beneficial to bone health [119]. Some elderly vegetarians and a few vegans may not consume sufficient protein for maintaining optimal bone health [114,120,121].

The impact of a vegetarian diet on bone health has many dimensions. Reports can vary considerably in study design, populations, and conclusions. Some find significantly lower BMD in vegetarians, especially vegans, which could increase fracture risk [122], while others see no difference in bone health, provided that calcium and vitamin D is adequate [123], and conclude that vegetarian food can provide a solid foundation for healthy bones and preventing fractures [124].

A large prospective UK study found that fish eaters and vegetarians had a higher risk of hip fractures compared to meat eaters, while vegans had a greater risk of total, hip, leg, and vertebral fractures [125]. Some of the differences may have been partly due to lower BMI and possibly lower intake of calcium and protein in the vegans.

A systematic review of some 20 studies involving 37,134 subjects found vegetarians and vegans had lower BMD at the femoral neck and lumbar spine compared to omnivores [126]. The effect was greater in vegans who also had higher fracture rates [127,128]. Another review concluded that the balance between protective factors in vegetarian and vegan diets and potential nutrient shortfalls may leave vegetarians, and especially vegans, at increased risk of bone loss and fractures [129]. Potential nutrient shortfalls can be remedied by appropriate food selections (including fortified foods) containing critical nutrients or by taking supplements. More research data on the bone health of vegans are needed before definitive recommendations can be made.

5. Eating Disorders

Previous use of a vegetarian or vegan diet apparently does not increase the risk of developing any eating disorder, such as anorexia nervosa, bulimia nervosa, and binge eating disorder [130,131]. Those with preexisting disordered eating tendencies may select vegetarian or partially vegetarian diets as a way to limit food intake in a socially acceptable fashion [130,132]. Semi-vegetarians appear to be at higher risk for developing eating disorders than vegetarians and vegans [130,133]. Those vegetarians whose motivation is weight control report more symptoms of disordered eating than do those with other motivations [134]. Commonly used assessment tools may incorrectly assess dietary restraint or eating disorder psychopathology in vegetarians [130].

6. Plant-Based Diets and the Gut Microbiome

The human gut microbiota is a highly complex community of some 10^{14} microorganisms. Diet has a significant impact upon the microbiota composition and function [135,136]. The microbiome has a profound impact on one's personal health and wellbeing [137]. Manipulating the gut microbiota has been viewed as a way to modulate the risk of chronic diseases such as obesity, T2D, cancer, and CVD [135,137].

Gut microbiota have a major role in the fermentation of nondigestible carbohydrates, namely resistant starch, soluble/insoluble dietary fiber, including plant wall polysaccharides and oligosaccharides. Fermentation of these nondigestible carbohydrates is associated with a higher abundance of microbes that produce butyrate and other short-chain fatty acids, which have an anti-inflammatory function, strengthen the intestinal barrier function, and improve overall gut health [138–141]. For example, the consumption of fiber-rich foods such as barley, wheat bran, brown rice, and other whole grains, as well as fructo-oligosaccharides and other prebiotics, are reported to increase butyrate-producing microbes [137,142–146]. Vegetarians would be expected to have an increased abundance of these microbes, as their fiber-rich diets are typically high in whole grains, fruits, vegetables, nuts, and legumes [107].

These plant foods also contain polyphenols—lignans, isoflavones, anthocyanins, and flavonols—in addition to other phytochemicals such as carotenoids and phytosterols [147–149]. These are metabolized into bioactive compounds by various microbes [150], some with health benefits and anti-inflammatory or antioxidant activity. Phytochemicals increase beneficial bacteria, including *Lactobacillus* and *Bifidobacterium*, which are the primary species present in probiotic supplements that are taken to improve gut health [151], in addition to some butyrate producers [150]. Among fiber-rich plant foods, nuts in particular (walnuts, almonds, pistachios) have been found to have prebiotic effects and are associated with increases in butyrate-producing microbes and other beneficial microbes [152]. Hence, the gut microbial composition is greatly influenced by dietary fiber as well as by polyphenols and other phytochemicals and their metabolites, all of which are more highly consumed by vegetarians.

Studies have supported the value of two so-called enterotypes, or clusters of microbes driven by distinct genera, in distinguishing dietary patterns. Accordingly, Bacteroides are associated with animal fat and high-protein diets [153–157], and *Prevotella* are associated with fiber-rich foods and carbohydrates, typical of a plant-based diet [158–160]. Higher abundance of *Prevotella* and other polysaccharide-degrading or potential butyrate-producing microbes has been seen particularly in agrarian cultures such as those in Tanzania, the Peruvian Amazon, and Burkina Faso, compared to U.S. or Western populations, reflecting the higher consumption of fiber-rich plant foods by these societies [160–162]. Hence, enterotypes may have some utility in distinguishing plant- and animal-based diets. Plant-based and high-fiber diets are also associated with increases in the Bacteroidetes phylum [160,162], or the Bacteroidetes/Firmicutes ratio, as well as microbial richness/diversity [142,155,160,162–164], in contrast to diets high in fat [165–169]. This is relevant in that various microbes from the Bacteroidetes phylum encode carbohydrate-active enzymes (CAZymes) necessary for degrading indigestible carbohydrates [139], and the Bacteroidetes:Firmicutes ratio may have implications for obesity and metabolic diseases, although the relationship is not clear as findings have been inconsistent [170,171].

Differences in gut microbial composition are not always observed in cross-sectional studies comparing vegans or other vegetarians with non-vegetarians [172]. In the AHS-2 cohort, only subtle differences were noted in the microbiome [173]. However, vast differences were discovered in the plasma metabolome, with vegans showing higher abundance of anti-inflammatory plant/polyphenol or microbial-related metabolites [174]. Non-vegetarians on the other hand may have higher abundance of amino acids and lipids conceivably associated with cardiometabolic phenotypes [174–176]. Intestinal microbiota convert choline and L-carnitine, derived from meat, fish, dairy, and eggs, into trimethylamine, which is oxidized by the liver to trimethylamine N-oxide (TMAO), a pro-inflammatory compound that has been associated with increased cardiometabolic risk [177–180]. Thus, it may be that microbial function is more relevant than composition, with metabolic profiles showing much greater differences, reflecting phenotypic changes.

There are physiological consequences of diet-induced shifts in the microbiome. Low consumption of plant-based foods may lead to increased penetration of the intestinal barrier, as a low-fiber diet triggers a shift from fiber-degrading to mucus-degrading bacteria [181]. This in turn could promote a hyperactive immune response, conceivably with the production of pro-inflammatory metabolites that fuel disease processes [182]. However, much remains to be understood about how vegetarian and plant-based dietary patterns impact the microbiome and associated metabolic responses to influence disease processes.

7. Plant-Based Diets and the Life Cycle

Vegetarian, including vegan, diets can satisfy the nutritional requirements of all stages of the life cycle. They can promote normal growth and development in infancy, childhood, and adolescence and meet the needs for energy and nutrients of these life cycle stages as well as those of pregnancy, lactation, and older adulthood.

7.1. *Pregnancy and Lactation*

Vegetarian diets can effectively meet energy and nutrient needs in pregnancy and lactation [183,184]. Several reviews, while noting the limited amount of information about vegetarian, including vegan, diets in pregnancy, have concluded that, with adequate nutrient intake, these diets are safe in pregnancy [183,185]. When food access is satisfactory, infant birth weights and the duration of gestation are similar in vegetarian and nonvegetarian pregnancy [186,187]. Some studies report that vegetarians are more likely to have infants who are small for their gestational age [188–190]. This finding may be due to lower mean pre-pregnancy BMI, lower weight gain, or inadequate weight gain in pregnancy. Well-nourished vegetarians produce nutritionally adequate breast milk that supports infant growth and development [191].

Health benefits of vegetarian diets in pregnancy include a lower risk of excessive weight gain and higher fiber and folate intakes [188,192,193]. Dietary patterns that are high in plant foods are associated with a reduced risk of gestational diabetes mellitus, hypertensive disorders of pregnancy, and preterm birth [194].

Nutrient requirements in vegetarian pregnancy and lactation generally do not differ from those for nonvegetarians [195]. Vegetarians may especially benefit from guidance on sources of iron, zinc, vitamin B12, iodine, and docosahexaenoic acid (DHA). Although iron absorption increases in pregnancy [196], iron needs are high, so iron-rich foods and low-dose iron supplements are recommended for all women [197,198]. The increased need for zinc can be met through a combination of increased intake and absorption [199].

Phytate's inhibitory effect on zinc absorption is markedly reduced in late pregnancy and early lactation [200]. In addition to the use of iodized salt, a 150 μg/d iodine supplement is recommended for all pregnant and lactating women [201].

During pregnancy, blood DHA concentrations are often lower in vegetarians than in nonvegetarians [202]; cord blood DHA is lower in infants of vegetarians [202]. Breast milk DHA concentrations of vegetarians and vegans are lower than worldwide averages [203]. DHA or omega-3 supplementation is associated with greater gestational duration and a reduced risk of preterm birth [204,205]. Supplemental DHA derived from microalgae should be used in vegetarian pregnancy and lactation [195].

Adequate vitamin B12 intake is especially important during periods of growth such as pregnancy and breastfeeding. Infants born to long-term vegan mothers and who are breastfed are at risk of B12 deficiency. This is especially true when the mother's diet is not well-supplemented. Symptoms of B12 deficiency in breastfed infants and small children fed a vegan diet include developmental delay or psychomotor regression, lethargy, anemia, neurological issues, and failure to thrive [206]. Pregnant and lactating vegetarians should consume reliable sources of vitamin B12, such as supplements and/or fortified foods, on a daily basis [195].

7.2. Infants, Children, and Adolescents

Vegetarian, including vegan, diets that are nutritionally adequate are appropriate for use in infancy, childhood, and adolescence and support normal growth [184,207,208]. Health benefits of vegetarian diets in childhood and adolescence include the potential for exposure to a wide variety of plant foods, lower risk for childhood obesity [209], and higher consumption of fruits and vegetables [210,211]. Vegan children appear to have lower intakes of total and saturated fat and cholesterol compared to non-vegan children [211]. A low-fat vegan diet has effectively treated children with obesity and elevated blood pressure [212].

Exclusive breastfeeding is recommended for infants for the first 6 months after birth, with breastfeeding continuing until at least 12 months of age [213]. If breastfeeding or exclusive breastfeeding is not possible, commercial infant formula should be used as the primary beverage for the first year. Plant milks, unmodified cow's milk, other milks, and homemade formulas should not be used to replace breast milk or formula during the first year. Standard practices should be used when introducing complementary foods to vegetarian infants. Vegetable proteins, such as pureed beans or tofu, are used in place of pureed meats. After the first year, if toddlers are growing normally and eating a variety of foods, fortified soy or pea protein milk or dairy milk can be started [195].

Several nutrients require special attention in the planning of nutritionally adequate diets for young vegetarians, including iron, zinc, iodine, and vitamin B12; calcium and vitamin D may also require attention, depending on dietary choices and other factors. Protein recommendations for vegan children may be somewhat higher than standard recommendations because of factors including protein digestibility and amino acid composition [195]. Protein needs of vegetarian or vegan children and adolescents are generally met when their diets contain adequate energy and a variety of plant protein sources. Deficiencies of iron and zinc are rarely seen in vegetarian children eating varied diets [207]. Zinc supplementation may be needed when complementary foods are introduced, if foods are mainly those with low zinc bioavailability [214]. Iron and zinc status in infants, children, and adolescents should be monitored, and fortified foods and/or supplements used as needed. Iodized salt is a reliable source of iodine for children and teens. If maternal vitamin B12 intake or status are inadequate, breastfed infants should be given supplemental vitamin B12 [206]. Vegetarian children and adolescents should use vitamin B12-fortified foods or supplements to supply adequate vitamin B12. Calcium sources for children and adolescents include fortified plant-based milks, green leafy vegetables, and dairy products.

7.3. Older Adults

Older adults generally have decreased energy requirements, although nutrient requirements are often similar to, or higher than, those of younger adults. The selection of nutrient-dense diets is especially important for older adults. Limited research indicates that nutrient intakes of older vegetarians are comparable to those of older non-vegetarians [195].

Recommendations for calcium, vitamin B6, and vitamin D are higher for older adults [215,216]. There is some evidence that protein needs increase as well [217]. Higher protein foods such as soy products (including tofu, soy beverage, soy yogurt alternative, etc.), legumes, nuts and seeds, and meat analogs should be used two to three times a day by older vegetarians. Vitamin B6 recommendations for all older adults are higher due to decreased absorption and alterations in metabolism [216]. Vegetarians generally have adequate intakes of vitamin B6. Sources include potatoes, bananas, fortified breakfast cereals, and spinach. Several factors increase older adults' risk for vitamin D insufficiency, including reduced dermal and renal synthesis [218,219], inadequate dietary intake, and limited sun exposure. Fortified foods and/or supplements may be needed for older adults to meet recommendations for calcium and vitamin D.

The main cause of vitamin B12 deficiency in older adults is impaired absorption of vitamin B12 from foods [220]. Absorption of purified vitamin B12 from fortified foods and supplements is not typically impaired, so recommendations call for older adults to use fortified foods and supplements as their primary sources of vitamin B12 [216].

8. Athletic Performance

Vegetarian diets can meet the needs of athletes at all levels, from recreational to elite athletes [221,222], and have been followed by athletes throughout history [223]. While a nutritionally adequate plant-based diet is thought to help optimize training and performance, due in part to its high carbohydrate [223–225] and high phytochemical content [225], limited evidence from well-controlled studies suggests that vegetarian diets neither enhance nor impair performance [225]. Additional research is needed to determine whether such diets enhance recovery and attenuate the oxidative damage and inflammation that occur with heavy training.

Nutrition recommendations for athletes should consider each athlete's training volume (intensity and frequency), sport, season, performance goals, and food preferences. Vegetarian diets that meet energy needs and contain a variety of plant-based protein sources, including soy foods, dried legumes, nuts, seeds, quinoa, and other grains, can provide adequate protein to support most training needs. There is some evidence that plant-derived proteins result in lower post-prandial muscle protein synthesis responses compared with equivalent amounts of animal-derived proteins [226]; this response may be improved by consuming blends of different plant-derived proteins [226]. Milk and eggs [227–229] can supplement plant-based sources for vegetarian athletes.

Depending on food preferences, athletes need to ensure they consume adequate amounts of the nutrients that are either found less abundantly in vegetarian foods or are less well absorbed from plants compared to animal sources. These nutrients include calcium, iron, zinc, iodine and vitamin B12. For example, female athletes and endurance athletes should ensure sufficient consumption of iron-rich plant foods along with dietary factors that enhance rather than inhibit iron absorption [230–232]. Female athletes with restricted intake and amenorrhea (i.e., low energy availability) [233] may require additional calcium (1500 mg/day along with 1500–2000 IU vitamin D) to optimize bone health [234]. Maintaining adequate vitamin D status is important for athletes due to its role in immune function, inflammatory modulation, physical performance, and overall health [235–238]. Vegetarian athletes may have lower blood and muscle creatinine and carnitine concentrations [239–242] compared to omnivores due to lower dietary intake. Athletes participating in resistance training and bouts of high-intensity exercise may benefit from creatine supplementation [243], but there is no recognized benefit to carnitine supplementation. Vegetarian athletes, like most others, may benefit from education about food choices to optimize health and peak performance [244].

9. Nutrients of Concern in a Plant-Based Diet

9.1. Calcium

In addition to its role in bone mineralization, calcium is required for blood clotting, muscle function, nerve transmission, hormone release, intracellular signaling, and regulating key enzymes [245]. Typically, vegans consume substantially less calcium than other vegetarians and omnivores [192,246]. When calcium intakes are low, the body can compensate somewhat by increasing the fractional calcium absorption [247] and decreasing urinary calcium excretion [215]. However, anyone, including all types of vegetarians, with inadequate calcium intake needs to consistently use calcium-fortified foods, such as fortified breakfast cereals, fortified fruit juices, and fortified plant-based beverages, or take a calcium supplement, to meet their calcium needs. Vegan diets in the UK have been associated with a clinically significant increased risk of fracture when the calcium intake was inadequate [248].

Phytic and oxalic acids in plant foods are both inhibitors of calcium absorption. The calcium absorption from oxalate-rich vegetables (spinach, Swiss chard) may be as low 5%; from beans, almonds, tahini, and figs 20–25%; from dairy products 32%; from soy products (tofu, fortified soy beverages), it is similar to dairy milk; and from low-oxalate vegetables (kale, Chinese cabbage, broccoli, bok choy, etc.) 50–60% [249–251]. Boiling can reduce oxalate content in green leafy vegetables [252]. A vegetarian diet, with its high intake of fruit and vegetables, is rich in anti-inflammatory phytonutrients, specifically carotenoids and flavonoids, and potassium and magnesium. Carotenoids and flavonoids are associated with an improved BMD and lower bone fractures [253–256].

Compared to a vegetarian diet, consuming an animal protein diet is associated with an increased loss of urinary calcium [257].

9.2. Iron

In addition to its ability to transfer oxygen by means of hemoglobin and myoglobin, iron functions as a co-factor for many important enzymes (such as myeloperoxidase, important for immune function) and has a role in thyroid hormone synthesis and amino acid metabolism [245]. Since heme iron is generally better absorbed (15–30%) than non-heme iron (typically 5–10%), omnivores are assumed to have better iron status. However, vegetarians who eat a varied and well-balanced diet do not appear to be at any greater risk of iron-deficiency anemia than omnivores [258,259]. Hemoglobin levels of the two dietary groups normally show no significant differences [259,260]. Additional studies of iron deficiency in vegetarians are needed before definitive conclusions can be reached. A varied diet that is rich in wholegrains, legumes, nuts, seeds, dried fruits, iron-fortified cereal products, and green leafy vegetables provides an adequate iron intake. Vegetarian diets generally contain as much or more iron than omnivore diets [92,195].

Non-heme iron absorption is significantly affected by several dietary components [261, 262]. Vitamin C, other organic acids (citric, malic, lactic, tartaric acids), and erythorbic acid (an antioxidant used in processed food) all enhance absorption [196,230,259,263,264]. Plant ferritin, found in soy and other legumes, is an easily absorbed source of iron (22–34%). While phytates (found in legumes, nuts, and whole grains) can inhibit non-heme iron absorption, their inhibitory effect is diminished by baking, soaking, leavening, and germination [184,265]. Furthermore, the overall long-term effect of enhancers and inhibitors of iron may be less important than once thought when the foods are eaten as part of a whole diet [266,267].

Absorption of non-heme iron is also inversely related to the body's iron status [196]. When stores are low and the need for iron increases, compensatory mechanisms facilitate greater absorption of iron. Absorption can be as low as 2–3% in people with good iron stores but as high as 14–23% in people with low iron stores [268].

Humans have a limited ability to excrete excess stored iron [258], so consuming large amounts of heme iron may be unhealthy due to its pro-oxidant nature. Consumption of heme iron has been associated with an increased risk of chronic diseases such as diabetes, metabolic syndrome, and colorectal cancer [269–272]. Vegetarians typically have lower iron stores (as reflected in lower serum ferritin levels), which may be an advantage as lower serum ferritin levels may be associated with improved insulin sensitivity and reduced risk of T2D [258,273].

Iron absorption from an omnivorous diet is claimed to be about 18%, whereas for a plant-based diet it is said to be about 10% [196]. Hence, the current Dietary Reference Intake (DRI) for iron for vegetarians has been set 1.8 times higher than that for non-vegetarians. This increased requirement is based on limited research, which has been unable to accurately measure adaptive absorption rates of non-heme iron in vegetarians [267,274]. Further research is needed to reassess the iron requirement recommended for vegetarians [232].

9.3. Zinc

Zinc acts as a coenzyme for multiple enzymes involved with growth, immunity, cognitive function, bone function, and regulation of gene expression [275,276]. Zinc deficiency causes stunted growth, poor appetite, dermatitis, alopecia, endocrine dysfunction, and impaired immunity [276]. Zinc deficiency is not any more commonly seen in vegetarians than in non-vegetarians [277]. Zinc intake and serum levels for adolescent and adult vegetarians in developed countries are the same or slightly lower than for omnivores, but within the normal range [214,231,275,278,279]. In developing countries, vegans and vegetarians are more likely to show marginal zinc status [278].

The bioavailability of zinc from plant foods may be reduced. However, zinc absorption and retention can be regulated by homeostatic mechanisms, adapting to lower intakes by reducing losses and increasing absorption [275]. During periods of high demand (pregnancy, infancy), absorption becomes more efficient [280].

Phytates in cereals and legumes lower absorption of zinc, but leavening, soaking, fermenting, or sprouting reduces the phytate levels and makes zinc more bioavailable [281]. Sulfur-containing amino acids and organic acids in a variety of plant foods will also enhance zinc absorption [279,282].

Vegetarian food sources for zinc include nuts, seeds, wholegrains, legumes, tofu, tempeh, and dairy products [283]. The use of supplements and fortified foods (such as fortified breakfast cereals) may be necessary for very restricted vegan diets [214,246].

9.4. Iodine

Iodine is essential for thyroid hormones, which regulate metabolic activity. Iodine is especially important in pregnancy for fetal development and during early childhood. Iodine deficiency in childhood can prevent children from attaining their full physical potential and intellectual capacity [284].

Major dietary sources of iodine include iodized salt, seafood, and dairy products [284]. The iodine content of seaweeds and dairy products can vary widely [195,285]. Sea salt, Himalayan salt, and the salt used in processed foods typically do not contain iodine [195]. Although foods such as soybeans, cruciferous vegetables, and sweet potatoes contain natural substances that interfere with iodine uptake by the thyroid, these foods have not been associated with thyroid dysfunction in healthy people, provided iodine intake is adequate [196,286].

Vegans who do not use iodized salt and/or sea vegetables may have low iodine intakes and may be at risk for iodine deficiency [287,288].

9.5. Vitamin B12

Vitamin B12 is required for red blood cell formation, DNA synthesis, homocysteine metabolism, and the myelination and function of the central nervous system [289]. Vitamin B12 deficiency is not uncommon among the elderly and unsupplemented vegans. It can manifest itself with consequential hematological and neurological changes. Typically, the mean dietary intake of vitamin B12 of vegans falls well below the DRI, while that of lacto-ovo-vegetarians may be marginal, depending on the use of dairy products [246,290,291]. Vegans must obtain their vitamin B12 either from regular use of vitamin B12-fortified foods, such as fortified plant-based beverages, fortified breakfast cereals, fortified vegetarian meat analogs, or from a regular vitamin B12 supplement. Unfortified plant foods such as fermented soy foods, leafy vegetables, seaweeds, mushrooms, and algae (including spirulina) do not contain significant amounts of active vitamin B12 to provide daily needs [292]. Furthermore, a number of medications can impair the absorption or utilization of B12. Vitamin B12 appears to be a cofactor involved in the production of nitric oxide [293], which would have important implications for vascular and immune health.

About 50% of dietary B12 is normally absorbed via ileal receptors, mediated by the intrinsic factor, a glycoprotein from the stomach. The ileal receptors become saturated with 1.5 to 2 µg of B12, limiting further absorption [216]. When ingesting large doses of supplemental B12, about 1% of the dose is absorbed by passive diffusion across the intestinal tract [216]. Daily needs can be adequately met in non-pregnant, non-lactating people by consuming a 500 µg B12 supplement at least three times a week. Vitamin B12 is well absorbed from either sub-lingual or chewable tablets. While the methylcobalamin supplement is touted as the more effective form of B12, its bioavailability is not superior to that of cyanocobalamin, which is the more stable and most commonly used form of B12 in fortified foods and many supplements [294,295].

A deficiency of vitamin B12 may take years to develop in adults, as most of the B12 secreted into the gut via the bile gets reabsorbed, thus conserving the body stores [216]. Therefore, a regular consumption of adequate B12 is important to avoid a sub-clinical deficiency that can go undetected for years. An elevated serum methylmalonic acid (MMA) level is a reliable indicator of B12 deficiency [216], while the serum B12 level is an insensitive indicator of B12 status. While serum B12 levels between 148 and 221 pmol/L (200–300 pg/mL) are considered borderline B12 deficiency [296], some individuals with B12 values in this range manifest neuropsychiatric problems and memory loss [297]. As a good preventative measure, all vegans should annually check their B12 status.

9.6. Vitamin D

Vitamin D facilitates calcium absorption from the gut, regulates bone mineralization, cell growth, and differentiation. Its other roles include reduction of inflammation and modulation of neuromuscular and immune function [298]. Because cutaneous production of vitamin D from sunlight exposure is not adequate (especially in the elderly, dark-skinned individuals, and heavy sunscreen users) to meet nutrition needs in populations living in high latitudes, especially during the winter months, regular food and supplement sources are necessary. Foods contain limited amounts of vitamin D, so supplements are often needed to meet needs. Depending upon one's age, geographical location, dietary preferences, and body weight, a daily supplemental dose of 10–50 µg (400 to 2000 IU) of vitamin D may be needed to achieve optimal serum levels of 25-hydroxyvitamin D (25(OH)D) year-round [299].

One study found no significant difference in serum 25(OH)D levels between vegetarians and non-vegetarians. Factors such as vitamin D supplementation, degree of skin pigmentation, and amount of sun exposure had a greater influence on serum 25(OH)D levels than did diet [300]. By contrast, in the large EPIC-Oxford study, plasma 25(OH)D levels in British vegetarians were 14.3% lower, and in vegans 27.5% lower, than in meat eaters [301].

Vitamin D intake by vegans tends to be substantially below that of lacto-ovo-vegetarians and omnivores [195]. Low serum 25(OH) D levels and reduced bone mass have been reported in vegans living in high latitudes who were not using vitamin D-fortified foods or supplements [302,303].

Fortified plant-based beverages, fortified orange juice, ready-to-eat breakfast cereals, and fortified margarines provide vitamin D for vegetarians. Modest levels of vitamin D are also obtained from mushrooms that have been exposed to ultraviolet light under controlled conditions [304]. Lacto-ovo-vegetarians also obtain vitamin D from fortified dairy products and eggs. Depending on sunlight exposure and dietary intake, supplements may be needed. For low daily doses, vitamin D2 appears to be as effective as vitamin D3 in maintaining circulating levels of serum 25(OH) D [305]. When given as a single large dose, vitamin D2 appears to be less effective than vitamin D3 for improving the vitamin D status [306].

With appropriate food and supplement choices, a vegetarian diet can be consistent with having an adequate vitamin D status and supporting a healthy BMD (bone mineral density) [129].

9.7. Omega-3 Fatty Acids

Omega-3 fatty acids (n-3) are associated with favorable cardiometabolic status [307]. The source of omega-3 for vegetarians is predominantly α-linolenic acid (ALA) [308]. Normally, only small amounts of ALA are converted to the longer-chain eicosapentaenoic acid (EPA), and to a less degree DHA, particularly if linoleic acid intake is high [308,309]. Conversion of ALA is also affected by health status, age, dietary composition, and gender [310]. Results from the EPIC-Norfolk cohort study revealed that omega-3 status differences were much smaller than dietary differences, with vegans and vegetarians showing a more efficient conversion of ALA to EPA and DHA [311]. Most studies indicate that plasma, serum, erythrocytes, adipose, and platelet levels of EPA and DHA are lower in vegetarians than omnivores [309], yet there is no evidence of adverse effects on heart health or cognitive function in vegetarians [312,313].

EPA has antithrombotic properties and confers cardiovascular protection [308,314], while DHA has been linked to eye and brain development and is important for ongoing visual, cognitive, cardiovascular health [308,315]. Omega-3 fatty acids may also help regulate gut microbiota and immunity and reduce the risk of inflammatory diseases [316–318]. ALA, EPA, and DHA intakes are all associated with a reduced risk of CVD [319].

The richest sources of ALA include flaxseed, hemp seed, walnuts, chia seeds, and their oils, with smaller amounts present in canola and soy oils, and green leafy vegetables [310]. Currently the National Academy of Medicine (Washington, DC, USA) has not established recommendations for EPA and DHA, while the European Food Safety Authority has recommended an intake of 250 mg/day for EPA and DHA [320]. To date, an adequate intake of ALA has been specified as 1.6 g for men and 1.1 g for women [321]. The ideal omega-6/omega-3 ratio for optimal health has not been defined, although various authors have debated the issue [321]. Improving the DHA status of an individual is generally regarded as desirable. For the vegetarian, a regular use of an algal DHA supplement would be an effective way to increase serum DHA levels [309,322].

The critical period of pregnancy and lactation requires a higher n-3 status (particularly DHA) [308,323,324]. Pregnant and breastfeeding women, and those at greater risk for poor ALA conversion, such as people with diabetes, older people, and premature infants, are most likely to benefit from DHA supplements derived from micro-algae [319,325]. Omega-3-rich eggs and DHA-fortified foods are also food sources of DHA for vegetarians.

9.8. Protein

Individuals following vegetarian diets generally consume more than adequate protein, particularly in western countries, although intakes are typically lower than those of omnivores [120]. Furthermore, as long as a variety of protein-rich foods are consumed, vegetarian diets are able to provide all of the indispensable amino acids [120,326]. While there is no need for different protein foods to be combined in one meal, a variety of plant foods should be included each day [326]. Most plant foods contain some protein, with the best sources being legumes, soy foods (including fortified soy milk, tofu, and tempeh), nuts, and seeds. Grains and vegetables also contain protein but in smaller amounts.

While the lower protein intake and quality of protein in a vegetarian diet is often cited as a concern, there is increasing evidence for the health benefits of consuming protein from plant sources rather than animal sources, and this may be one of the reasons why vegetarians have a lower risk of obesity and chronic diseases [327].

Those consuming omnivorous diets in western countries tend to get 1.5 to 2 times the recommended protein intake, and such high protein intakes can have a variety of deleterious effects, such as increased calcium excretion and reduced insulin sensitivity [328,329].

10. Guidelines for Health Professionals

Significant health benefits are associated with vegetarian, including vegan, diets. Plant-based diets, even if not completely vegetarian, also offer significant health benefits. Health professionals should discuss the benefits of vegetarian and near-vegetarian diets with their clients and provide supportive, reliable, evidence-based information and resources. If the practitioner is unfamiliar with vegetarian nutrition, clients should be referred to other health professionals with expertise in this area, such as registered dietitians.

Health professionals are ethically obligated to respect vegetarian dietary patterns and to provide information so that clients are aware of their nutritional needs, sources of nutrients, and any dietary modifications needed to meet their individual situation. The client's food preferences should be determined and respected. This may include religious or cultural factors that influence one's food choices.

Health professionals who work with vegetarians and those interested in vegetarian diets should be familiar with current research on vegetarian nutrition as well as with vegetarian foods and food preparation. There are a number of excellent books and other resources available for health professionals to acquaint themselves with evidence-based data [195,222,330,331]. Individualized counseling materials should be developed that feature vegetarian foods.

Some traditional cultures have plant-based traditions. When working with clients from these cultures, professionals should focus on the retention of healthy traditional practices, with modification of other practices to promote more healthful diets instead of promoting the eating patterns of the dominant culture [332].

It is incumbent on any health professional providing counsel regarding dietary choices to remember it is not what a diet is called, but what foods an individual consumes on a regular basis that determines the adequacy of a diet.

11. Conclusions

Plant-based diets continue to grow in popularity. Currently, there is a vibrant interest in the sustainability of diets and a growing awareness of the need to focus on both human health and the health of the planet in formulating dietary guidelines. Plant-based diets are more sustainable than diets based on animal products, since they use fewer natural resources and produce fewer GHG emissions. Vegetarian and vegan diets provide protection against a number of common chronic diseases, such as CVD, obesity, T2D, and certain types of cancer. The consumption of a plant-based diet rich in fiber and phytochemicals not only provides disease-preventing benefits but also has a substantial impact on the composition and function of the gut microbiome, which in turn influences overall health.

Both a vegetarian and vegan diet are appropriate for all stages of the life cycle, including pregnancy and lactation, all stages of childhood, the elderly, and for athletes. When appropriately planned, a plant-based diet (consisting substantially of minimally processed foods) can be nutritionally adequate. Vegetarians and especially vegans should consume a well-balanced diet and regularly use fortified foods and/or supplements. Special attention should be paid to calcium, iron, vitamin D, and vitamin B12. A deficiency may be exacerbated when supplements are not utilized and when food choices are limited or self-restricting.

Health professionals who work with vegetarians and those interested in vegetarian diets should be familiar with current research on vegetarian nutrition and be able to provide information so that clients are aware of their nutritional needs, sources of nutrients, and any dietary modifications needed to meet their individual situation. The health professional should be sensitive to the client's food preferences and respect any religious or cultural factors that influence their food choices.

Author Contributions: Conceptualization, W.J.C.; writing—original draft preparation, W.J.C., A.R.M., U.F., K.M., F.L.M., A.V.S., E.H.H., C.E.H., P.J., E.L.-M., M.O.; writing—review and editing, W.J.C., A.R.M., U.F. All authors have read and agreed to the published version of the manuscript.

Funding: This research received no external funding.

Institutional Review Board Statement: Not applicable.

Informed Consent Statement: Not applicable.

Data Availability Statement: Not applicable.

Acknowledgments: U.F. acknowledges support from the Spanish Ministry of Science and Innovation and State Research Agency through the "Centro de Excelencia Severo Ochoa 2019–2023" Program (CEX2018-000806-S) and support from the Generalitat de Catalunya through the CERCA Program. C.H. wishes to acknowledge the research assistance of Chassitie Char.

Conflicts of Interest: The authors declare no conflict of interest.

References

1. Kamiński, M.; Skonieczna-Żydecka, K.; Nowak, J.K.; Stachowska, E. Global and local diet popularity rankings, their secular trends, and seasonal variation in Google Trends data. *Nutrition* **2020**, *79–80*, 110759. [CrossRef]
2. Shridhar, K.; Dhillon, P.K.; Bowen, L.; Kinra, S.; Bharathi, A.V.; Prabhakaran, D.; Reddy, K.S.; Ebrahim, S. Nutritional profile of Indian vegetarian diets–the Indian Migration Study (IMS). *Nutr. J.* **2014**, *13*, 55. [CrossRef] [PubMed]
3. Natrajan, B.; Jacob, S. "Provincialising" vegetarianism. Putting Indian food habits in their place. *Econ. Polit. Wkly.* **2018**, *53*, 54–64.
4. The Vegetarian Resource Group. How Many Adults in the U.S. Are Vegan? How Many Adults Eat Vegetarian When Eating Out? *VRG Blog* **2020**. Available online: https://www.vrg.org/blog/2020/08/07/how-many-adults-in-the-u-s-are-vegan-how-many-adults-eat-vegetarian-when-eating-out-asks-the-vegetarian-resource-group-in-a-national-poll/ (accessed on 18 August 2021).
5. Stahler, C.; Mangels, R. How Many Youth in the U.S. are Vegan? How Many Teens Eat Vegetarian When Eating Out? The Vegetarian Resource Group Asks in a YouGov Poll. *Veg. Resour. Group.* **2021**. Available online: https://www.vrg.org/nutshell/Yougov2021youthteenwriteup.pdf (accessed on 18 August 2021).
6. Fortune Business Insights. Dairy Alternatives Market Size Worth $25.12 billion by 2026; Rising Adoption of Vegan Diet to Propel Growth. 2020. Available online: https://www.globenewswire.com/news-release/2020/02/19/1986821/0/en/Dairy-Alternatives-Market-Size-Worth-25-12-Billion-by-2026-Rising-Adoption-of-Vegan-Diet-to-Propel-Growth-says-Fortune-Business-Insights.html (accessed on 20 August 2021).
7. Plant Based Food Association. 2020 Retail Data Sales Announcement. 2021. Available online: https://www.plantbasedfoods.org/2020-retail-sales-data-announcement/ (accessed on 18 August 2021).
8. U.S. Department of Agriculture; U.S. Department of Health and Human Services. *Dietary Guidelines for Americans, 2020–2025*, 9th ed.; 2020. Available online: https://www.dietaryguidelines.gov/sites/default/files/2020-12/Dietary_Guidelines_for_Americans_2020-2025.pdf (accessed on 18 August 2021).
9. Mangels, R. Vegetarian diets. Trends in acceptance and perception. In *Vegetarian Nutrition and Wellness*, 1st ed.; Craig, W., Ed.; CRC Press: Boca Raton, FL, USA, 2018; pp. 1–9.
10. Kukura, J. Berkeley Passes 'Vegan Monday' Requirement. Available online: https://www.sfweekly.com/news/berkeley-passes-vegan-monday-requirement/ (accessed on 11 November 2021).
11. Storz, M.A. Should Plant-Based Hospital Meals be the Law? An American Experience. *Hosp. Pract.* **2020**, *48*, 241–243. [CrossRef] [PubMed]
12. Regan, T. Dished: Meatless Meals Gain Steam in Senior Living Dining Programs. Available online: https://seniorhousingnews.com/2018/03/16/dished-meatless-meals-gain-steam-senior-living-dining-programs/ (accessed on 11 November 2021).
13. Clune, S.; Crossin, E.; Verghese, K. Systematic review of greenhouse gas emissions for different fresh food categories. *J. Clean. Prod.* **2017**, *140*, 766–783. [CrossRef]
14. Poore, J.; Nemecek, T. Reducing food's environmental impacts through producers and consumers. *Science* **2018**, *360*, 987–992. [CrossRef] [PubMed]
15. Clark, M.; Tilman, D. Comparative analysis of environmental impacts of agricultural production systems, agricultural input efficiency, and food choice. *Environ. Res. Lett.* **2017**, *12*, 064016. [CrossRef]

16. Santo, R.E.; Kim, B.F.; Goldman, S.E.; Dutkiewicz, J.; Biehl, E.M.B.; Bloem, M.W.; Neff, R.A.; Nachman, K.E. Considering Plant-Based Meat Substitutes and Cell-Based Meats: A Public Health and Food Systems Perspective. *Front. Sustain. Food Syst.* **2020**, *4*, 134. [CrossRef]
17. Tilman, D.; Clark, M. Global diets link environmental sustainability and human health. *Nature* **2014**, *515*, 518–522. [CrossRef]
18. Ritchie, H.; Roser, M. Environmental Impacts of Food Production. Our World in Data. 2020. Available online: https://ourworldindata.org/environmental-impacts-of-food (accessed on 7 July 2021).
19. European Comision. Our Oceans, Seas and Coast. Descriptor 5: Eutrophication. Available online: https://ec.europa.eu/environment/marine/good-environmental-status/descriptor-5/index_en.html (accessed on 10 November 2021).
20. Fresán, U.; Sabaté, J. Vegetarian Diets: Planetary Health and Its Alignment with Human Health. *Adv. Nutr.* **2019**, *10*, S380–S388. [CrossRef]
21. Aleksandrowicz, L.; Green, R.; Joy, E.J.; Smith, P.; Haines, A. The Impacts of Dietary Change on Greenhouse Gas Emissions, Land Use, Water Use, and Health: A Systematic Review. *PLoS ONE* **2016**, *11*, e0165797. [CrossRef]
22. Springmann, M.; Godfray, H.C.J.; Rayner, M.; Scarborough, P. Analysis and valuation of the health and climate change cobenefits of dietary change. *Proc. Natl. Acad. Sci. USA* **2016**, *113*, 4146–4151. [CrossRef] [PubMed]
23. Springmann, M.; Wiebe, K.; Mason-D'Croz, D.; Sulser, T.B.; Rayner, M.; Scarborough, P. Health and nutritional aspects of sustainable diet strategies and their association with environmental impacts: A global modelling analysis with country-level detail. *Lancet Planet. Health* **2018**, *2*, e451–e461. [CrossRef]
24. Virani, S.S.; Alonso, A.; Aparicio, H.J.; Benjamin, E.J.; Bittencourt, M.S.; Callaway, C.W.; Carson, A.P.; Chamberlain, A.M.; Cheng, S.; Delling, F.N.; et al. Heart Disease and Stroke Statistics. 2021 Update. *Circulation* **2021**, *143*, e254–e743. [CrossRef] [PubMed]
25. World Health Organization. World Health Statistics 2021: Monitoring Health for the SDGs, Sustainable Development Goals. 2021. Available online: https://www.who.int/publications/i/item/9789240027053 (accessed on 13 August 2021).
26. Arnett, D.K.; Blumenthal, R.S.; Albert, M.A.; Buroker, A.B.; Goldberger, Z.D.; Hahn, E.J.; Himmelfarb, C.D.; Khera, A.; Lloyd-Jones, D.; McEvoy, J.W.; et al. 2019 ACC/AHA Guideline on the Primary Prevention of Cardiovascular Disease: A Report of the American College of Cardiology/American Heart Association Task Force on Clinical Practice Guidelines. *J. Am. Coll. Cardiol.* **2019**, *74*, e177–e232. [CrossRef]
27. Grundy, S.M.; Stone, N.J. 2018 Cholesterol Clinical Practice Guidelines: Synopsis of the 2018 American Heart Association/American College of Cardiology/Multisociety Cholesterol Guideline. *Ann. Intern. Med.* **2019**, *170*, 779–783. [CrossRef]
28. Rizzo, N.S.; Sabate, J.; Jaceldo-Siegl, K.; Fraser, G.E. Vegetarian dietary patterns are associated with a lower risk of metabolic syndrome: The adventist health study 2. *Diabetes Care* **2011**, *34*, 1225–1227. [CrossRef]
29. Matsumoto, S.; Beeson, W.L.; Shavlik, D.J.; Siapco, G.; Jaceldo-Siegl, K.; Fraser, G.; Knutsen, S.F. Association between vegetarian diets and cardiovascular risk factors in non-Hispanic white participants of the Adventist Health Study-2. *J. Nutr. Sci.* **2019**, *8*, e6. [CrossRef]
30. Bradbury, K.E.; Crowe, F.L.; Appleby, P.N.; Schmidt, J.A.; Travis, R.C.; Key, T.J. Serum concentrations of cholesterol, apolipoprotein A-I and apolipoprotein B in a total of 1694 meat-eaters, fish-eaters, vegetarians and vegans. *Eur. J. Clin. Nutr.* **2015**, *69*, 1180. [CrossRef]
31. Shah, B.; Newman, J.D.; Woolf, K.; Ganguzza, L.; Guo, Y.; Allen, N.; Zhong, J.; Fisher, E.A.; Slater, J. Anti-Inflammatory Effects of a Vegan Diet Versus the American Heart Association-Recommended Diet in Coronary Artery Disease Trial. *J. Am. Heart Assoc.* **2018**, *7*, e011367. [CrossRef]
32. Yokoyama, Y.; Nishimura, K.; Barnard, N.D.; Takegami, M.; Watanabe, M.; Sekikawa, A.; Okamura, T.; Miyamoto, Y. Vegetarian diets and blood pressure: A meta-analysis. *JAMA Intern. Med.* **2014**, *174*, 577–587. [CrossRef] [PubMed]
33. Barnard, N.D.; Alwarith, J.; Rembert, E.; Brandon, L.; Nguyen, M.; Goergen, A.; Horne, T.; do Nascimento, G.F.; Lakkadi, K.; Tura, A.; et al. A Mediterranean Diet and Low-Fat Vegan Diet to Improve Body Weight and Cardiometabolic Risk Factors: A Randomized, Cross-over Trial. *J. Am. Coll. Nutr.* **2021**, 1–13. [CrossRef]
34. Ferdowsian, H.R.; Barnard, N.D.; Hoover, V.J.; Katcher, H.I.; Levin, S.M.; Green, A.A.; Cohen, J.L. A multicomponent intervention reduces body weight and cardiovascular risk at a GEICO corporate site. *Am. J. Health Promot.* **2010**, *24*, 384–387. [CrossRef]
35. Wang, F.; Zheng, J.; Yang, B.; Jiang, J.; Fu, Y.; Li, D. Effects of Vegetarian Diets on Blood Lipids: A Systematic Review and Meta-Analysis of Randomized Controlled Trials. *J. Am. Heart Assoc.* **2015**, *4*, e002408. [CrossRef]
36. Barnard, N.D.; Levin, S.M.; Yokoyama, Y. A systematic review and meta-analysis of changes in body weight in clinical trials of vegetarian diets. *J. Acad. Nutr. Diet* **2015**, *115*, 954–969. [CrossRef] [PubMed]
37. Kahleova, H.; Dort, S.; Holubkov, R.; Barnard, N.D. A Plant-Based High-Carbohydrate, Low-Fat Diet in Overweight Individuals in a 16-Week Randomized Clinical Trial: The Role of Carbohydrates. *Nutrients* **2018**, *10*, 1302. [CrossRef]
38. Viguiliouk, E.; Kendall, C.W.; Kahleova, H.; Rahelic, D.; Salas-Salvado, J.; Choo, V.L.; Mejia, S.B.; Stewart, S.E.; Leiter, L.A.; Jenkins, D.J.; et al. Effect of vegetarian dietary patterns on cardiometabolic risk factors in diabetes: A systematic review and meta-analysis of randomized controlled trials. *Clin. Nutr.* **2019**, *38*, 1133–1145. [CrossRef] [PubMed]
39. Ornish, D.; Scherwitz, L.W.; Billings, J.H.; Brown, S.E.; Gould, K.L.; Merritt, T.A.; Sparler, S.; Armstrong, W.T.; Ports, T.A.; Kirkeeide, R.L.; et al. Intensive lifestyle changes for reversal of coronary heart disease. *JAMA* **1998**, *280*, 2001–2007. [CrossRef]
40. Esselstyn, C.B., Jr.; Gendy, G.; Doyle, J.; Golubic, M.; Roizen, M.F. A way to reverse CAD? *J. Fam. Pract.* **2014**, *63*, 356–364b.
41. Tonstad, S.; Butler, T.; Yan, R.; Fraser, G.E. Type of vegetarian diet, body weight, and prevalence of type 2 diabetes. *Diabetes Care* **2009**, *32*, 791–796. [CrossRef]

42. Crowe, F.L.; Appleby, P.N.; Travis, R.C.; Key, T.J. Risk of hospitalization or death from ischemic heart disease among British vegetarians and nonvegetarians: Results from the EPIC-Oxford cohort study. *Am. J. Clin. Nutr.* **2013**, *97*, 597–603. [CrossRef] [PubMed]
43. Tong, T.Y.N.; Appleby, P.N.; Bradbury, K.E.; Perez-Cornago, A.; Travis, R.C.; Clarke, R.; Key, T.J. Risks of ischaemic heart disease and stroke in meat eaters, fish eaters, and vegetarians diaover 18 years of follow-up: Results from the prospective EPIC-Oxford study. *BMJ* **2019**, *366*, l4897. [CrossRef]
44. Al-Shaar, L.; Satija, A.; Wang, D.D.; Rimm, E.B.; Smith-Warner, S.A.; Stampfer, M.J.; Hu, F.B.; Willett, W.C. Red meat intake and risk of coronary heart disease among US men: Prospective cohort study. *BMJ* **2020**, *371*, m4141. [CrossRef] [PubMed]
45. Huang, J.; Liao, L.M.; Weinstein, S.J.; Sinha, R.; Graubard, B.I.; Albanes, D. Association Between Plant and Animal Protein Intake and Overall and Cause-Specific Mortality. *JAMA Intern. Med.* **2020**, *180*, 1173–1184. [CrossRef]
46. Kim, H.; Caulfield, L.E.; Rebholz, C.M. Healthy Plant-Based Diets Are Associated with Lower Risk of All-Cause Mortality in US Adults. *J. Nutr.* **2018**, *148*, 624–631. [CrossRef]
47. Glenn, A.J.; Viguiliouk, E.; Seider, M.; Boucher, B.A.; Khan, T.A.; Blanco Mejia, S.; Jenkins, D.J.A.; Kahleova, H.; Rahelic, D.; Salas-Salvado, J.; et al. Relation of Vegetarian Dietary Patterns With Major Cardiovascular Outcomes: A Systematic Review and Meta-Analysis of Prospective Cohort Studies. *Front. Nutr.* **2019**, *6*, 80. [CrossRef]
48. Dinu, M.; Abbate, R.; Gensini, G.F.; Casini, A.; Sofi, F. Vegetarian, vegan diets and multiple health outcomes: A systematic review with meta-analysis of observational studies. *Crit. Rev. Food Sci. Nutr.* **2017**, *57*, 3640–3649. [CrossRef] [PubMed]
49. Tonstad, S.; Stewart, K.; Oda, K.; Batech, M.; Herring, R.P.; Fraser, G.E. Vegetarian diets and incidence of diabetes in the Adventist Health Study-2. *Nutr. Metab. Cardiovasc. Dis.* **2013**, *23*, 292–299. [CrossRef]
50. Satija, A.; Bhupathiraju, S.N.; Rimm, E.B.; Spiegelman, D.; Chiuve, S.E.; Borgi, L.; Willett, W.C.; Manson, J.E.; Sun, Q.; Hu, F.B. Plant-Based Dietary Patterns and Incidence of Type 2 Diabetes in US Men and Women: Results from Three Prospective Cohort Studies. *PLoS Med.* **2016**, *13*, e1002039. [CrossRef]
51. Koloverou, E.; Panagiotakos, D.B.; Georgousopoulou, E.N.; Grekas, A.; Christou, A.; Chatzigeorgiou, M.; Chrysohoou, C.; Tousoulis, D.; Stefanadis, C.; Pitsavos, C.; et al. Dietary Patterns and 10-year (2002-2012) Incidence of Type 2 Diabetes: Results from the ATTICA Cohort Study. *Rev. Diabet Stud.* **2016**, *13*, 246–256. [CrossRef]
52. Chen, Z.; Zuurmond, M.G.; van der Schaft, N.; Nano, J.; Wijnhoven, H.A.H.; Ikram, M.A.; Franco, O.H.; Voortman, T. Plant versus animal based diets and insulin resistance, prediabetes and type 2 diabetes: The Rotterdam Study. *Eur. J. Epidemiol.* **2018**, *33*, 883–893. [CrossRef] [PubMed]
53. Papier, K.; Appleby, P.N.; Fensom, G.K.; Knuppel, A.; Perez-Cornago, A.; Schmidt, J.A.; Tong, T.Y.N.; Key, T.J. Vegetarian diets and risk of hospitalisation or death with diabetes in British adults: Results from the EPIC-Oxford study. *Nutr. Diabetes* **2019**, *9*, 7. [CrossRef] [PubMed]
54. Lee, Y.; Park, K. Adherence to a Vegetarian Diet and Diabetes Risk: A Systematic Review and Meta-Analysis of Observational Studies. *Nutrients* **2017**, *9*, 603. [CrossRef] [PubMed]
55. Pollakova, D.; Andreadi, A.; Pacifici, F.; Della-Morte, D.; Lauro, D.; Tubili, C. The Impact of Vegan Diet in the Prevention and Treatment of Type 2 Diabetes: A Systematic Review. *Nutrients* **2021**, *13*, 2123. [CrossRef]
56. Toumpanakis, A.; Turnbull, T.; Alba-Barba, I. Effectiveness of plant-based diets in promoting well-being in the management of type 2 diabetes: A systematic review. *BMJ Open Diabetes Res. Care* **2018**, *6*, e000534. [CrossRef]
57. Yokoyama, Y.; Barnard, N.D.; Levin, S.M.; Watanabe, M. Vegetarian diets and glycemic control in diabetes: A systematic review and meta-analysis. *Cardiovasc. Diagn. Ther.* **2014**, *4*, 373–382. [CrossRef]
58. de Carvalho, G.B.; Dias-Vasconcelos, N.L.; Santos, R.K.F.; Brandao-Lima, P.N.; da Silva, D.G.; Pires, L.V. Effect of different dietary patterns on glycemic control in individuals with type 2 diabetes mellitus: A systematic review. *Crit. Rev. Food Sci. Nutr.* **2020**, *60*, 1999–2010. [CrossRef]
59. McMacken, M.; Shah, S. A plant-based diet for the prevention and treatment of type 2 diabetes. *J. Geriatr. Cardiol.* **2017**, *14*, 342–354. [CrossRef]
60. Carter, P.; Gray, L.J.; Troughton, J.; Khunti, K.; Davies, M.J. Fruit and vegetable intake and incidence of type 2 diabetes mellitus: Systematic review and meta-analysis. *BMJ* **2010**, *341*, c4229. [CrossRef]
61. Du, H.; Li, L.; Bennett, D.; Guo, Y.; Turnbull, I.; Yang, L.; Bragg, F.; Bian, Z.; Chen, Y.; Chen, J.; et al. Fresh fruit consumption in relation to incident diabetes and diabetic vascular complications: A 7-y prospective study of 0.5 million Chinese adults. *PLoS Med.* **2017**, *14*, e1002279. [CrossRef]
62. Muraki, I.; Imamura, F.; Manson, J.E.; Hu, F.B.; Willett, W.C.; van Dam, R.M.; Sun, Q. Fruit consumption and risk of type 2 diabetes: Results from three prospective longitudinal cohort studies. *BMJ* **2013**, *347*, f5001. [CrossRef] [PubMed]
63. Bondonno, N.P.; Davey, R.J.; Murray, K.; Radavelli-Bagatini, S.; Bondonno, C.P.; Blekkenhorst, L.C.; Sim, M.; Magliano, D.J.; Daly, R.M.; Shaw, J.E.; et al. Associations Between Fruit Intake and Risk of Diabetes in the AusDiab Cohort. *J. Clin. Endocrinol. Metab.* **2021**, *106*, e4097–e4108. [CrossRef]
64. Feskens, E.J.; Sluik, D.; van Woudenbergh, G.J. Meat consumption, diabetes, and its complications. *Curr. Diabetes Rep.* **2013**, *13*, 298–306. [CrossRef]
65. Kunutsor, S.K.; Apekey, T.A.; Walley, J.; Kain, K. Ferritin levels and risk of type 2 diabetes mellitus: An updated systematic review and meta-analysis of prospective evidence. *Diabetes Metab. Res. Rev.* **2013**, *29*, 308–318. [CrossRef] [PubMed]

66. Ye, J.; Yu, Q.; Mai, W.; Liang, P.; Liu, X.; Wang, Y. Dietary protein intake and subsequent risk of type 2 diabetes: A dose-response meta-analysis of prospective cohort studies. *Acta Diabetol.* **2019**, *56*, 851–870. [CrossRef]
67. Viguiliouk, E.; Stewart, S.E.; Jayalath, V.H.; Ng, A.P.; Mirrahimi, A.; de Souza, R.J.; Hanley, A.J.; Bazinet, R.P.; Blanco Mejia, S.; Leiter, L.A.; et al. Effect of Replacing Animal Protein with Plant Protein on Glycemic Control in Diabetes: A Systematic Review and Meta-Analysis of Randomized Controlled Trials. *Nutrients* **2015**, *7*, 9804–9824. [CrossRef]
68. American Diabetes Association. 8. Obesity Management for the Treatment of Type 2 Diabetes: Standards of Medical Care in Diabetes—2021. *Diabetes Care* **2021**, *44*, S100–S110. [CrossRef]
69. Clarys, P.; Deliens, T.; Huybrechts, I.; Deriemaeker, P.; Vanaelst, B.; De Keyzer, W.; Hebbelinck, M.; Mullie, P. Comparison of nutritional quality of the vegan, vegetarian, semi-vegetarian, pesco-vegetarian and omnivorous diet. *Nutrients* **2014**, *6*, 1318–1332. [CrossRef]
70. Aune, D.; Keum, N.; Giovannucci, E.; Fadnes, L.T.; Boffetta, P.; Greenwood, D.C.; Tonstad, S.; Vatten, L.J.; Riboli, E.; Norat, T. Nut consumption and risk of cardiovascular disease, total cancer, all-cause and cause-specific mortality: A systematic review and dose-response meta-analysis of prospective studies. *BMC Med.* **2016**, *14*, 207. [CrossRef]
71. Grosso, G.; Yang, J.; Marventano, S.; Micek, A.; Galvano, F.; Kales, S.N. Nut consumption on all-cause, cardiovascular, and cancer mortality risk: A systematic review and meta-analysis of epidemiologic studies. *Am. J. Clin. Nutr.* **2015**, *101*, 783–793. [CrossRef] [PubMed]
72. Aune, D.; Giovannucci, E.; Boffetta, P.; Fadnes, L.T.; Keum, N.; Norat, T.; Greenwood, D.C.; Riboli, E.; Vatten, L.J.; Tonstad, S. Fruit and vegetable intake and the risk of cardiovascular disease, total cancer and all-cause mortality-a systematic review and dose-response meta-analysis of prospective studies. *Int. J. Epidemiol.* **2017**, *46*, 1029–1056. [CrossRef]
73. Chen, G.C.; Tong, X.; Xu, J.Y.; Han, S.F.; Wan, Z.X.; Qin, J.B.; Qin, L.Q. Whole-grain intake and total, cardiovascular, and cancer mortality: A systematic review and meta-analysis of prospective studies. *Am. J. Clin. Nutr.* **2016**, *104*, 164–172. [CrossRef] [PubMed]
74. Aune, D.; De Stefani, E.; Ronco, A.; Boffetta, P.; Deneo-Pellegrini, H.; Acosta, G.; Mendilaharsu, M. Legume intake and the risk of cancer: A multisite case-control study in Uruguay. *Cancer Causes Control* **2009**, *20*, 1605–1615. [CrossRef] [PubMed]
75. Naujokat, C.; McKee, D.L. The "Big Five" Phytochemicals Targeting Cancer Stem Cells: Curcumin, EGCG, Sulforaphane, Resveratrol and Genistein. *Curr. Med. Chem.* **2021**, *28*, 4321–4342. [CrossRef]
76. Hosseini, A.; Ghorbani, A. Cancer therapy with phytochemicals: Evidence from clinical studies. *Avicenna J. Phytomed.* **2015**, *5*, 84–97.
77. Wolk, A. Potential health hazards of eating red meat. *J. Intern. Med.* **2017**, *281*, 106–122. [CrossRef] [PubMed]
78. Diallo, A.; Deschasaux, M.; Latino-Martel, P.; Hercberg, S.; Galan, P.; Fassier, P.; Allès, B.; Guéraud, F.; Pierre, F.H.; Touvier, M. Red and processed meat intake and cancer risk: Results from the prospective NutriNet-Santé cohort study. *Int. J. Cancer* **2018**, *142*, 230–237. [CrossRef]
79. Sinha, R.; Cross, A.J.; Graubard, B.I.; Leitzmann, M.F.; Schatzkin, A. Meat intake and mortality: A prospective study of over half a million people. *Arch. Intern. Med.* **2009**, *169*, 562–571. [CrossRef]
80. Tantamango-Bartley, Y.; Jaceldo-Siegl, K.; Fan, J.; Fraser, G. Vegetarian diets and the incidence of cancer in a low-risk population. *Cancer Epidemiol. Biomark. Prev.* **2013**, *22*, 286–294. [CrossRef]
81. Tantamango-Bartley, Y.; Knutsen, S.F.; Knutsen, R.; Jacobsen, B.K.; Fan, J.; Beeson, W.L.; Sabate, J.; Hadley, D.; Jaceldo-Siegl, K.; Penniecook, J.; et al. Are strict vegetarians protected against prostate cancer? *Am. J. Clin. Nutr.* **2016**, *103*, 153–160. [CrossRef]
82. Penniecook-Sawyers, J.A.; Jaceldo-Siegl, K.; Fan, J.; Beeson, L.; Knutsen, S.; Herring, P.; Fraser, G.E. Vegetarian dietary patterns and the risk of breast cancer in a low-risk population. *Br. J. Nutr.* **2016**, *115*, 1790–1797. [CrossRef] [PubMed]
83. Orlich, M.J.; Singh, P.N.; Sabate, J.; Fan, J.; Sveen, L.; Bennett, H.; Knutsen, S.F.; Beeson, W.L.; Jaceldo-Siegl, K.; Butler, T.L.; et al. Vegetarian dietary patterns and the risk of colorectal cancers. *JAMA Intern. Med.* **2015**, *175*, 767–776. [CrossRef]
84. Key, T.J.; Appleby, P.N.; Crowe, F.L.; Bradbury, K.E.; Schmidt, J.A.; Travis, R.C. Cancer in British vegetarians: Updated analyses of 4998 incident cancers in a cohort of 32,491 meat eaters, 8612 fish eaters, 18,298 vegetarians, and 2246 vegans. *Am. J. Clin. Nutr.* **2014**, *100* (Suppl. S1), 378s–385s. [CrossRef]
85. Cade, J.E.; Taylor, E.F.; Burley, V.J.; Greenwood, D.C. Common Dietary Patterns and Risk of Breast Cancer: Analysis From the United Kingdom Women's Cohort Study. *Nutr. Cancer* **2010**, *62*, 300–306. [CrossRef] [PubMed]
86. Fryar, C.D.; Carroll, M.D.; Afful, J. Prevalence of Overweight, Obesity, and Severe Obesity among Adults Aged 20 and Over: United States, 1960–1962 through 2017–2018. NCHS Health E-Stats. 2020. Available online: https://www.cdc.gov/nchs/data/hestat/obesity-adult-17-18/obesity-adult.htm (accessed on 15 August 2021).
87. World Health Organization. Obesity and Overweight. Key Facts. Available online: https://www.who.int/news-room/fact-sheets/detail/obesity-and-overweight (accessed on 29 September 2021).
88. Orlich, M.J.; Fraser, G.E. Vegetarian diets in the Adventist Health Study 2: A review of initial published findings. *Am. J. Clin. Nutr.* **2014**, *100* (Suppl. S1), 353S–358S. [CrossRef]
89. Spencer, E.A.; Appleby, P.N.; Davey, G.K.; Key, T.J. Diet and body mass index in 38000 EPIC-Oxford meat-eaters, fish-eaters, vegetarians and vegans. *Int. J. Obes. Relat. Metab. Disord.* **2003**, *27*, 728–734. [CrossRef]
90. Gomez-Donoso, C.; Martinez-Gonzalez, M.A.; Martinez, J.A.; Gea, A.; Sanz-Serrano, J.; Perez-Cueto, F.J.A.; Bes-Rastrollo, M. A Provegetarian Food Pattern Emphasizing Preference for Healthy Plant-Derived Foods Reduces the Risk of Overweight/Obesity in the SUN Cohort. *Nutrients* **2019**, *11*, 1553. [CrossRef] [PubMed]

91. Rosell, M.; Appleby, P.; Spencer, E.; Key, T. Weight gain over 5 years in 21,966 meat-eating, fish-eating, vegetarian, and vegan men and women in EPIC-Oxford. *Int. J. Obes.* **2006**, *30*, 1389–1396. [CrossRef] [PubMed]
92. Davey, G.K.; Spencer, E.A.; Appleby, P.N.; Allen, N.E.; Knox, K.H.; Key, T.J. EPIC-Oxford: Lifestyle characteristics and nutrient intakes in a cohort of 33,883 meat-eaters and 31,546 non meat-eaters in the UK. *Public Health Nutr.* **2003**, *6*, 259–269. [CrossRef]
93. Gogga, P.; Śliwińska, A.; Aleksandrowicz-Wrona, E.; Małgorzewicz, S. Association between different types of plant-based diets and leptin levels in healthy volunteers. *Acta Biochim. Pol.* **2019**, *66*, 77–82. [CrossRef]
94. Kahleova, H.; Petersen, K.F.; Shulman, G.I.; Alwarith, J.; Rembert, E.; Tura, A.; Hill, M.; Holubkov, R.; Barnard, N.D. Effect of a Low-Fat Vegan Diet on Body Weight, Insulin Sensitivity, Postprandial Metabolism, and Intramyocellular and Hepatocellular Lipid Levels in Overweight Adults: A Randomized Clinical Trial. *JAMA Netw. Open* **2020**, *3*, e2025454. [CrossRef]
95. Kahleova, H.; Hlozkova, A.; Fleeman, R.; Fletcher, K.; Holubkov, R.; Barnard, N.D. Fat Quantity and Quality, as Part of a Low-Fat, Vegan Diet, Are Associated with Changes in Body Composition, Insulin Resistance, and Insulin Secretion. A 16-Week Randomized Controlled Trial. *Nutrients* **2019**, *11*, 615. [CrossRef]
96. Barnard, N.D.; Levin, S.M.; Gloede, L.; Flores, R. Turning the Waiting Room into a Classroom: Weekly Classes Using a Vegan or a Portion-Controlled Eating Plan Improve Diabetes Control in a Randomized Translational Study. *J. Acad. Nutr. Diet.* **2018**, *118*, 1072–1079. [CrossRef]
97. Turner-McGrievy, G.M.; Davidson, C.R.; Wingard, E.E.; Wilcox, S.; Frongillo, E.A. Comparative effectiveness of plant-based diets for weight loss: A randomized controlled trial of five different diets. *Nutrition* **2015**, *31*, 350–358. [CrossRef] [PubMed]
98. Mishra, S.; Xu, J.; Agarwal, U.; Gonzales, J.; Levin, S.; Barnard, N.D. A multicenter randomized controlled trial of a plant-based nutrition program to reduce body weight and cardiovascular risk in the corporate setting: The GEICO study. *Eur. J. Clin. Nutr.* **2013**, *67*, 718–724. [CrossRef]
99. Kahleova, H.; Fleeman, R.; Hlozkova, A.; Holubkov, R.; Barnard, N.D. A plant-based diet in overweight individuals in a 16-week randomized clinical trial: Metabolic benefits of plant protein. *Nutr. Diabetes* **2018**, *8*, 58. [CrossRef] [PubMed]
100. Djekic, D.; Shi, L.; Brolin, H.; Carlsson, F.; Sarnqvist, C.; Savolainen, O.; Cao, Y.; Backhed, F.; Tremaroli, V.; Landberg, R.; et al. Effects of a Vegetarian Diet on Cardiometabolic Risk Factors, Gut Microbiota, and Plasma Metabolome in Subjects With Ischemic Heart Disease: A Randomized, Crossover Study. *J. Am. Heart Assoc.* **2020**, *9*, e016518. [CrossRef]
101. Wright, N.; Wilson, L.; Smith, M.; Duncan, B.; McHugh, P. The BROAD study: A randomised controlled trial using a whole food plant-based diet in the community for obesity, ischaemic heart disease or diabetes. *Nutr. Diabetes* **2017**, *7*, e256. [CrossRef] [PubMed]
102. Academy of Nutrition and Dietetics. Adult Weight Management Guideline—Realistic Goal Weight Setting 2014. Academy of Nutrition and Dietetics Evidence Analysis Library. Available online: https://www.andeal.org/template.cfm?template=guide_summary&key=4325 (accessed on 30 September 2021).
103. Shahavandi, M.; Djafari, F.; Shahinfar, H.; Davarzani, S.; Babaei, N.; Ebaditabar, M.; Djafarian, K.; Clark, C.C.T.; Shab-Bidar, S. The association of plant-based dietary patterns with visceral adiposity, lipid accumulation product, and triglyceride-glucose index in Iranian adults. *Complement. Ther. Med.* **2020**, *53*, 102531. [CrossRef]
104. Ratjen, I.; Morze, J.; Enderle, J.; Both, M.; Borggrefe, J.; Muller, H.P.; Kassubek, J.; Koch, M.; Lieb, W. Adherence to a plant-based diet in relation to adipose tissue volumes and liver fat content. *Am. J. Clin. Nutr.* **2020**, *112*, 354–363. [CrossRef]
105. Mayra, S.; Ugarte, N.; Johnston, C.S. Health Biomarkers in Adults Are More Closely Linked to Diet Quality Attributes Than to Plant-Based Diet Categorization. *Nutrients* **2019**, *11*, 1427. [CrossRef]
106. Sofi, F.; Dinu, M.; Pagliai, G.; Cesari, F.; Gori, A.M.; Sereni, A.; Becatti, M.; Fiorillo, C.; Marcucci, R.; Casini, A. Low-Calorie Vegetarian Versus Mediterranean Diets for Reducing Body Weight and Improving Cardiovascular Risk Profile: CARDIVEG Study (Cardiovascular Prevention With Vegetarian Diet). *Circulation* **2018**, *137*, 1103–1113. [CrossRef] [PubMed]
107. Orlich, M.J.; Jaceldo-Siegl, K.; Sabaté, J.; Fan, J.; Singh, P.N.; Fraser, G.E. Patterns of food consumption among vegetarians and non-vegetarians. *Br. J. Nutr.* **2014**, *112*, 1644–1653. [CrossRef]
108. Klementova, M.; Thieme, L.; Haluzik, M.; Pavlovicova, R.; Hill, M.; Pelikanova, T.; Kahleova, H. A Plant-Based Meal Increases Gastrointestinal Hormones and Satiety More Than an Energy- and Macronutrient-Matched Processed-Meat Meal in T2D, Obese, and Healthy Men: A Three-Group Randomized Crossover Study. *Nutrients* **2019**, *11*, 157. [CrossRef]
109. Tomova, A.; Bukovsky, I.; Rembert, E.; Yonas, W.; Alwarith, J.; Barnard, N.D.; Kahleova, H. The Effects of Vegetarian and Vegan Diets on Gut Microbiota. *Front. Nutr.* **2019**, *6*, 47. [CrossRef]
110. Weaver, C.M.; Gordon, C.M.; Janz, K.F.; Kalkwarf, H.J.; Lappe, J.M.; Lewis, R.; O'Karma, M.; Wallace, T.C.; Zemel, B.S. The National Osteoporosis Foundation's position statement on peak bone mass development and lifestyle factors: A systematic review and implementation recommendations. *Osteoporos Int.* **2016**, *27*, 1281–1386. [CrossRef]
111. Brondani, J.E.; Comim, F.V.; Flores, L.M.; Martini, L.A.; Premaor, M.O. Fruit and vegetable intake and bones: A systematic review and meta-analysis. *PLoS ONE* **2019**, *14*, e0217223. [CrossRef]
112. Benetou, V.; Orfanos, P.; Feskanich, D.; Michaëlsson, K.; Pettersson-Kymmer, U.; Eriksson, S.; Grodstein, F.; Wolk, A.; Bellavia, A.; Ahmed, L.A.; et al. Fruit and Vegetable Intake and Hip Fracture Incidence in Older Men and Women: The CHANCES Project. *J. Bone Miner. Res.* **2016**, *31*, 1743–1752. [CrossRef] [PubMed]
113. Qiu, R.; Cao, W.T.; Tian, H.Y.; He, J.; Chen, G.D.; Chen, Y.M. Greater Intake of Fruit and Vegetables Is Associated with Greater Bone Mineral Density and Lower Osteoporosis Risk in Middle-Aged and Elderly Adults. *PLoS ONE* **2017**, *12*, e0168906. [CrossRef] [PubMed]

114. Papier, K.; Tong, T.Y.; Appleby, P.N.; Bradbury, K.E.; Fensom, G.K.; Knuppel, A.; Perez-Cornago, A.; Schmidt, J.A.; Travis, R.C.; Key, T.J. Comparison of Major Protein-Source Foods and Other Food Groups in Meat-Eaters and Non-Meat-Eaters in the EPIC-Oxford Cohort. *Nutrients* **2019**, *11*, 824. [CrossRef]
115. Blekkenhorst, L.C.; Hodgson, J.M.; Lewis, J.R.; Devine, A.; Woodman, R.J.; Lim, W.H.; Wong, G.; Zhu, K.; Bondonno, C.P.; Ward, N.C.; et al. Vegetable and Fruit Intake and Fracture-Related Hospitalisations: A Prospective Study of Older Women. *Nutrients* **2017**, *9*, 511. [CrossRef]
116. Sim, M.; Blekkenhorst, L.C.; Lewis, J.R.; Bondonno, C.P.; Devine, A.; Zhu, K.; Woodman, R.J.; Prince, R.L.; Hodgson, J.M. Vegetable and fruit intake and injurious falls risk in older women: A prospective cohort study. *Br. J. Nutr.* **2018**, *120*, 925–934. [CrossRef] [PubMed]
117. Shams-White, M.M.; Chung, M.; Du, M.; Fu, Z.; Insogna, K.L.; Karlsen, M.C.; LeBoff, M.S.; Shapses, S.A.; Sackey, J.; Wallace, T.C.; et al. Dietary protein and bone health: A systematic review and meta-analysis from the National Osteoporosis Foundation. *Am. J. Clin. Nutr.* **2017**, *105*, 1528–1543. [CrossRef] [PubMed]
118. Shams-White, M.M.; Chung, M.; Fu, Z.; Insogna, K.L.; Karlsen, M.C.; LeBoff, M.S.; Shapses, S.A.; Sackey, J.; Shi, J.; Wallace, T.C.; et al. Animal versus plant protein and adult bone health: A systematic review and meta-analysis from the National Osteoporosis Foundation. *PLoS ONE* **2018**, *13*, e0192459. [CrossRef]
119. Burckhardt, P. The role of low acid load in vegetarian diet on bone health: A narrative review. *Swiss Med. Wkly.* **2016**, *146*, w14277. [CrossRef]
120. Mariotti, F.; Gardner, C.D. Dietary Protein and Amino Acids in Vegetarian Diets—A Review. *Nutrients* **2019**, *11*, 2661. [CrossRef]
121. Sobiecki, J.G.; Appleby, P.N.; Bradbury, K.E.; Key, T.J. High compliance with dietary recommendations in a cohort of meat eaters, fish eaters, vegetarians, and vegans: Results from the European Prospective Investigation into Cancer and Nutrition-Oxford study. *Nutr. Res.* **2016**, *36*, 464–477. [CrossRef]
122. Veronese, N.; Reginster, J.Y. The effects of calorie restriction, intermittent fasting and vegetarian diets on bone health. *Aging Clin. Exp. Res.* **2019**, *31*, 753–758. [CrossRef]
123. Hsu, E. Plant-based diets and bone health: Sorting through the evidence. *Curr. Opin. Endocrinol. Diabetes Obes.* **2020**, *27*, 248–252. [CrossRef]
124. Chuang, T.L.; Lin, C.H.; Wang, Y.F. Effects of vegetarian diet on bone mineral density. *Tzu Chi Med. J.* **2021**, *33*, 128–134. [CrossRef]
125. Tong, T.Y.N.; Appleby, P.N.; Armstrong, M.E.G.; Fensom, G.K.; Knuppel, A.; Papier, K.; Perez-Cornago, A.; Travis, R.C.; Key, T.J. Vegetarian and vegan diets and risks of total and site-specific fractures: Results from the prospective EPIC-Oxford study. *BMC Med.* **2020**, *18*, 353. [CrossRef]
126. Iguacel, I.; Miguel-Berges, M.L.; Gomez-Bruton, A.; Moreno, L.A.; Julian, C. Veganism, vegetarianism, bone mineral density, and fracture risk: A systematic review and meta-analysis. *Nutr. Rev.* **2019**, *77*, 1–18. [CrossRef]
127. Appleby, P.N.; Key, T.J.A. Letter: Veganism, vegetarianism, bone mineral density, and fracture risk: A systematic review and meta-analysis. *Nutr. Rev.* **2019**, *77*, 451. [CrossRef]
128. Iguacel, I.; Miguel-Berges, M.L.; Gómez-Bruton, A.; Moreno, L.A.; Julian, C. Authors' Reply: Veganism, vegetarianism, bone mineral density, and fracture risk: A systematic review and meta-analysis. *Nutr. Rev.* **2019**, *77*, 452–453. [CrossRef] [PubMed]
129. Tucker, K.L. Vegetarian diets and bone status. *Am. J. Clin. Nutr.* **2014**, *100* (Suppl. S1), 329S–335S. [CrossRef]
130. Timko, C.A.; Hormes, J.M.; Chubski, J. Will the real vegetarian please stand up? An investigation of dietary restraint and eating disorder symptoms in vegetarians versus non-vegetarians. *Appetite* **2012**, *58*, 982–990. [CrossRef]
131. Heiss, S.; Walker, D.C.; Anderson, D.A.; Morison, J.N.; Hormes, J.M. Vegetarians and omnivores with diagnosed eating disorders exhibit no difference in symptomology: A retrospective clinical chart review. *Eat. Weight Disord.* **2021**, *26*, 1007–1012. [CrossRef] [PubMed]
132. Martins, Y.; Pliner, P.; O'Connor, R. Restrained Eating Among Vegetarians: Does a Vegetarian Eating Style Mask Concerns About Weight? *Appetite* **1999**, *32*, 145–154. [CrossRef] [PubMed]
133. Heiss, S.; Coffino, J.A.; Hormes, J.M. What does the ORTO-15 measure? Assessing the construct validity of a common orthorexia nervosa questionnaire in a meat avoiding sample. *Appetite* **2019**, *135*, 93–99. [CrossRef]
134. Zickgraf, H.F.; Hazzard, V.M.; O'Connor, S.M.; Simone, M.; Williams-Kerver, G.A.; Anderson, L.M.; Lipson, S.K. Examining vegetarianism, weight motivations, and eating disorder psychopathology among college students. *Int. J. Eat. Disord.* **2020**, *53*, 1506–1514. [CrossRef]
135. Hills, R.D., Jr.; Pontefract, B.A.; Mishcon, H.R.; Black, C.A.; Sutton, S.C.; Theberge, C.R. Gut Microbiome: Profound Implications for Diet and Disease. *Nutrients* **2019**, *11*, 1613. [CrossRef] [PubMed]
136. Mills, S.; Stanton, C.; Lane, J.A.; Smith, G.J.; Ross, R.P. Precision Nutrition and the Microbiome, Part I: Current State of the Science. *Nutrients* **2019**, *11*, 923. [CrossRef]
137. Singh, R.K.; Chang, H.W.; Yan, D.; Lee, K.M.; Ucmak, D.; Wong, K.; Abrouk, M.; Farahnik, B.; Nakamura, M.; Zhu, T.H.; et al. Influence of diet on the gut microbiome and implications for human health. *J. Transl. Med.* **2017**, *15*, 73. [CrossRef]
138. DeMartino, P.; Cockburn, D.W. Resistant starch: Impact on the gut microbiome and health. *Curr. Opin. Biotechnol.* **2020**, *61*, 66–71. [CrossRef]
139. Flint, H.J.; Scott, K.P.; Duncan, S.H.; Louis, P.; Forano, E. Microbial degradation of complex carbohydrates in the gut. *Gut Microbes* **2012**, *3*, 289–306. [CrossRef] [PubMed]

140. Liu, P.; Wang, Y.; Yang, G.; Zhang, Q.; Meng, L.; Xin, Y.; Jiang, X. The role of short-chain fatty acids in intestinal barrier function, inflammation, oxidative stress, and colonic carcinogenesis. *Pharmacol. Res.* **2021**, *165*, 105420. [CrossRef] [PubMed]
141. Fu, X.; Liu, Z.; Zhu, C.; Mou, H.; Kong, Q. Nondigestible carbohydrates, butyrate, and butyrate-producing bacteria. *Crit. Rev. Food Sci. Nutr.* **2019**, *59*, S130–S152. [CrossRef]
142. Martínez, I.; Lattimer, J.M.; Hubach, K.L.; Case, J.A.; Yang, J.; Weber, C.G.; Louk, J.A.; Rose, D.J.; Kyureghian, G.; Peterson, D.A.; et al. Gut microbiome composition is linked to whole grain-induced immunological improvements. *ISME J.* **2013**, *7*, 269–280. [CrossRef]
143. Ferrocino, I.; Di Cagno, R.; De Angelis, M.; Turroni, S.; Vannini, L.; Bancalari, E.; Rantsiou, K.; Cardinali, G.; Neviani, E.; Cocolin, L. Fecal Microbiota in Healthy Subjects Following Omnivore, Vegetarian and Vegan Diets: Culturable Populations and rRNA DGGE Profiling. *PLoS ONE* **2015**, *10*, e0128669. [CrossRef]
144. Matijasic, B.B.; Obermajer, T.; Lipoglavsek, L.; Grabnar, I.; Avgustin, G.; Rogelj, I. Association of dietary type with fecal microbiota in vegetarians and omnivores in Slovenia. *Eur. J. Nutr.* **2014**, *53*, 1051–1064. [CrossRef]
145. O'Keefe, S.J.; Li, J.V.; Lahti, L.; Ou, J.; Carbonero, F.; Mohammed, K.; Posma, J.M.; Kinross, J.; Wahl, E.; Ruder, E.; et al. Fat, fibre and cancer risk in African Americans and rural Africans. *Nat. Commun.* **2015**, *6*, 6342. [CrossRef]
146. So, D.; Whelan, K.; Rossi, M.; Morrison, M.; Holtmann, G.; Kelly, J.T.; Shanahan, E.R.; Staudacher, H.M.; Campbell, K.L. Dietary fiber intervention on gut microbiota composition in healthy adults: A systematic review and meta-analysis. *Am. J. Clin. Nutr.* **2018**, *107*, 965–983. [CrossRef]
147. Cuevas-Tena, M.; Alegría, A.; Lagarda, M.J. Relationship Between Dietary Sterols and Gut Microbiota: A Review. *Eur. J. Lipid Sci. Technol.* **2018**, *120*, 1800054. [CrossRef]
148. Dingeo, G.; Brito, A.; Samouda, H.; Iddir, M.; La Frano, M.R.; Bohn, T. Phytochemicals as modifiers of gut microbial communities. *Food Funct.* **2020**, *11*, 8444–8471. [CrossRef] [PubMed]
149. Wiese, M.; Bashmakov, Y.; Chalyk, N.; Nielsen, D.S.; Krych, L.; Kot, W.; Klochkov, V.; Pristensky, D.; Bandaletova, T.; Chernyshova, M.; et al. Prebiotic Effect of Lycopene and Dark Chocolate on Gut Microbiome with Systemic Changes in Liver Metabolism, Skeletal Muscles and Skin in Moderately Obese Persons. *Biomed. Res. Int.* **2019**, *2019*, 4625279. [CrossRef] [PubMed]
150. Corrêa, T.A.F.; Rogero, M.M.; Hassimotto, N.M.A.; Lajolo, F.M. The Two-Way Polyphenols-Microbiota Interactions and Their Effects on Obesity and Related Metabolic Diseases. *Front. Nutr.* **2019**, *6*, 188. [CrossRef]
151. Azad, M.A.K.; Sarker, M.; Li, T.; Yin, J. Probiotic Species in the Modulation of Gut Microbiota: An Overview. *Biomed. Res. Int.* **2018**, *2018*, 9478630. [CrossRef]
152. Muralidharan, J.; Galie, S.; Hernandez-Alonso, P.; Bullo, M.; Salas-Salvado, J. Plant-Based Fat, Dietary Patterns Rich in Vegetable Fat and Gut Microbiota Modulation. *Front. Nutr.* **2019**, *6*, 157. [CrossRef]
153. David, L.A.; Maurice, C.F.; Carmody, R.N.; Gootenberg, D.B.; Button, J.E.; Wolfe, B.E.; Ling, A.V.; Devlin, A.S.; Varma, Y.; Fischbach, M.A.; et al. Diet rapidly and reproducibly alters the human gut microbiome. *Nature* **2014**, *505*, 559–563. [CrossRef]
154. Klimenko, N.S.; Tyakht, A.V.; Popenko, A.S.; Vasiliev, A.S.; Altukhov, I.A.; Ischenko, D.S.; Shashkova, T.I.; Efimova, D.A.; Nikogosov, D.A.; Osipenko, D.A.; et al. Microbiome Responses to an Uncontrolled Short-Term Diet Intervention in the Frame of the Citizen Science Project. *Nutrients* **2018**, *10*, 576. [CrossRef]
155. Lin, A.; Bik, E.M.; Costello, E.K.; Dethlefsen, L.; Haque, R.; Relman, D.A.; Singh, U. Distinct distal gut microbiome diversity and composition in healthy children from Bangladesh and the United States. *PLoS ONE* **2013**, *8*, e53838. [CrossRef] [PubMed]
156. Ruengsomwong, S.; La-Ongkham, O.; Jiang, J.; Wannissorn, B.; Nakayama, J.; Nitisinprasert, S. Microbial Community of Healthy Thai Vegetarians and Non-Vegetarians, Their Core Gut Microbiota, and Pathogen Risk. *J. Microbiol. Biotechnol.* **2016**, *26*, 1723–1735. [CrossRef]
157. Zimmer, J.; Lange, B.; Frick, J.S.; Sauer, H.; Zimmermann, K.; Schwiertz, A.; Rusch, K.; Klosterhalfen, S.; Enck, P. A vegan or vegetarian diet substantially alters the human colonic faecal microbiota. *Eur. J. Clin. Nutr.* **2012**, *66*, 53–60. [CrossRef]
158. Wu, G.D.; Chen, J.; Hoffmann, C.; Bittinger, K.; Chen, Y.Y.; Keilbaugh, S.A.; Bewtra, M.; Knights, D.; Walters, W.A.; Knight, R.; et al. Linking long-term dietary patterns with gut microbial enterotypes. *Science* **2011**, *334*, 105–108. [CrossRef]
159. Franco-de-Moraes, A.C.; de Almeida-Pititto, B.; da Rocha Fernandes, G.; Gomes, E.P.; da Costa Pereira, A.; Ferreira, S.R.G. Worse inflammatory profile in omnivores than in vegetarians associates with the gut microbiota composition. *Diabetol. Metab. Syndr.* **2017**, *9*, 62. [CrossRef]
160. De Filippo, C.; Di Paola, M.; Ramazzotti, M.; Albanese, D.; Pieraccini, G.; Banci, E.; Miglietta, F.; Cavalieri, D.; Lionetti, P. Diet, Environments, and Gut Microbiota. A Preliminary Investigation in Children Living in Rural and Urban Burkina Faso and Italy. *Front. Microbiol.* **2017**, *8*, 1979. [CrossRef] [PubMed]
161. Obregon-Tito, A.J.; Tito, R.Y.; Metcalf, J.; Sankaranarayanan, K.; Clemente, J.C.; Ursell, L.K.; Zech Xu, Z.; Van Treuren, W.; Knight, R.; Gaffney, P.M.; et al. Subsistence strategies in traditional societies distinguish gut microbiomes. *Nat. Commun.* **2015**, *6*, 6505. [CrossRef] [PubMed]
162. Schnorr, S.L.; Candela, M.; Rampelli, S.; Centanni, M.; Consolandi, C.; Basaglia, G.; Turroni, S.; Biagi, E.; Peano, C.; Severgnini, M.; et al. Gut microbiome of the Hadza hunter-gatherers. *Nat. Commun.* **2014**, *5*, 3654. [CrossRef]
163. Fava, F.; Gitau, R.; Griffin, B.A.; Gibson, G.R.; Tuohy, K.M.; Lovegrove, J.A. The type and quantity of dietary fat and carbohydrate alter faecal microbiome and short-chain fatty acid excretion in a metabolic syndrome 'at-risk' population. *Int. J. Obes.* **2013**, *37*, 216–223. [CrossRef]

164. Losasso, C.; Eckert, E.M.; Mastrorilli, E.; Villiger, J.; Mancin, M.; Patuzzi, I.; Di Cesare, A.; Cibin, V.; Barrucci, F.; Pernthaler, J.; et al. Assessing the Influence of Vegan, Vegetarian and Omnivore Oriented Westernized Dietary Styles on Human Gut Microbiota: A Cross Sectional Study. *Front. Microbiol.* **2018**, *9*, 317. [CrossRef] [PubMed]
165. Abulizi, N.; Quin, C.; Brown, K.; Chan, Y.K.; Gill, S.K.; Gibson, D.L. Gut Mucosal Proteins and Bacteriome Are Shaped by the Saturation Index of Dietary Lipids. *Nutrients* **2019**, *11*, 418. [CrossRef]
166. De Wit, N.; Derrien, M.; Bosch-Vermeulen, H.; Oosterink, E.; Keshtkar, S.; Duval, C.; van der Meer, R. Saturated fat stimulates obesity and hepatic steatosis and affects gut microbiota composition by an enhanced overflow of dietary fat to the distal intestine. *Am. J. Physiol. Gastrointest. Liver Physiol.* **2012**, *303*, G589–G599. [CrossRef]
167. Lecomte, V.; Kaakoush, N.O.; Maloney, C.A.; Raipuria, M.; Huinao, K.D.; Mitchell, H.M.; Morris, M.J. Changes in gut microbiota in rats fed a high fat diet correlate with obesity-associated metabolic parameters. *PLoS ONE* **2015**, *10*, e0126931. [CrossRef] [PubMed]
168. Wolters, M.; Ahrens, J.; Romaní-Pérez, M.; Watkins, C.; Sanz, Y.; Benítez-Páez, A.; Stanton, C.; Günther, K. Dietary fat, the gut microbiota, and metabolic health—A systematic review conducted within the MyNewGut project. *Clin. Nutr.* **2019**, *38*, 2504–2520. [CrossRef] [PubMed]
169. Orlich, M.J.; Siapco, G.; Jung, S. 24—Vegetarian Diets and the Microbiome. In *Vegetarian and Plant-Based Diets in Health and Disease Prevention*; Mariotti, F., Ed.; Academic Press: London, UK, 2017; pp. 429–461.
170. Koliada, A.; Syzenko, G.; Moseiko, V.; Budovska, L.; Puchkov, K.; Perederiy, V.; Gavalko, Y.; Dorofeyev, A.; Romanenko, M.; Tkach, S.; et al. Association between body mass index and Firmicutes/Bacteroidetes ratio in an adult Ukrainian population. *BMC Microbiol.* **2017**, *17*, 120. [CrossRef] [PubMed]
171. Magne, F.; Gotteland, M.; Gauthier, L.; Zazueta, A.; Pesoa, S.; Navarrete, P.; Balamurugan, R. The Firmicutes/Bacteroidetes Ratio: A Relevant Marker of Gut Dysbiosis in Obese Patients? *Nutrients* **2020**, *12*, 1474. [CrossRef]
172. Wu, G.D.; Compher, C.; Chen, E.Z.; Smith, S.A.; Shah, R.D.; Bittinger, K.; Chehoud, C.; Albenberg, L.G.; Nessel, L.; Gilroy, E.; et al. Comparative metabolomics in vegans and omnivores reveal constraints on diet-dependent gut microbiota metabolite production. *Gut* **2016**, *65*, 63–72. [CrossRef] [PubMed]
173. Orlich, M.; Zhou, L.; Vogtmann, E.; Wan, Y.; Hua, X.; Shi, J.; Siapco, G.; Knight, R.; Fraser, G.; Sinha, R. *Vegetarian Diets and Fecal and Oral Microbiome in Adventist Health Study—2*; Loma Linda University: Loma Linda, CA, USA, in press.
174. Miles, F.L.; Orlich, M.J.; Mashchak, A.; Chandler, P.D.; Duerksen-Hughes, P.; Fraser, G.E. *The Biology of Veganism: Plasma Metabolomics Profiles Reveals Distinct Profiles of Vegans and Non-Vegetarians in the Adventist Health Study-2 Cohort*; Loma Linda University: Loma Linda, CA, USA, in press.
175. Schmidt, J.A.; Rinaldi, S.; Ferrari, P.; Carayol, M.; Achaintre, D.; Scalbert, A.; Cross, A.J.; Gunter, M.J.; Fensom, G.K.; Appleby, P.N.; et al. Metabolic profiles of male meat eaters, fish eaters, vegetarians, and vegans from the EPIC-Oxford cohort. *Am. J. Clin. Nutr.* **2015**, *102*, 1518–1526. [CrossRef] [PubMed]
176. Wang, N.C.; Bagheri, M.; Olszewski, T.; Friese, K.A.; Smith, H.M.; Robles, M.E.; Wang, C.; Brooks, A.; Bordenstein, S.R.; Ferguson, J.F.; et al. New-onset vegetarian diet shows differences in fatty acid metabolites in European American and African American women. *Nutr. Metab. Cardiovasc. Dis.* **2021**, *31*, 2436–2448. [CrossRef]
177. Tang, W.H.W.; Wang, Z.; Kennedy, D.J.; Wu, Y.; Buffa, J.A.; Agatisa-Boyle, B.; Li, X.S.; Levison, B.S.; Hazen, S.L. Gut Microbiota-Dependent Trimethylamine N-oxide (TMAO) Pathway Contributes to Both Development of Renal Insufficiency and Mortality Risk in Chronic Kidney Disease. *Circ. Res.* **2015**, *116*, 448–455. [CrossRef] [PubMed]
178. Tang, W.H.; Wang, Z.; Levison, B.S.; Koeth, R.A.; Britt, E.B.; Fu, X.; Wu, Y.; Hazen, S.L. Intestinal microbial metabolism of phosphatidylcholine and cardiovascular risk. *NEJM* **2013**, *368*, 1575–1584. [CrossRef] [PubMed]
179. Fu, B.C.; Hullar, M.A.J.; Randolph, T.W.; Franke, A.A.; Monroe, K.R.; Cheng, I.; Wilkens, L.R.; Shepherd, J.A.; Madeleine, M.M.; Le Marchand, L.; et al. Associations of plasma trimethylamine N-oxide, choline, carnitine, and betaine with inflammatory and cardiometabolic risk biomarkers and the fecal microbiome in the Multiethnic Cohort Adiposity Phenotype Study. *Am. J. Clin. Nutr.* **2020**, *111*, 1226–1234. [CrossRef]
180. Koeth, R.A.; Wang, Z.; Levison, B.S.; Buffa, J.A.; Org, E.; Sheehy, B.T.; Britt, E.B.; Fu, X.; Wu, Y.; Li, L.; et al. Intestinal microbiota metabolism of L-carnitine, a nutrient in red meat, promotes atherosclerosis. *Nat. Med.* **2013**, *19*, 576–585. [CrossRef] [PubMed]
181. Desai, M.S.; Seekatz, A.M.; Koropatkin, N.M.; Kamada, N.; Hickey, C.A.; Wolter, M.; Pudlo, N.A.; Kitamoto, S.; Terrapon, N.; Muller, A.; et al. A Dietary Fiber-Deprived Gut Microbiota Degrades the Colonic Mucus Barrier and Enhances Pathogen Susceptibility. *Cell* **2016**, *167*, 1339–1353. [CrossRef]
182. Toor, D.; Wsson, M.K.; Kumar, P.; Karthikeyan, G.; Kaushik, N.K.; Goel, C.; Singh, S.; Kumar, A.; Prakash, H. Dysbiosis Disrupts Gut Immune Homeostasis and Promotes Gastric Diseases. *Int. J. Mol. Sci.* **2019**, *20*, 2432. [CrossRef] [PubMed]
183. Sebastiani, G.; Herranz Barbero, A.; Borras-Novell, C.; Alsina Casanova, M.; Aldecoa-Bilbao, V.; Andreu-Fernandez, V.; Pascual-Tutusaus, M.; Ferrero Martinez, S.; Gomez Roig, M.D.; Garcia-Algar, O. The Effects of Vegetarian and Vegan Diet during Pregnancy on the Health of Mothers and Offspring. *Nutrients* **2019**, *11*, 557. [CrossRef]
184. Agnoli, C.; Baroni, L.; Bertini, I.; Ciappellano, S.; Fabbri, A.; Papa, M.; Pellegrini, N.; Sbarbati, R.; Scarino, M.L.; Siani, V.; et al. Position paper on vegetarian diets from the working group of the Italian Society of Human Nutrition. *Nutr. Metab. Cardiovasc. Dis.* **2017**, *27*, 1037–1052. [CrossRef]
185. Piccoli, G.B.; Clari, R.; Vigotti, F.N.; Leone, F.; Attini, R.; Cabiddu, G.; Mauro, G.; Castelluccia, N.; Colombi, N.; Capizzi, I.; et al. Vegan-vegetarian diets in pregnancy: Danger or panacea? A systematic narrative review. *BJOG* **2015**, *122*, 623–633. [CrossRef]

186. Zulyniak, M.A.; de Souza, R.J.; Shaikh, M.; Desai, D.; Lefebvre, D.L.; Gupta, M.; Wilson, J.; Wahi, G.; Subbarao, P.; Becker, A.B.; et al. Does the impact of a plant-based diet during pregnancy on birth weight differ by ethnicity? A dietary pattern analysis from a prospective Canadian birth cohort alliance. *BMJ Open* **2017**, *7*, e017753. [CrossRef]
187. Tan, C.; Zhao, Y.; Wang, S. Is a vegetarian diet safe to follow during pregnancy? A systematic review and meta-analysis of observational studies. *Crit. Rev. Food Sci. Nutr.* **2019**, *59*, 2586–2596. [CrossRef] [PubMed]
188. Kesary, Y.; Avital, K.; Hiersch, L. Maternal plant-based diet during gestation and pregnancy outcomes. *Arch. Gynecol. Obstet.* **2020**, *302*, 887–898. [CrossRef]
189. Yisahak, S.F.; Hinkle, S.N.; Mumford, S.L.; Li, M.; Andriessen, V.C.; Grantz, K.L.; Zhang, C.; Grewal, J. Vegetarian diets during pregnancy, and maternal and neonatal outcomes. *Int. J. Epidemiol.* **2021**, *50*, 165–178. [CrossRef]
190. Avnon, T.; Paz Dubinsky, E.; Lavie, I.; Ben-Mayor Bashi, T.; Anbar, R.; Yogev, Y. The impact of a vegan diet on pregnancy outcomes. *J. Perinatol.* **2021**, *41*, 1129–1133. [CrossRef]
191. Karcz, K.; Krolak-Olejnik, B. Vegan or vegetarian diet and breast milk composition—A systematic review. *Crit. Rev. Food Sci. Nutr.* **2021**, *61*, 1081–1098. [CrossRef] [PubMed]
192. Rizzo, N.S.; Jaceldo-Siegl, K.; Sabate, J.; Fraser, G.E. Nutrient profiles of vegetarian and nonvegetarian dietary patterns. *J. Acad. Nutr. Diet.* **2013**, *113*, 1610–1619. [CrossRef]
193. Koebnick, C.; Heins, U.A.; Hoffmann, I.; Dagnelie, P.C.; Leitzmann, C. Folate status during pregnancy in women is improved by long-term high vegetable intake compared with the average western diet. *J. Nutr.* **2001**, *131*, 733–739. [CrossRef] [PubMed]
194. Raghavan, R.; Dreibelbis, C.; Kingshipp, B.L.; Wong, Y.P.; Abrams, B.; Gernand, A.D.; Rasmussen, K.M.; Siega-Riz, A.M.; Stang, J.; Casavale, K.O.; et al. Dietary patterns before and during pregnancy and maternal outcomes: A systematic review. *Am. J. Clin. Nutr.* **2019**, *109*, 705s–728s. [CrossRef]
195. Mangels, R.; Messina, V.; Messina, M. *The Dietitian's Guide to Vegetarian Diets*, 4th ed.; Jones & Bartlett: Burlington, MA, USA, 2021.
196. Institute of Medicine (US) Panel on Micronutrients. *Dietary Reference Intakes for Vitamin A, Vitamin K, Arsenic, Boron, Chromium, Copper, Iodine, Iron, Manganese, Molybdenum, Nickel, Silicon, Vanadium, and Zinc*; National Academies Press: Washington, WA, USA, 2001.
197. Centers for Disease Control and Prevention. Recommendations to prevent and control iron deficiency in the United States. *MMWR Recomm. Rep.* **1998**, *47*, 1–29.
198. World Health Organization. WHO Recommendations on Antenatal care for a Positive Pregnancy Experience. 2016. Available online: https://www.who.int/publications/i/item/9789241549912 (accessed on 11 October 2021).
199. King, J.C. Determinants of maternal zinc status during pregnancy. *Am. J. Clin. Nutr.* **2000**, *71*, 1334S–1343S. [CrossRef]
200. Hambidge, K.M.; Miller, L.V.; Mazariegos, M.; Westcott, J.; Solomons, N.W.; Raboy, V.; Kemp, J.F.; Das, A.; Goco, N.; Hartwell, T.; et al. Upregulation of Zinc Absorption Matches Increases in Physiologic Requirements for Zinc in Women Consuming High- or Moderate-Phytate Diets during Late Pregnancy and Early Lactation. *J. Nutr.* **2017**, *147*, 1079–1085. [CrossRef]
201. Stagnaro-Green, A.; Abalovich, M.; Alexander, E.; Azizi, F.; Mestman, J.; Negro, R.; Nixon, A.; Pearce, E.N.; Soldin, O.P.; Sullivan, S.; et al. Guidelines of the American Thyroid Association for the diagnosis and management of thyroid disease during pregnancy and postpartum. *Thyroid* **2011**, *21*, 1081–1125. [CrossRef]
202. Lakin, V.; Haggarty, P.; Abramovich, D.R.; Ashton, J.; Moffat, C.F.; McNeill, G.; Danielian, P.J.; Grubb, D. Dietary intake and tissue concentration of fatty acids in omnivore, vegetarian and diabetic pregnancy. *Prostaglandins Leukot Essent Fat. Acids* **1998**, *59*, 209–220. [CrossRef]
203. Perrin, M.T.; Pawlak, R.; Dean, L.L.; Christis, A.; Friend, L. A cross-sectional study of fatty acids and brain-derived neurotrophic factor (BDNF) in human milk from lactating women following vegan, vegetarian, and omnivore diets. *Eur. J. Nutr.* **2019**, *58*, 2401–2410. [CrossRef]
204. Middleton, P.; Gomersall, J.C.; Gould, J.F.; Shepherd, E.; Olsen, S.F.; Makrides, M. Omega-3 fatty acid addition during pregnancy. *Cochrane Database Syst. Rev.* **2018**, *11*, CD003402. [CrossRef]
205. Kar, S.; Wong, M.; Rogozinska, E.; Thangaratinam, S. Effects of omega-3 fatty acids in prevention of early preterm delivery: A systematic review and meta-analysis of randomized studies. *Eur. J. Obstet. Gynecol. Reprod. Biol.* **2016**, *198*, 40–46. [CrossRef] [PubMed]
206. Dror, D.K.; Allen, L.H. Effect of vitamin B12 deficiency on neurodevelopment in infants: Current knowledge and possible mechanisms. *Nutr. Rev.* **2008**, *66*, 250–255. [CrossRef] [PubMed]
207. American Academy of Pediatrics Committee on Nutrition. Nutritional aspects of vegetarian diets. In *Pediatric Nutrition*, 8th ed.; Kleinman, R.E., Greer, F.R., Eds.; American Academy of Pediatrics: Itasca, IL, USA, 2020; pp. 293–319.
208. Amit, M. Vegetarian diets in children and adolescents. *Paediatr. Child. Health* **2010**, *15*, 303–314.
209. Sabate, J.; Wien, M. Vegetarian diets and childhood obesity prevention. *Am. J. Clin. Nutr.* **2010**, *91*, 1525S–1529S. [CrossRef]
210. Yen, C.E.; Yen, C.H.; Huang, M.C.; Cheng, C.H.; Huang, Y.C. Dietary intake and nutritional status of vegetarian and omnivorous preschool children and their parents in Taiwan. *Nutr. Res.* **2008**, *28*, 430–436. [CrossRef]
211. Larsson, C.L.; Johansson, G.K. Dietary intake and nutritional status of young vegans and omnivores in Sweden. *Am. J. Clin. Nutr.* **2002**, *76*, 100–106. [CrossRef]
212. Macknin, M.; Kong, T.; Weier, A.; Worley, S.; Tang, A.S.; Alkhouri, N.; Golubic, M. Plant-based, no-added-fat or American Heart Association diets: Impact on cardiovascular risk in obese children with hypercholesterolemia and their parents. *J. Pediatr.* **2015**, *166*, 953–959. [CrossRef] [PubMed]

213. Gartner, L.M.; Morton, J.; Lawrence, R.A.; Naylor, A.J.; O'Hare, D.; Schanler, R.J.; Eidelman, A.I. Breastfeeding and the use of human milk. *Pediatrics* **2012**, *129*, e827–e841. [CrossRef]
214. Gibson, R.S.; Heath, A.L.; Szymlek-Gay, E.A. Is iron and zinc nutrition a concern for vegetarian infants and young children in industrialized countries? *Am. J. Clin. Nutr.* **2014**, *100* (Suppl. S1), 459s–468s. [CrossRef] [PubMed]
215. Institute of Medicine Committee to Review Dietary Reference Intakes for Vitamin D and Calcium. The National Academies Collection: Reports funded by National Institutes of Health. In *Dietary Reference Intakes for Calcium and Vitamin D*; Ross, A.C., Taylor, C.L., Yaktine, A.L., Del Valle, H.B., Eds.; National Academy of Sciences: Washington, WA, USA, 2011.
216. Institute of Medicine (US) Standing Committee on the Scientific Evaluation of Dietary Reference Intakes and its Panel on Folate, Other B Vitamins, and Choline. Dietary Reference Intakes for Thiamin, Riboflavin, Niacin, Vitamin B6, Folate, Vitamin B12, Pantothenic Acid, Biotin, and Choline. In *The National Academies Collection: Reports Funded by National Institutes of Health*; Ross, A.C., Taylor, C.L., Yaktine, A.L., Del Valle, H.B., Eds.; National Academy of Sciences: Washington, WA, USA, 2011.
217. Deutz, N.E.; Bauer, J.M.; Barazzoni, R.; Biolo, G.; Boirie, Y.; Bosy-Westphal, A.; Cederholm, T.; Cruz-Jentoft, A.; Krznaric, Z.; Nair, K.S.; et al. Protein intake and exercise for optimal muscle function with aging: Recommendations from the ESPEN Expert Group. *Clin. Nutr.* **2014**, *33*, 929–936. [CrossRef]
218. MacLaughlin, J.; Holick, M.F. Aging decreases the capacity of human skin to produce vitamin D3. *J. Clin. Investig.* **1985**, *76*, 1536–1538. [CrossRef]
219. Tsai, K.S.; Heath, H., 3rd; Kumar, R.; Riggs, B.L. Impaired vitamin D metabolism with aging in women. Possible role in pathogenesis of senile osteoporosis. *J. Clin. Investig.* **1984**, *73*, 1668–1672. [CrossRef]
220. Allen, L.H. How common is vitamin B-12 deficiency? *Am. J. Clin. Nutr.* **2009**, *89*, 693s–696s. [CrossRef]
221. Thomas, D.T.; Erdman, K.A.; Burke, L.M. American College of Sports Medicine Joint Position Statement. Nutrition and Athletic Performance. *Med. Sci. Sports Exerc.* **2016**, *48*, 543–568. [CrossRef]
222. Melina, V.; Craig, W.; Levin, S. Position of the Academy of Nutrition and Dietetics: Vegetarian Diets. *J. Acad. Nutr. Diet.* **2016**, *116*, 1970–1980. [CrossRef]
223. Nieman, D.C. Vegetarian dietary practices and endurance performance. *Am. J. Clin. Nutr.* **1988**, *48*, 754–761. [CrossRef] [PubMed]
224. Nieman, D.C. Physical fitness and vegetarian diets: Is there a relation? *Am. J. Clin. Nutr.* **1999**, *70*, 570s–575s. [CrossRef]
225. Craddock, J.C.; Probst, Y.C.; Peoples, G.E. Vegetarian and Omnivorous Nutrition—Comparing Physical Performance. *Int. J. Sport Nutr. Exerc. Metab.* **2016**, *26*, 212–220. [CrossRef]
226. Pinckaers, P.J.M.; Trommelen, J.; Snijders, T.; van Loon, L.J.C. The Anabolic Response to Plant-Based Protein Ingestion. *Sports Med.* **2021**, *51* (Suppl. 1), 59–74. [CrossRef] [PubMed]
227. Pinckaers, P.J.M.; Kouw, I.W.K.; Hendriks, F.K.; van Kranenburg, J.M.X.; de Groot, L.; Verdijk, L.B.; Snijders, T.; van Loon, L.J.C. No differences in muscle protein synthesis rates following ingestion of wheat protein, milk protein, and their protein blend in healthy, young males. *Br. J. Nutr.* **2021**, *126*, 1832–1842. [CrossRef]
228. Wilkinson, S.B.; Tarnopolsky, M.A.; Macdonald, M.J.; Macdonald, J.R.; Armstrong, D.; Phillips, S.M. Consumption of fluid skim milk promotes greater muscle protein accretion after resistance exercise than does consumption of an isonitrogenous and isoenergetic soy-protein beverage. *Am. J. Clin. Nutr.* **2007**, *85*, 1031–1040. [CrossRef] [PubMed]
229. Abou Sawan, S.; van Vliet, S.; West, D.W.D.; Beals, J.W.; Paluska, S.A.; Burd, N.A.; Moore, D.R. Whole egg, but not egg white, ingestion induces mTOR colocalization with the lysosome after resistance exercise. *Am. J. Physiol. Cell Physiol.* **2018**, *315*, C537–C543. [CrossRef]
230. Hallberg, L.; Hulthén, L. Prediction of dietary iron absorption: An algorithm for calculating absorption and bioavailability of dietary iron. *Am. J. Clin. Nutr.* **2000**, *71*, 1147–1160. [CrossRef] [PubMed]
231. Medeiros, D.M. *Dietary Reference Intakes: The Essential Guide to Nutrient Requirements*; Otten, J.J., Hellwig, J.P., Meyers, L.D., Eds.; The National Academies Press: Washington, WA, USA, 2006.
232. Saunders, A.V.; Craig, W.J.; Baines, S.K.; Posen, J.S. Iron and vegetarian diets. *Med. J. Aust.* **2013**, *199*, S11–S16. [CrossRef]
233. Cialdella-Kam, L.; Kulpins, D.; Manore, M.M. Vegetarian, Gluten-Free, and Energy Restricted Diets in Female Athletes. *Sports* **2016**, *4*, 50. [CrossRef]
234. Thomas, D.T.; Erdman, K.A.; Burke, L.M. Position of the Academy of Nutrition and Dietetics, Dietitians of Canada, and the American College of Sports Medicine: Nutrition and Athletic Performance. *J. Acad. Nutr. Diet.* **2016**, *116*, 501–528. [CrossRef]
235. Larson-Meyer, D.E.; Willis, K.S. Vitamin D and athletes. *Curr. Sports. Med. Rep.* **2010**, *9*, 220–226. [CrossRef]
236. He, C.S.; Aw Yong, X.H.; Walsh, N.P.; Gleeson, M. Is there an optimal vitamin D status for immunity in athletes and military personnel? *Exerc. Immunol. Rev.* **2016**, *22*, 42–64. [PubMed]
237. Owens, D.J.; Allison, R.; Close, G.L. Vitamin D and the Athlete: Current Perspectives and New Challenges. *Sports Med.* **2018**, *48*, 3–16. [CrossRef]
238. Yoon, S.; Kwon, O.; Kim, J. Vitamin D in athletes: Focus on physical performance and musculoskeletal injuries. *Phys. Act. Nutr.* **2021**, *25*, 20–25. [CrossRef] [PubMed]
239. Delanghe, J.; De Slypere, J.P.; De Buyzere, M.; Robbrecht, J.; Wieme, R.; Vermeulen, A. Normal reference values for creatine, creatinine, and carnitine are lower in vegetarians. *Clin. Chem.* **1989**, *35*, 1802–1803. [CrossRef] [PubMed]
240. Shomrat, A.; Weinstein, Y.; Katz, A. Effect of creatine feeding on maximal exercise performance in vegetarians. *Eur. J. Appl. Physiol.* **2000**, *82*, 321–325. [CrossRef]

241. Lukaszuk, J.M.; Robertson, R.J.; Arch, J.E.; Moore, G.E.; Yaw, K.M.; Kelley, D.E.; Rubin, J.T.; Moyna, N.M. Effect of creatine supplementation and a lacto-ovo-vegetarian diet on muscle creatine concentration. *Int. J. Sport Nutr. Exerc. Metab.* **2002**, *12*, 336–348. [CrossRef]
242. Burke, D.G.; Chilibeck, P.D.; Parise, G.; Candow, D.G.; Mahoney, D.; Tarnopolsky, M. Effect of creatine and weight training on muscle creatine and performance in vegetarians. *Med. Sci. Sports Exerc.* **2003**, *35*, 1946–1955. [CrossRef]
243. Peeling, P.; Binnie, M.J.; Goods, P.S.R.; Sim, M.; Burke, L.M. Evidence-Based Supplements for the Enhancement of Athletic Performance. *Int. J. Sport Nutr. Exerc. Metab.* **2018**, *28*, 178–187. [CrossRef] [PubMed]
244. Larson-Meyer, D.E.; Ruscigno, M. Plant-Based Sports Nutrition: Expert Fueling Strategies for Training, Recovery, and Performance. *Med. Sci. Sports Exerc.* **2020**, *52*, 1639. [CrossRef]
245. Gropper, S.S.; Smith, J.L.; Carr, T.P. *Advanced Nutrition and Human Metabolism*, 7th ed.; Cengage Learning: Boston, MA, USA, 2017.
246. Craig, W.J. Nutrition concerns and health effects of vegetarian diets. *Nutr. Clin. Pract.* **2010**, *25*, 613–620. [CrossRef]
247. Dawson-Hughes, B.; Harris, S.; Kramich, C.; Dallal, G.; Rasmussen, H.M. Calcium retention and hormone levels in black and white women on high—And low-calcium diets. *J. Bone Miner. Res.* **1993**, *8*, 779–787. [CrossRef]
248. Appleby, P.; Roddam, A.; Allen, N.; Key, T. Comparative fracture risk in vegetarians and nonvegetarians in EPIC-Oxford. *Eur. J. Clin. Nutr.* **2007**, *61*, 1400–1406. [CrossRef] [PubMed]
249. Weaver, C.M.; Proulx, W.R.; Heaney, R. Choices for achieving adequate dietary calcium with a vegetarian diet. *Am. J. Clin. Nutr.* **1999**, *70*, 543s–548s. [CrossRef]
250. Heaney, R.P.; Weaver, C.M.; Fitzsimmons, M.L. Soybean phytate content: Effect on calcium absorption. *Am. J. Clin. Nutr.* **1991**, *53*, 745–747. [CrossRef]
251. Zhao, Y.; Martin, B.R.; Weaver, C.M. Calcium bioavailability of calcium carbonate fortified soymilk is equivalent to cow's milk in young women. *J. Nutr.* **2005**, *135*, 2379–2382. [CrossRef] [PubMed]
252. Chai, W.; Liebman, M. Effect of different cooking methods on vegetable oxalate content. *J. Agric. Food Chem.* **2005**, *53*, 3027–3030. [CrossRef]
253. Sahni, S.; Hannan, M.T.; Blumberg, J.; Cupples, L.A.; Kiel, D.P.; Tucker, K.L. Protective effect of total carotenoid and lycopene intake on the risk of hip fracture: A 17-year follow-up from the Framingham Osteoporosis Study. *J. Bone Miner. Res.* **2009**, *24*, 1086–1094. [CrossRef] [PubMed]
254. Wattanapenpaiboon, N.; Lukito, W.; Wahlqvist, M.L.; Strauss, B.J.G. Dietary carotenoid intake as a predictor of bone mineral density. *Asia Pac. J. Clin. Nutr.* **2003**, *12*, 467–473. [PubMed]
255. Sugiura, M.; Nakamura, M.; Ogawa, K.; Ikoma, Y.; Ando, F.; Yano, M. Bone mineral density in post-menopausal female subjects is associated with serum antioxidant carotenoids. *Osteoporos Int.* **2008**, *19*, 211–219. [CrossRef]
256. Zhang, Z.Q.; He, L.P.; Liu, Y.H.; Liu, J.; Su, Y.X.; Chen, Y.M. Association between dietary intake of flavonoid and bone mineral density in middle aged and elderly Chinese women and men. *Osteoporos Int.* **2014**, *25*, 2417–2425. [CrossRef] [PubMed]
257. Ball, D.; Maughan, R.J. Blood and Urine Acid-Base Status of Premenopausal Omnivorous and Vegetarian Women. *Br. J. Nutr.* **1997**, *78*, 683–693. [CrossRef]
258. Haider, L.M.; Schwingshackl, L.; Hoffmann, G.; Ekmekcioglu, C. The effect of vegetarian diets on iron status in adults: A systematic review and meta-analysis. *Crit. Rev. Food Sci. Nutr.* **2018**, *58*, 1359–1374. [CrossRef]
259. Craig, W.J. Iron status of vegetarians. *Am. J. Clin. Nutr.* **1994**, *59*, 1233S–1237S. [CrossRef] [PubMed]
260. Obeid, R.; Geisel, J.; Schorr, H.; Hübner, U.; Herrmann, W. The impact of vegetarianism on some haematological parameters. *Eur. J. Haematol.* **2002**, *69*, 275–279. [CrossRef]
261. Monsen, E.R. Iron nutrition and absorption: Dietary factors which impact iron bioavailability. *J. Am. Diet. Assoc.* **1988**, *88*, 786–790. [CrossRef]
262. Hallberg, L. Bioavailability of dietary iron in man. *Annu. Rev. Nutr.* **1981**, *1*, 123–147. [CrossRef]
263. Gillooly, M.; Bothwell, T.H.; Torrance, J.D.; MacPhail, A.P.; Derman, D.P.; Bezwoda, W.R.; Mills, W.; Charlton, R.W.; Mayet, F. The effects of organic acids, phytates and polyphenols on the absorption of iron from vegetables. *Br. J. Nutr.* **1983**, *49*, 331–342. [CrossRef] [PubMed]
264. Fidler, M.C.; Davidsson, L.; Zeder, C.; Hurrell, R.F. Erythorbic acid is a potent enhancer of nonheme-iron absorption. *Am. J. Clin. Nutr.* **2004**, *79*, 99–102. [CrossRef]
265. Davidsson, L. Approaches to improve iron bioavailability from complementary foods. *J. Nutr.* **2003**, *133*, 1560s–1562s. [CrossRef]
266. Calder, P.C. Scientific Advisory Committee on Nutrition (SACN). In *Iron and Health*; The Stationery Office: Norwich, UK, 2010.
267. Armah, S.M.; Carriquiry, A.; Sullivan, D.; Cook, J.D.; Reddy, M.B. A complete diet-based algorithm for predicting nonheme iron absorption in adults. *J. Nutr.* **2013**, *143*, 1136–1140. [CrossRef]
268. Collings, R.; Harvey, L.J.; Hooper, L.; Hurst, R.; Brown, T.J.; Ansett, J.; King, M.; Fairweather-Tait, S.J. The absorption of iron from whole diets: A systematic review. *Am. J. Clin. Nutr.* **2013**, *98*, 65–81. [CrossRef] [PubMed]
269. White, D.L.; Collinson, A. Red meat, dietary heme iron, and risk of type 2 diabetes: The involvement of advanced lipoxidation endproducts. *Adv. Nutr.* **2013**, *4*, 403–411. [CrossRef]
270. Qi, L.; van Dam, R.M.; Rexrode, K.; Hu, F.B. Heme iron from diet as a risk factor for coronary heart disease in women with type 2 diabetes. *Diabetes Care* **2007**, *30*, 101–106. [CrossRef]

271. Cross, A.J.; Ferrucci, L.M.; Risch, A.; Graubard, B.I.; Ward, M.H.; Park, Y.; Hollenbeck, A.R.; Schatzkin, A.; Sinha, R. A Large Prospective Study of Meat Consumption and Colorectal Cancer Risk: An Investigation of Potential Mechanisms Underlying this Association. *Cancer Res.* **2010**, *70*, 2406–2414. [CrossRef]
272. Agarwal, U. Rethinking Red Meat as a Prevention Strategy for Iron Deficiency. *Infant Child. Adolesc. Nutr.* **2013**, *5*, 231–235. [CrossRef]
273. Park, S.K.; Ryoo, J.H.; Kim, M.G.; Shin, J.Y. Association of serum ferritin and the development of metabolic syndrome in middle-aged Korean men: A 5-year follow-up study. *Diabetes Care* **2012**, *35*, 2521–2526. [CrossRef]
274. Hunt, J.R.; Roughead, Z.K. Nonheme-iron absorption, fecal ferritin excretion, and blood indexes of iron status in women consuming controlled lactoovovegetarian diets for 8 week. *Am. J. Clin. Nutr.* **1999**, *69*, 944–952. [CrossRef]
275. King, J.C. Zinc: An essential but elusive nutrient. *Am. J. Clin. Nutr.* **2011**, *94*, 679S–684S. [CrossRef] [PubMed]
276. Huang, L.; Drake, V.J.; Ho, E. Zinc. *Adv. Nutr.* **2015**, *6*, 224–226. [CrossRef] [PubMed]
277. Hunt, J.R. Bioavailability of iron, zinc, and other trace minerals from vegetarian diets. *Am. J. Clin. Nutr.* **2003**, *78*, 633s–639s. [CrossRef] [PubMed]
278. Foster, M.; Chu, A.; Petocz, P.; Samman, S. Effect of vegetarian diets on zinc status: A systematic review and meta-analysis of studies in humans. *J. Sci. Food Agric.* **2013**, *93*, 2362–2371. [CrossRef]
279. Foster, M.; Samman, S. Vegetarian diets across the lifecycle: Impact on zinc intake and status. *Adv. Food Nutr. Res.* **2015**, *74*, 93–131. [CrossRef]
280. Krebs, N.F.; Hambidge, K.M. Zinc requirements and zinc intakes of breast-fed infants. *Am. J. Clin. Nutr.* **1986**, *43*, 288–292. [CrossRef]
281. Gibson, R.S.; Perlas, L.; Hotz, C. Improving the bioavailability of nutrients in plant foods at the household level. *Proc. Nutr. Soc.* **2006**, *65*, 160–168. [CrossRef]
282. Lönnerdal, B. Dietary factors influencing zinc absorption. *J. Nutr.* **2000**, *130*, 1378s–1383s. [CrossRef]
283. Saunders, A.V.; Craig, W.J.; Baines, S.K. Zinc and vegetarian diets. *Med. J. Aust.* **2013**, *199*, S17–S21. [CrossRef]
284. National Institute of Health. ODS. Iodine. Available online: https://ods.od.nih.gov/factsheets/Iodine-Consumer/ (accessed on 26 September 2021).
285. Teas, J.; Pino, S.; Critchley, A.; Braverman, L.E. Variability of iodine content in common commercially available edible seaweeds. *Thyroid* **2004**, *14*, 836–841. [CrossRef] [PubMed]
286. Messina, M.; Redmond, G. Effects of soy protein and soybean isoflavones on thyroid function in healthy adults and hypothyroid patients: A review of the relevant literature. *Thyroid* **2006**, *16*, 249–258. [CrossRef]
287. Groufh-Jacobsen, S.; Hess, S.Y.; Aakre, I.; Folven Gjengedal, E.L.; Blandhoel Pettersen, K.; Henjum, S. Vegans, Vegetarians and Pescatarians Are at Risk of Iodine Deficiency in Norway. *Nutrients* **2020**, *12*, 3555. [CrossRef] [PubMed]
288. Leung, A.M.; Lamar, A.; He, X.; Braverman, L.E.; Pearce, E.N. Iodine status and thyroid function of Boston-area vegetarians and vegans. *J. Clin. Endocrinol. Metab.* **2011**, *96*, E1303–E1307. [CrossRef]
289. National Institutes of Health. Office of Dietary Supplements. Vitamin B12. Available online: https://ods.od.nih.gov/factsheets/VitaminB12-HealthProfessional/ (accessed on 11 November 2021).
290. Rizzo, G.; Laganà, A.S.; Rapisarda, A.M.; La Ferrera, G.M.; Buscema, M.; Rossetti, P.; Nigro, A.; Muscia, V.; Valenti, G.; Sapia, F.; et al. Vitamin B12 among Vegetarians: Status, Assessment and Supplementation. *Nutrients* **2016**, *8*, 767. [CrossRef] [PubMed]
291. Gilsing, A.M.; Crowe, F.L.; Lloyd-Wright, Z.; Sanders, T.A.; Appleby, P.N.; Allen, N.E.; Key, T.J. Serum concentrations of vitamin B12 and folate in British male omnivores, vegetarians and vegans: Results from a cross-sectional analysis of the EPIC-Oxford cohort study. *Eur. J. Clin. Nutr.* **2010**, *64*, 933–939. [CrossRef]
292. Watanabe, F.; Yabuta, Y.; Bito, T.; Teng, F. Vitamin B_{12}-containing plant food sources for vegetarians. *Nutrients* **2014**, *6*, 1861–1873. [CrossRef] [PubMed]
293. Wheatley, C. The return of the Scarlet Pimpernel: Cobalamin in inflammation II—Cobalamins can both selectively promote all three nitric oxide synthases (NOS), particularly iNOS and eNOS, and, as needed, selectively inhibit iNOS and nNOS. *J. Nutr. Environ. Med.* **2007**, *16*, 181–211. [CrossRef]
294. Obeid, R.; Fedosov, S.N.; Nexo, E. Cobalamin coenzyme forms are not likely to be superior to cyano- and hydroxyl-cobalamin in prevention or treatment of cobalamin deficiency. *Mol. Nutr. Food Res.* **2015**, *59*, 1364–1372. [CrossRef]
295. Paul, C.; Brady, D.M. Comparative Bioavailability and Utilization of Particular Forms of B(12) Supplements With Potential to Mitigate B(12)-related Genetic Polymorphisms. *Integr. Med.* **2017**, *16*, 42–49.
296. Chatthanawaree, W. Biomarkers of cobalamin (vitamin B12) deficiency and its application. *J. Nutr. Health Aging* **2011**, *15*, 227–231. [CrossRef] [PubMed]
297. Beck, W.S. Neuropsychiatric consequences of cobalamin deficiency. *Adv. Intern. Med.* **1991**, *36*, 33–56.
298. National Institutes of Health. Office of Dietary Supplements. Vitamin D. Available online: https://ods.od.nih.gov/factsheets/VitaminD-HealthProfessional/ (accessed on 12 November 2021).
299. Pludowski, P.; Holick, M.F.; Grant, W.B.; Konstantynowicz, J.; Mascarenhas, M.R.; Haq, A.; Povoroznyuk, V.; Balatska, N.; Barbosa, A.P.; Karonova, T.; et al. Vitamin D supplementation guidelines. *J. Steroid Biochem. Mol. Biol.* **2018**, *175*, 125–135. [CrossRef]
300. Chan, J.; Jaceldo-Siegl, K.; Fraser, G.E. Serum 25-hydroxyvitamin D status of vegetarians, partial vegetarians, and nonvegetarians: The Adventist Health Study-2. *Am. J. Clin. Nutr.* **2009**, *89*, 1686S–1692S. [CrossRef] [PubMed]

301. Crowe, F.L.; Steur, M.; Allen, N.E.; Appleby, P.N.; Travis, R.C.; Key, T.J. Plasma concentrations of 25-hydroxyvitamin D in meat eaters, fish eaters, vegetarians and vegans: Results from the EPIC–Oxford study. *Public Health Nutr.* **2011**, *14*, 340–346. [CrossRef] [PubMed]
302. Outila, T.A.; Kärkkäinen, M.U.; Seppänen, R.H.; Lamberg-Allardt, C.J. Dietary intake of vitamin D in premenopausal, healthy vegans was insufficient to maintain concentrations of serum 25-hydroxyvitamin D and intact parathyroid hormone within normal ranges during the winter in Finland. *J. Am. Diet. Assoc.* **2000**, *100*, 434–441. [CrossRef]
303. Elorinne, A.-L.; Alfthan, G.; Erlund, I.; Kivimäki, H.; Paju, A.; Salminen, I.; Turpeinen, U.; Voutilainen, S.; Laakso, J. Food and Nutrient Intake and Nutritional Status of Finnish Vegans and Non-Vegetarians. *PLoS ONE* **2016**, *11*, e0148235. [CrossRef] [PubMed]
304. Kamweru, P.K.; Tindibale, E.L. Vitamin D and Vitamin D from Ultraviolet-Irradiated Mushrooms (Review). *Int. J. Med. Mushrooms* **2016**, *18*, 205–214. [CrossRef]
305. Holick, M.F.; Biancuzzo, R.M.; Chen, T.C.; Klein, E.K.; Young, A.; Bibuld, D.; Reitz, R.; Salameh, W.; Ameri, A.; Tannenbaum, A.D. Vitamin D2 is as effective as vitamin D3 in maintaining circulating concentrations of 25-hydroxyvitamin D. *J. Clin. Endocrinol. Metab.* **2008**, *93*, 677–681. [CrossRef]
306. Tripkovic, L.; Lambert, H.; Hart, K.; Smith, C.P.; Bucca, G.; Penson, S.; Chope, G.; Hyppönen, E.; Berry, J.; Vieth, R.; et al. Comparison of vitamin D2 and vitamin D3 supplementation in raising serum 25-hydroxyvitamin D status: A systematic review and meta-analysis. *Am. J. Clin. Nutr.* **2012**, *95*, 1357–1364. [CrossRef]
307. Santos, H.O.; Price, J.C.; Bueno, A.A. Beyond Fish Oil Supplementation: The Effects of Alternative Plant Sources of Omega-3 Polyunsaturated Fatty Acids upon Lipid Indexes and Cardiometabolic Biomarkers—An Overview. *Nutrients* **2020**, *12*, 3159. [CrossRef]
308. Lane, K.; Derbyshire, E.; Li, W.; Brennan, C. Bioavailability and Potential Uses of Vegetarian Sources of Omega-3 Fatty Acids: A Review of the Literature. *Crit. Rev. Food Sci. Nutr.* **2014**, *54*, 572–579. [CrossRef]
309. Burns-Whitmore, B.; Froyen, E.; Heskey, C.; Parker, T.; San Pablo, G. Alpha-Linolenic and Linoleic Fatty Acids in the Vegan Diet: Do They Require Dietary Reference Intake/Adequate Intake Special Consideration? *Nutrients* **2019**, *11*, 2365. [CrossRef]
310. Saunders, A.V.; Davis, B.C.; Garg, M.L. Omega-3 polyunsaturated fatty acids and vegetarian diets. *Med. J. Aust.* **2013**, *199*, S22–S26. [CrossRef]
311. Welch, A.A.; Shakya-Shrestha, S.; Lentjes, M.A.; Wareham, N.J.; Khaw, K.T. Dietary intake and status of n-3 polyunsaturated fatty acids in a population of fish-eating and non-fish-eating meat-eaters, vegetarians, and vegans and the product-precursor ratio [corrected] of α-linolenic acid to long-chain n-3 polyunsaturated fatty acids: Results from the EPIC-Norfolk cohort. *Am. J. Clin. Nutr.* **2010**, *92*, 1040–1051. [CrossRef]
312. Rocha, J.P.; Laster, J.; Parag, B.; Shah, N.U. Multiple Health Benefits and Minimal Risks Associated with Vegetarian Diets. *Curr. Nutr. Rep.* **2019**, *8*, 374–381. [CrossRef] [PubMed]
313. Sanders, T.A. Plant compared with marine n-3 fatty acid effects on cardiovascular risk factors and outcomes: What is the verdict? *Am. J. Clin. Nutr.* **2014**, *100* (Suppl. S1), 453s–458s. [CrossRef] [PubMed]
314. Metcalf, R.G.; James, M.J.; Gibson, R.A.; Edwards, J.R.; Stubberfield, J.; Stuklis, R.; Roberts-Thomson, K.; Young, G.D.; Cleland, L.G. Effects of fish-oil supplementation on myocardial fatty acids in humans. *Am. J. Clin. Nutr.* **2007**, *85*, 1222–1228. [CrossRef] [PubMed]
315. Arterburn, L.M.; Oken, H.A.; Hoffman, J.P.; Bailey-Hall, E.; Chung, G.; Rom, D.; Hamersley, J.; McCarthy, D. Bioequivalence of Docosahexaenoic acid from different algal oils in capsules and in a DHA-fortified food. *Lipids* **2007**, *42*, 1011–1024. [CrossRef]
316. Rousseau, G. Microbiota, a New Playground for the Omega-3 Polyunsaturated Fatty Acids in Cardiovascular Diseases. *Mar. Drugs* **2021**, *19*, 54. [CrossRef]
317. Fu, Y.; Wang, Y.; Gao, H.; Li, D.; Jiang, R.; Ge, L.; Tong, C.; Xu, K. Associations among Dietary Omega-3 Polyunsaturated Fatty Acids, the Gut Microbiota, and Intestinal Immunity. *Mediat. Inflamm.* **2021**, *2021*, 8879227. [CrossRef]
318. Hutchinson, A.N.; Tingö, L.; Brummer, R.J. The Potential Effects of Probiotics and ω-3 Fatty Acids on Chronic Low-Grade Inflammation. *Nutrients* **2020**, *12*, 2402. [CrossRef]
319. Davis, B.C.; Kris-Etherton, P.M. Achieving optimal essential fatty acid status in vegetarians: Current knowledge and practical implications. *Am. J. Clin. Nutr.* **2003**, *78*, 640s–646s. [CrossRef]
320. European Food Safety Authority. Labelling reference intake values for n-3 and n-6 polyunsaturated fatty acids. *EFSA J.* **2009**, *7*, 1176. [CrossRef]
321. National Institutes of Health. Omega-3 Fatty Acids. Available online: https://ods.od.nih.gov/factsheets/Omega3FattyAcids-HealthProfessional/ (accessed on 26 September 2021).
322. Lane, K.E.; Wilson, M.; Hellon, T.G.; Davies, I.G. Bioavailability and conversion of plant based sources of omega-3 fatty acids—A scoping review to update supplementation options for vegetarians and vegans. *Crit. Rev. Food Sci. Nutr.* **2021**, 1–16. [CrossRef] [PubMed]
323. Carlson, S.E.; Colombo, J.; Gajewski, B.J.; Gustafson, K.M.; Mundy, D.; Yeast, J.; Georgieff, M.K.; Markley, L.A.; Kerling, E.H.; Shaddy, D.J. DHA supplementation and pregnancy outcomes. *Am. J. Clin. Nutr.* **2013**, *97*, 808–815. [CrossRef] [PubMed]
324. Pistollato, F.; Sumalla Cano, S.; Elio, I.; Masias Vergara, M.; Giampieri, F.; Battino, M. Plant-Based and Plant-Rich Diet Patterns during Gestation: Beneficial Effects and Possible Shortcomings. *Adv. Nutr.* **2015**, *6*, 581–591. [CrossRef] [PubMed]

325. Sanders, T.A. Essential fatty acid requirements of vegetarians in pregnancy, lactation, and infancy. *Am. J. Clin. Nutr.* **1999**, *70*, 555s–559s. [CrossRef] [PubMed]
326. Young, V.R.; Pellett, P.L. Plant proteins in relation to human protein and amino acid nutrition. *Am. J. Clin. Nutr.* **1994**, *59*, 1203S–1212S. [CrossRef]
327. Katz, D.L.; Doughty, K.N.; Geagan, K.; Jenkins, D.A.; Gardner, C.D. Perspective: The Public Health Case for Modernizing the Definition of Protein Quality. *Adv. Nutr.* **2019**, *10*, 755–764. [CrossRef]
328. Metges, C.C.; Barth, C.A. Metabolic Consequences of a High Dietary-Protein Intake in Adulthood: Assessment of the Available Evidence. *J. Nutr.* **2000**, *130*, 886–889. [CrossRef]
329. Mittendorfer, B.; Klein, S.; Fontana, L. A word of caution against excessive protein intake. *Nat. Reviews. Endocrinol.* **2020**, *16*, 59–66. [CrossRef]
330. Craig, W.J. *Vegetarian Nutrition and Wellness*; CRC Press: Boca Raton, FL, USA; Taylor and Francis Group: Boca Raton, FL, USA, 2018.
331. The Vegetarian Resource Group. Available online: https://www.vrg.org/ (accessed on 4 October 2021).
332. Deng, F.; Zhang, A.; Chan, C. Acculturation, Dietary Acceptability, and Diabetes Management among Chinese in North America. *Front. Endocrinol.* **2013**, *4*, 108. [CrossRef]

nutrients

MDPI

Article

Designing Nutritionally Adequate and Climate-Friendly Diets for Omnivorous, Pescatarian, Vegetarian and Vegan Adolescents in Sweden Using Linear Optimization

Patricia Eustachio Colombo [1,*], Liselotte Schäfer Elinder [1,2], Anna Karin Lindroos [3,4] and Alexandr Parlesak [5,6]

1 Department of Global Public Health, Karolinska Institutet, 113 65 Stockholm, Sweden; liselotte.schafer-elinder@sll.se
2 Centre for Epidemiology and Social Medicine, Region Stockholm, 113 65 Stockholm, Sweden
3 The Swedish Food Agency, 751 26 Uppsala, Sweden; annakarin.lindroos@slv.se
4 Department of Internal Medicine and Clinical Nutrition, The Sahlgrenska Academy, University of Gothenburg, 405 30 Gothenburg, Sweden
5 Baden-Wuerttemberg Cooperative State University, 74076 Heilbronn, Germany; alexandr.parlesak@heilbronn.dhbw.de
6 Department of Nutrition, Exercise and Sports, University of Copenhagen, 1165 Copenhagen, Denmark
* Correspondence: patricia.eustachio.colombo@ki.se

Abstract: Low-carbon diets can counteract climate change and promote health if they are nutritionally adequate, affordable and culturally acceptable. This study aimed at developing sustainable diets and to compare these with the EAT-Lancet diet. The Swedish national dietary survey Riksmaten Adolescents 2016–2017 was used as the baseline. Diets were optimized using linear programming for four dietary patterns: omnivores, pescatarians, vegetarians and vegans. The deviation from the baseline Riksmaten diet was minimized for all optimized diets while fulfilling nutrient and climate footprint constraints. Constraining the diet-related carbon dioxide equivalents of omnivores to 1.57 kg/day resulted in a diet associated with a reduction of meat, dairy products, and processed foods and an increase in potatoes, pulses, eggs and seafood. Climate-friendly, nutritionally adequate diets for pescatarians, vegetarians and vegans contained fewer foods and included considerable amounts of fortified dairy and meat substitutes. The optimized diets did not align very well with the food-group pattern of the EAT-Lancet diet. These findings suggest how to design future diets that are climate-friendly, nutritionally adequate, affordable, and culturally acceptable for Swedish adolescents with different dietary patterns. The discrepancies with the EAT diet indicate that the cultural dietary context is likely to play an important role in characterizing sustainable diets for specific populations.

Keywords: planetary health; Paris agreement; linear programming; nutrition; greenhouse gas emission; alternative diets; sustainability

Citation: Eustachio Colombo, P.; Elinder, L.S.; Lindroos, A.K.; Parlesak, A. Designing Nutritionally Adequate and Climate-Friendly Diets for Omnivorous, Pescatarian, Vegetarian and Vegan Adolescents in Sweden Using Linear Optimization. *Nutrients* **2021**, *13*, 2507. https://doi.org/10.3390/nu13082507

Academic Editors: Winston Craig and Ujué Fresán

Received: 9 June 2021
Accepted: 20 July 2021
Published: 22 July 2021

1. Introduction

All regions around the world are facing severe consequences of global warming [1], resulting in adverse effects on human health and the economy [2]. So far, more than 95% of parties to the United Nations Framework Convention on Climate Change (UNFCCC) have ratified the Paris Agreement, which commits governments to pursue actions to keep the increase in global average temperatures below 1.5 °C above pre-industrial levels and thus prevent dramatic climate change [3]. To reach this goal, environmental, social, and economic aspects of sustainability have to be considered. In the aftermath of the ratification of the Paris Agreement in October 2016 [4], Sweden adopted a climate policy framework [4] with the long-term goal of becoming a net-zero carbon economy by 2045 [4].

Food production contributes globally to about 25–30% of all anthropogenic greenhouse gas emissions (GHGE), through altered land use, storage, transport, packaging, processing, retail, and preparation for consumption [1]. Hence, successful transition into a society that produces less GHGE requires changes at all levels of the food chain. In a market economy, consumer demand is one of the most relevant ways to achieve these changes [5]. Such changes would require a shift towards more plant-based diets, which are generally less GHGE intensive [6–10]. As in other countries [11,12], the motivation to switch to environmentally friendly diets is more pronounced in younger than in older people in Sweden. According to the Swedish Youth Barometer, about a third of all young people are currently consuming more plant-based diets for environmental reasons [13]. However, guidance is needed that can guarantee nutritional adequacy when initiating major dietary changes.

Promoting diets that omit entire food groups such as vegan diets can lead to nutritional deficiencies such as inadequate intakes of calcium, vitamin B12, vitamin D and iron [14], as well as a too-high intake of sugar [14,15]. The choice of foods to replace meat has also been shown to be questionable from a climate perspective as, on a per calorie basis, the substitution of meat products with increased fruit and vegetables can result in higher or similar environmental impacts [6–8]. Thus, consumers who want to change their diet to be more climate friendly, yet nutritionally adequate, face a challenge when having to combine foods to meet all these demands. Average dietary intakes of Swedish adolescents are far from meeting the dietary guidelines that aim at preventing chronic disease [16,17]. Therefore, any suggestions on future sustainable diets for adolescents need to consider health-promoting aspects at the same time [18].

A frequently suggested approach to reduce the environmental impact from food is to avoid specific food categories such as meat (pescatarian diet), meat and fish (vegetarian diet), or any animal product (vegan diet), as these diets are associated with lower GHGE [19]. However, deficiencies in in the supply of some nutrients may affect the nutritional status of vegetarians and vegans negatively [20,21]. In 2019, the EAT-Lancet Commission suggested a healthy reference diet, based on studies of dietary patterns and health outcomes, that also had been evaluated against different environmental aspects [5]. The authors of the report called on all countries to make national adaptations to this generic diet. However, this diet neither has been fully controlled for nutritional adequacy, nor for specific cultural acceptability or affordability.

A comprehensive way to fulfil a broad range of criteria simultaneously is by optimization analysis through linear programming (LP) [22]. Using this methodology, diets that are nutritionally adequate, while at the same being reduced in GHGE and limited in cost, can be developed [22,23]. Additionally, this methodology has been shown to be successful for meeting cultural acceptability by minimizing the deviation from reported dietary patterns of the population [22,24–26].

The aim of the present study was to apply LP in designing nutritionally adequate and culturally acceptable diets with significantly reduced GHGE based on the current diet of adolescents in Sweden. We optimized the diet for four patterns, which varied based on their inclusion of animal foods (omnivores, pescatarians, vegetarians and vegan). The optimized diets were set to meet the maximum tolerable diet-related GHGE limit defined to keep the increase in global average temperatures below 1.5 °C above pre-industrial levels, as calculated by the World Wildlife Fund based on targets of the Intergovernmental Panel on Climate Change (IPCC) [27]. We also compared the optimized diets to the proposed EAT-Lancet diet [5].

2. Materials and Methods

2.1. Design and Dietary Data

This was a modeling study using linear programming to design nutritionally adequate and climate-friendly diets for omnivorous, pescatarian, vegetarian and vegan adolescents in Sweden. Dietary data were derived from the national dietary survey Riksmaten Adolescents 2016–2017, which is a school-based cross-sectional dietary survey of 3099 pupils from 130 schools including grades 5, 8 and 11 [28]. Consumed foods and their amounts were recorded using a validated, web-based 24-h recall method (RiksmatenFlex) on two non-consecutive days with the option to choose from 778 foods, of which 725 foods were recorded at least once [29]. The sample consisted of 55% girls and the participants were evenly distributed between the three grades: 34% pupils were between 10 and 11 years old in grade 5, 34% pupils were between 14 and 15 years in grade 8, and 32% pupils were between 17 and 18 years in grade 11. A more detailed description of the survey, methodology, data acquisition and evaluation can be found elsewhere [28].

2.2. Intake of Energy and Nutrients

Energy and nutrient intakes of the edible parts of foods as eaten (e.g., cooked rice) were automatically calculated through linkage with the Swedish Food Agency's Food composition database version "Riksmaten Adolescents 2016–2017". Added sugars are defined as all refined sugars added to foods during cooking and manufacturing, not including honey and unsweetened fruit juices (NNR 2012, EFSA) [30].

For optimization purposes, the reported intake of each food item (g/day) was standardized to 2410 kcal, i.e., the estimated energy requirement for a reference pupil/child as indicated in the Nordic Nutrition Recommendations 2012 [31]. The energy requirement was weighted according to the different age and sex groups in the study sample (see Table S1 and Section 2.7 for more details). The energy-proportional shares of each food for the reference pupil were calculated for modeling purposes and represented the pupils' baseline food consumption. The reference energy intake 2410 kcal was also used as the pre-set daily energy constraint of all optimized diets.

2.3. Cost of Foods

The price of each food was searched for through the webpage "Matpriskollen" [32], which compares the prices of foods among twelve of Sweden's largest food retailers. Based on the different available prices for a food item (including low budget, conventional and organic varieties), an average price was calculated for each food item.

2.4. Greenhouse Gas Emissions (GHGE) of Foods

The carbon dioxide equivalents (CO_2eq) of foods were obtained from the Climate Database from Research Institutes of Sweden (RISE) [33], which is linked to the Swedish Food Agency's Food composition database. It contains 2129 foods and reflects typical Swedish food supply/purchasing patterns. The Climate database builds on life cycle analyses [34,35], covering the GHGE of food production from resource extraction (cradle) to the factory gate. It contains values for carbon dioxide (CO_2), methane (CH_4), and nitrous oxide (N_2O) that have been weighted in line with their respective global-warming potential over a 100-year period using factors recommended by the IPCC [36]. The combined emissions from the greenhouse gases from each food item yields a single value measured as kg of CO_2eq per kg of food. We used the CO_2eq-values which corresponded to the environmental impact of a food in its edible (e.g., boiled pasta) form.

2.5. Grouping of Foods

For analytical and descriptive purposes, foods were grouped into 22 food categories, based on the categorizations used in the RISE Climate Database (Bread; Cereals, other (including, e.g., pasta and rice); Nuts and seeds; Fruits and berries (including smoothies); Potatoes; Vegetables (e.g., tomatoes, cucumber, lettuce, bell pepper, carrots and a few vegetable-based dishes); Pulses (beans, lentils, peas and chickpeas); Meat substitutes; Dairy substitutes; Dairy, other (e.g., milk); Dairy, solid (including cheese, curd and yoghurt); Eggs; Pasta and rice dishes with meat/fish (e.g., composite dishes like pasta Bolognese); Poultry; Red/processed meat (e.g., beef, pork, including offal and meat-based dishes); Seafood (including fish, mussels and crabs); Oils; Fats, solid (e.g., butter, margarine); Drinks w/o milk; Sugar and sweets (including chocolate); Seasonings and sauces, and; Other (e.g., seeds, salt, sugar, jams).

The baseline and optimized diets were also re-grouped in order to be comparable to the EAT-Lancet Commission's food categorization used in the published report (Figure 1) [5]. This report applied the following categories: Whole grains (rice, wheat, corn and other); Tubers or starchy vegetables (e.g., potatoes); Vegetables; Fruits; Dairy foods (whole milk or equivalents, including butter); Beef, lamb and pork; Chicken and other poultry; Eggs; Fish; Legumes; Nuts; Added fats (unsaturated oils and saturated oils); and Added sugars.

2.6. Linear Optimization

Linear programming (LP) has successfully been applied to optimize goal determinants of diets while considering complex patterns of different constraints [22,37]. Briefly, LP is the application of an algorithm for maximizing or minimizing a given linear objective function (the variable to be optimized) subjected to a set of linear constraints (conditions to be met) on a list of decision variables (amount of each food item) [38]. A solution is found when all conditions can be met. If conditions are too strictly chosen, no solution is possible. Constraints that set the limit for the objective function's ability of being minimized or maximized (e.g., those being met by exactly 100% with regards to its applied limit) are called "active constraints" [39]. Linear optimization was performed with the CBC (COIN-OR Branch and Cut) Solver algorithm, which is part of the Excel® 2016 software add-in OpenSolver, V. 2.9.0 [40].

2.7. Nutritional Adequacy of Optimized Diets

Dietary reference values (DRVs) based on the Nordic Nutrition Recommendations 2012 [31], covering the nutritional needs of 97.5% of the population, were used as obligatory constraints for all solutions provided (Table S1). These constraints comprised the daily estimated energy requirements (EER), the recommended intake ranges for macronutrients, and the recommended intakes (RIs) for micronutrients [31]. The upper level for the salt intake was set to 6 g/day and the minimum value of fiber intake to 26 g/day [31]. In cases where the DRVs differed depending on age and/or sex, the nutritional constraints were weighted according to the DRVs and population size of the different age and sex groups in the study sample. All optimized diets met the DRVs for a reference pupil. Active nutrient constraints were identified for each solution (Table S2). As the bioavailability of iron is generally lower in vegetarian diets, an iron constraint of 1.8 times the RI provided by the Nordic Nutrition Guidelines was set for the "Veg", the "Veg+" and the "Plant" models [41].

2.8. Total GHGE of the Baseline and Optimized Diet

The overall GHGE of the baseline food intake and the optimized diets was calculated as the sum of the products of the corresponding food weights and their specific CO_2eq values as recorded in the Climate Database [33]. Based on the latest IPCC report [42], the World Wildlife Fund (WWF) has estimated that the GHGE from an individual's diet should amount to a maximum of 11 kg CO_2eq/week in order to keep global temperature increase below a 1.5 degrees, compared to preindustrial levels [27]. Hence, the GHGE upper limit for the daily diet was set to 1571 g CO_2eq in all optimizations (see Section 2.11).

2.9. Total Cost of Baseline and Optimized Diet

From the total edible weight of each food item in the diets, the raw weight was calculated and multiplied by the specific cost to obtain the total cost of the baseline and the optimized diets, respectively.

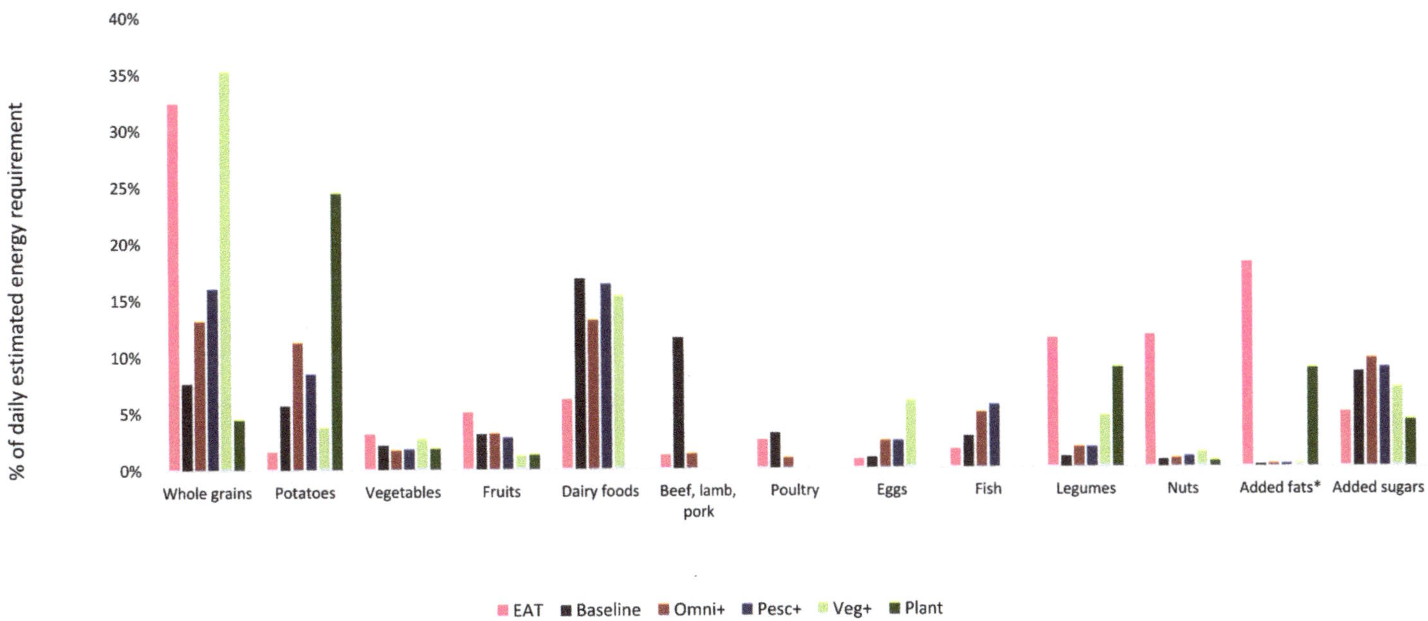

Figure 1. Comparison between the EAT-Lancet diet and (baseline and optimized) diets of Swedish Adolescents. Columns represent the percent of the daily estimated energy requirement for different food groups in the EAT-Lancet diet, the observed (baseline) diet, and in the four main optimized diets ("Omni+", "Pesc+", "Veg+", "Plant"). Food categories used in this comparison were based on the ones used for the EAT-Lancet diet [5]; * Added fats exclude dairy-based fats (such as butter), which are included in "Dairy foods".

2.10. Deviation from Baseline Diet

As the objective function of the LP model, we chose the minimization of the total relative deviation (TRD) from the baseline diet [26,43]. The minimized TRD from baseline was used as a proxy for cultural acceptability of the optimized diet solutions. The TRD is the (total) sum of the absolute (non-negative) values of the relative deviations (RDs) of the weight of a food in the optimized food supply from the reported intake of this food (Equation (1)).

$$RD_i = \frac{M_i - m_i}{m_i} \quad (1)$$

In Equation (1), i indicates the running index of the food, M its mass in the optimized diet, and m the reported intake of that food. The TRD from all N food items in the model was calculated as the total sum of the absolute values of RDs:

$$TRD = \sum_{i=1}^{N} abs(RD_i) \quad (2)$$

Since TRD is not a linear function and thus cannot be part of the linear equation system which LP builds upon, the non-negative values of RDi: RD1→RDN, with N being the number of foods included into the optimization, were created as described and applied previously [26,43]. In brief, the constraints applied to achieve the optimized absolute RD values were set so that the optimized values were greater or equal to both the actual negative and the positive RD value, which resulted in the optimized RD value being equal to the positive RD value, irrespective of whether the deviation was negative (reduced in comparison to the reported intake) or positive (increased). The decision variables were submitted to the following constraints (Formula (3)):

$$abs(RD_i) \geq (mi - Mi)/mi \text{ and } abs(RD_i) \geq -(mi - Mi)/mi \quad (3)$$

Thus, for each standardized difference, its absolute (positive) value was selected because RDi, by definition, has to be greater than or equal to both the relative difference and its negative value.

To be able to control for unacceptably high amounts of individual food items in the optimized FBs, a maximum relative deviation of single foods from baseline was introduced, which had to be adapted during the optimization of each diet to reach a feasible solution (see also Section 2.11).

The average relative deviation (ARD) from the baseline food consumption was used as a proxy of similarity between the baseline and the optimized food consumption and was calculated by dividing the TRD by the total number of food items included in the model (N), as given in Formula (4):

$$ARD = TRD/N \quad (4)$$

2.11. Models

The baseline food consumption was optimized following a strategy described previously [26]. For each of the dietary patterns, besides the vegan diet, the optimization was run without (Omni, Pesc, Veg, and Plant) and with (Omni+, Pesc+ and Veg+) the CO_2eq constraint of 1.571 g per day [27]. Because the total CO_2eq of the vegan diet (when modeled without a CO_2eq constraint) was already below the WWF threshold, only a "Plant" diet was modeled. Hence, since the CO_2eq constraint was not active in the vegan diet, a "Plant+" diet would have been identical to the "Plant" diet". Hence, seven different LP models were applied (Table 1), which all had the minimization of the total relative deviation (TRD) from the baseline food consumption as the objective function. DRVs were implemented as obligatory constraints in all models (Table S1).

Table 1. Names and characteristics of all applied models.

Model Number and Acronym	Objective Function (Minimum)	Foods Available	Nutritional Constraints	CO_2eq Constraint	Acceptability Constraint Max RD [b] for Food Items
1. Omni	TRD [a] from baseline diet	All food items	Meet all DRVs [c]	Not applied	+200%
2. Omni+				Max. 1571 g CO_2eq	+200%
3. Pesc		No meat or poultry products		Not applied	+200%
4. Pesc+				Max. 1571 g CO_2eq	+200%
5. Veg		No red/processed meat, poultry meat, or seafood products	Meet all DRVs, minimum iron intake constraint 1.8 × the DRV of omnivores	Not applied	+600%
6. Veg+				Max. 1571 g CO_2eq	+600%
7. Plant		No animal products		Not applied [d]	+5000%

[a] Total relative deviation. [b] Relative deviation from baseline food consumption. [c] Estimated energy requirements (EERs), recommended intake ranges for macronutrients, recommended intakes (RIs) for micronutrients [31]. [d] A CO_2eq constraint was not needed since the "Plant" model (without a CO_2eq constraint) resulted in a total CO_2eq below 1571 g CO_2eq.

Model 1 ("Omni") was run with nutritional constraints only, without constraining the GHGE (Table 1). In Model 2 ("Omni+"), the indicated CO_2eq constraint was imposed. Consequently, Model 3 ("Pesc", not CO_2eq-constrained) and Model 4 ("Pesc+", with CO_2eq constraint), representing a pescatarian diet, were set up as per Models 1 and 2, but without red/processed meat and poultry meat products (=constrained to zero). Omitting specific food categories such as meat for the pescatarians increased other food groups to achieve isocaloric diets. This in turn required to increase the tolerated maximum relative deviations of single foods from baseline (right column in Table 1). In Models 5 and 6, representing an ovo-vegetarian diet ("Veg", not CO_2eq-constrained and "Veg+", with CO_2eq constraint, respectively) red/processed meat, poultry meat and seafood products were excluded. In the seventh model, representing a vegan diet ("Plant"), all animal products were made unavailable to the model. To avoid extreme deviations of single foods, the absolute RDs of individual food items were limited as much as possible until no feasible solution could be provided by the linear programming algorithm. This corresponded to +200% for Models 1–4, 600% for the vegetarian models 5 and 6, and 5000% for Model 7.

3. Results

The baseline GHGE based on the average food intake of an adolescent was 4.48 kg CO_2eq/day (Table 2). This diet was lower than recommended in dietary fiber (90% coverage of DRV), polyunsaturated fatty acids (89% of DRV), vitamin D (83% of DRV), iron (89% of DRV), contained too much saturated fatty acids (135% of upper DRV) and sodium (157% of upper DRV) (Table S2).

Table 2. Cost, average relative deviation (ARD), min/max relative deviation (RD) values, CO_2eq values, and the number of foods removed, reduced or increased in the optimized diets for omnivores, pescatarians, vegetarians, and vegans compared with the baseline consumption of Swedish adolescents.

Diet [a]	CO_2eq Constraint (g/Day)	Max RD Set (%)	CO_2eq (g/Day)	ARD (%)	Cost (SEK [b])	FB Weight	of Foods Available [c]	of Foods Unavailable [c]	of Foods Removed by Optimization	of Foods Removed in Total	of Foods with Reduced Amount	of Foods with Increased Amount
Baseline	none	-	4481	0.0	77.24	2130	725	0	0	0	0	0
Omni	none	200	2729	12.8	60.71	2018	725	0	47	47	319	359
Omni+	1571	200	1571	21.1	61.73	1843	725	0	81	81	300	344
Pesc	none	200	1861	29.4	53.14	2144	596	129	13	142	265	314
Pesc+	1571	200	1571	31.4	51.43	1925	596	129	17	146	272	306
Veg	none	600	1682	72.0	61.07	1916	550	175	74	249	214	262
Veg+	1571	600	1571	73.0	59.14	1793	550	175	77	252	209	264
Plant	none	5000	1227	118	57.28	2034	334	391	21	412	145	168

[a] All optimized diets meet the dietary recommended values (DRVs). [b] Swedish Krona (1 SEK = approximately 0.10 Euro). [c] Availability based on type of diet (e.g., all red meat was made unavailable in the "Pesc" and "Pesc+" models).

In the four optimized diets, GHGE was reduced by 39–73% (Table 2). The lowest reduction in GHGE was achieved for omnivores ("Omni", −39%) and the greatest reduction was observed in the vegan model ("Plant", −73%) (Table 2). The ARD of the models ranged from 12.8% in the nutritionally adequate diet for omnivores ("Omni") to 118% for the plant-based diet ("Plant"). Compared to baseline, the diet cost was reduced by approximately 20–30% in all optimized diets, with the pescatarian diet being the most affordable (approximately EUR 5/person/day) (Table 2).

Based on the exclusion of food groups when moving from an omnivorous to a plant-based diet, fewer foods were part of the modeled diets. For example, the "Omni+" model included the majority of the original foods (644 out of 725 foods), while the "Plant" diet contained only 313 foods (Table 2).

All optimized diets constrained to meet both nutritional and climate targets had a lower share of animal-based foods (Table 3, Figure 1). The "Omni+" diet contained 91% less Red/processed meat, 73% less Poultry, 65% less Pasta and rice dishes with meat/fish, and about half of the Solid dairy (mainly cheese) compared to the baseline diet (Table 3). However, considerable increases in other animal foods such as Eggs (+158%) and Seafood (+55%) were observed in the optimized "Omni+" diet (Table 3). In the pescatarian, vegetarian and vegan models, the categories Red/processed meat and Poultry were removed entirely (Figure 1). The "Pesc+" diet compensated for the absence of Red/processed meat and Poultry by increasing the share of Seafood (+72%) and Eggs (+158%).

Table 3. Baseline intakes of food groups among Swedish adolescents and relative changes in food groups after optimization of different dietary models.

Model Name	Baseline Diet (g/Day)	Omni (% Change)	Omni+ (% Change)	Pesc (% Change)	Pesc+ (% Change)	Veg (% Change)	Veg+ (% Change)	Plant (% Change)
Model		1	2	3	4	5	6	7
CO_2eq limit (1571 g) applied		no	yes	no	yes	no	yes	no
Bread	85.7	89.1	102	94.5	156	160	160	−59.3
Cereals. other	218	−30.4	−26.5	7.0	−15.0	15.7	20.2	41.2
Nuts and seeds	4.2	16.7	3.0	16.7	24.3	55.0	55.0	−24.4
Fruits and berries	121	9.3	0.0	−21.2	−6.7	−44.8	−56.3	−55.1
Potatoes	121	19.1	97.4	7.1	33.6	−14.5	−36.7	309
Vegetables	104	−2.6	−13.0	−6.9	−6.3	188	121	−3.3
Pulses	21.7	37.4	82.4	82.3	82.3	348	348	1125
Meat substitutes	5.7	0.0	0.0	121	40.4	440	439	1165
Dairy substitutes	9.2	0.0	0.0	0.0	0.0	0.0	0.0	4867
Dairy, other	490	−6.1	−19.8	51.3	2.3	12.1	1.6	−100
Dairy, solid	25.7	−52.0	−54.8	−52.0	−52.0	−80.0	−88.8	−100
Eggs	13.0	132	158	139	158	533	533	−100
Pasta and rice dishes with meat/fish	111	−73.5	−65.9	−99.0	−99.0	−100	−100	−100
Poultry	44.2	−10.3	−73.1	−100	−100	−100	−100	−100
Red/processed meat	161	−63.5	−90.8	−100	−100	−100	−100	−100
Seafood	44.8	32.2	55.0	62.5	71.9	−100	−100	−100
Oils	0.1	0.0	0.0	0.0	0.0	0.0	0.0	−21.6
Fats. solid	10.7	45.2	51.4	15.4	73.3	82.3	83.4	215
Drinks w/o milk	425	10.9	−27.0	−13.4	−18.7	−73.0	−69.8	−58.1
Sugar and sweets	35.7	49.1	34.9	49.2	24.1	−20.2	15.2	−53.6
Seasonings and sauces	79.1	−40.0	−30.8	−19.2	−33.1	−66.2	−60.3	−80.6
Other	0.2	0.0	0.0	0.0	174	521	521	−100

The optimized diets also differed with respect to the amount and type of plant-based foods (Figure 1, Table 3). Pulses increased in all models, with the greatest changes seen in the "Plant" diet, where it increased more than ten-fold compared to baseline (Table 3). In contrast, the amount of Vegetables only increased in the vegetarian ("Veg" and "Veg+") diet (Figure 1, Table 3). The amount of Potatoes increased in all optimized diets with the exception of the "Veg" and "Veg+" diets, the "Plant" diet showing the largest increase (+309%) (Table 3). Fruits remained almost unchanged (+9% in the "Omni" diet) or was reduced by up to 56% in the rest of the optimized diets (Figure 1, Table 3). Cereals such as pasta and oats increased in the models containing little or no animal products ("Veg", "Veg+", and "Plant"), and decreased in the optimized diets for omnivore and pescatarians (Table 3). Bread increased in all models with the exception of the "Plant" model.

The more the baseline dietary pattern was restricted, the more meat substitutes were included in the modeled diet (Table 3). For example, the "Omni+" solution contained the same amount of meat substitutes as the baseline diet, while the optimized "Plant" diet experienced a more than ten-fold increase in these foods compared to baseline. Overall, the LP algorithm was able to meet both nutritional and climate objectives without increasing the amount of Dairy substitutes with the exception of the "Plant" diet, where their amount increased by more than 50-fold, from roughly 9 g/day (baseline) to about 460 g/day (Table 3).

The active nutritional constraints of all models are shown in Table S2. Iron, selenium, and vitamin D were active lower-threshold constraints, while sodium was an upper-threshold active constraint in all models. Calcium was an active lower-threshold constraint in the models "Omni", "Omni+" and "Pesc+", but not in the "Pesc", the "Veg" and the "Plant" models, which contained relatively high amounts of calcium-fortified dairy substitutes. Added sugars actively constrained the "Omni+" diet only. Achieving a minimum amount of polyunsaturated fatty acids was also an active constraint in the diets "Veg+", "Pesc+", and "Plant". Vitamin A acted as an active constraint in all models except in the "Veg" diet.

When comparing the omnivorous EAT-Lancet diet [5] to our optimized models, pronounced differences were observed (Figure 1). Overall, the EAT-Lancet diet's amounts were higher in Whole grain foods, Vegetables, Fruits, Legumes, Nuts, and Added fats, but lower in Potatoes, Dairy foods, Eggs, Fish and Added sugars than that provided by the optimized diets. The "Omni+" diet matched the EAT-Lancet diet with respect to red (beef, lamb, pork) meat. Naturally, the "Pesc+", "Veg+" and" "Plant" diets did not match the suggested amounts of red or poultry meat in the EAT-Lancet diet. Similarly, the "Veg+" and "Plant" diets were below the maximum limit on Fish. The "Veg+" diet aligned to the EAT-Lancet diet in terms of "Whole grains", whereas the "Plant" diet was the only diet mirroring the target for "Added sugars". The average relative deviation for all food groups between the EAT diet and the optimized diets (i.e., the sum of absolute relative deviations divided by the number of food groups compared) was 134, 136, 127, and 181 percent for the "Omni+", "Pesc+", "Veg+" and "Plant" diets, respectively (Figure 1).

4. Discussion

In this study, we demonstrated that nutritionally adequate diets, which align with the maximum tolerable diet-related GHGE limit defined to keep the increase in global average temperatures below 1.5 °C above pre-industrial levels, can be achieved for four different dietary patterns. Simply modifying the current diet of Swedish adolescents to meet DRV values resulted in a 39% decrease in GHGE, which was mainly achieved by a pronounced reduction in solid dairy foods (cheese and curd) and meat. Relative to the baseline diet, the GHGE in the nutritionally adequate pescatarian model ("Pesc") was reduced by 59%, by 62% in the vegetarian model ("Veg") and by 73% in the vegan ("Plant") model. The amount of CO_2eq in the baseline diet of the adolescents was 4.5 kg/day, a value that is comparable to the ~5 kg CO_2eq/day previously reported for adults [15]. This means that in order to reach the threshold of 1.57 kg CO_2eq/day proposed by the WWF [27], the GHGE had to be reduced by 65% [26,43]. Only the optimized, nutritionally adequate vegan diet ("Plant") dropped below the IPCC/WWF threshold without further active restriction of the model's GHGE. The exclusion of food groups in the pescatarian, vegetarian and vegan diets along with constraining the GHGE increased the deviation from the baseline diet, especially for the optimized vegetarian and vegan models as compared to the omnivoric or pescatarian solutions. The optimized diets, despite being nutritionally adequate and reaching the recommended GHGE level, did not align very well with the food-group pattern of the EAT-Lancet diet [5].

Constraining the reported food intake to meet the DRVs alone resulted in a marked reduction of GHGE, which is in line with previous findings [26,44]. However, the 39% reduction in GHGE achieved in the "Omni" diet is surprisingly high compared to previous studies in UK adults where the reduction was 17% [44]. This can be explained mainly by the DRV-enforced reduction of saturated fatty acids and sodium as well as the increased inclusion of foods that are rich in fiber and polyunsaturated fatty acids. These changes increase the share of plant-based foods with a low climate impact at the expense of animal-based foods, the consumption of which is comparably high in this sub-population [16].

The climate-friendly and nutritionally adequate food profile for omnivores ("Omni+"), which mimics the dietary habits of Swedish adolescents the best, showed a more pronounced trend towards reduction of meat, poultry, and solid dairy than the non-GHGE-constrained alternative ("Omni"). This reduction was compensated by an increase in the amounts of less GHGE-intense animal products such as eggs, but a major part of the substitution was based on an increased inclusion of pulses, potatoes, and bread. Table 4 summarizes the optimized solution of the "Omni+" diet. Others have also calculated climate-friendly diets for the general population [5,45], but without ensuring nutritional adequacy.

Table 4. Quantities of food groups for an omnivorous diet with 2410 kcal, generating a maximum of 1571 g of CO_2eq/day, based on the "Omni+" model.

- About 180 g of (whole grain) bread and approximately 160 g of other cereals (rice, pasta, etc.) per day
- At least 40 g of pulses per day
- At least 230 g of potatoes per day
- Around 220 g of fruits and vegetables per day
- About one egg per day
- One portion of fish and other seafood (~150 g) every second day, every third portion being oily fish
- Around one portion (~190 g) of meat, meat dishes and poultry per week (preferably pork, poultry, and offal such as liver and blood products rather than beef)
- Not more than 400 g of dairy products and about one slice of cheese (15 g) per day
- A handful of nuts and seeds per week (~30 g)

In the pescatarian model ("Pesc+"), the optimized solution is very similar to that of the omnivore diet ("Omni+"), except that meat and meat products are replaced by moderately increased amounts of fish, meat substitutes, and dairy products (Table S3). Both the omnivorous and the pescatarian diets include increased amounts of fish compared to the baseline diet. Presently, a large part of the fish consumed originates from marine capture fisheries [46], which explains the low CO_2eq-value of this micronutrient- and protein-rich commodity. However, 96 of the world's fish stocks are either moderately or fully exploited, or over-fished [47]. Farmed fish such as salmon has GHGE values comparable to or even higher than that of poultry, pork and dairy and can in addition be a source of eutrophication [48]. If a high proportion of the population follows the recommendation to increase the intake of farmed or captured fish, the biodiversity of certain fish types should be considered in addition to their production-related climate impacts.

The climate-friendly solution for vegetarians includes considerably increased amounts of dairy and meat substitutes (which are mostly mycoprotein-, pea- or soy-based products), pulses, bread, potatoes, and some vegetables to compensate for excluding meat and fish (Table S4). Vegetarian diets have been recommended as a principal approach to reduce the climate impact of the diet, though again, these are not based on calculations that ensure full nutritional adequacy [49–51] and may increase the risk of micronutrient deficiencies. For example, one third of Swedish female adolescents have low iron stores [17]. Excluding meat and fish from the diet may result in lower iron intakes as well as in a diet with a lower iron bioavailability. Haem iron, found in meat, is more readily absorbed than non-haem varieties. Furthermore, meat and fish enhances absorption of iron from plant-based foods [31]. Absence of haem iron in the diet may affect iron status negatively in vulnerable populations and highlights the need for reliable guidance on what to replace meat with and how to combine foods to increase bioavailability [52]. Therefore, in the optimized diets building on the Veg, Veg+, and Plant models, a higher minimum threshold of iron was set as recommended by the US Institute of Medicine [41]. The high bioavailability of ferritin-bound iron in legumes may also help to overcome this shortcoming [53].

Excluding all animal products in the "Plant" model resulted in a considerable inclusion of (mostly fortified) meat and dairy substitutes along with an increased intake of pulses, potatoes and non-dairy fats (Table S5). Although plant-based foods are considered to have a low bioavailability of iron, calcium, vitamin D and B12 and although the minimum threshold was raised for iron, all applied DRV values were covered by the optimized solution for vegans. Besides iron, a sufficient supply of calcium and vitamin B12 was also guaranteed even for the vegetarians and vegans. This was primarily achieved due to the high fortification of dairy replacements with these micronutrients. These results mirror a recent optimization study on Dutch eating habits, where the optimized diet for vegans met DRVs for vitamin B12 and calcium only through the inclusion of sufficiently high amounts of fortified soy milk [54]. This raises the question as to whether fortification or, alternatively, supplementation are acceptable ways forward to reduce diet-related GHGE. More studies on replacement food, fortification, and health outcomes are clearly needed. Furthermore, the production of meat and dairy replacements raises concerns about other environmental indicators. For example, plant-based milk replacements may contribute to water scarcity, deforestation and biodiversity loss [55], although this may vary depending on type of product and country. Further investigations are needed to fully understand how the "Veg+"- and "Plant" diets would impact the full range of health and environmental indicators in the context under study.

As is evident from Figure 1, the optimized "Plant" diet contained the lowest amount of whole grains and the highest amount of potatoes. Furthermore, the amount of vegetables (excluding legumes), fruits and nuts was comparably low. This food pattern differs somewhat from other recommendations on plant-based diets. For example, recent recommendations on plant-based diets for adolescents [56] emphasize the inclusion of whole grains, legumes, nuts and seeds, vegetables, and fruits to the diet. These differences are likely to result from the fact that environmental aspects have so far insufficiently been considered in the development of food-based recommendations. Studies show that the increased inclusion of fruit and vegetables in the diet, although beneficial from a health point of view, can lead to higher environmental impacts [6–8,57], or be less effective in reducing them [58]. Furthermore, diets optimized to meet nutritional constraints only [59,60] have been shown to have higher climate footprints. On the other hand, self-selected, plant-based diets with lower climate footprints have been shown to lead to the overconsumption of refined sugars [14,15]. This stands in contrast to the optimized "Plant" model, that had the lowest amounts of added sugars. In summary, these findings add to the challenges in defining the sustainability of diets. It is, therefore, advisable to use a holistic approach such as linear programming (that consider both health and environmental priorities) in the definition of food-based recommendations for different dietary patterns.

Our findings reveal that neither the baseline nor the optimized diets of Swedish adolescents align with the EAT-Lancet Commission's dietary recommendation for a sustainable diet [5]. This could be due to three reasons: (1) we optimized for similarity to the reported food consumption patterns of Swedish adolescents to achieve a high cultural acceptability instead of using the EAT-Lancet diet as the reference; (2) our models were all constrained to ensure the fulfilment of 27 DRVs, which the EAT-Lancet diet was not; (3) the EAT-Lancet diet considered additional dimensions of sustainability such as blue water footprint, land use change and animal welfare, which were not considered in the study at hand. In contrast to the EAT-Lancet diet, the models "Omni" and "Omni+" include significant amounts of dairy, fish, and eggs. Another difference is the much higher amount of potatoes and a markedly lower amount of legumes in the optimized diets as compared to the EAT-Lancet diet. Potatoes, commonly consumed in the Swedish adolescent population, are a dominant and nutritious staple-crop in Sweden considered to be healthy [61]. Although all optimized diets diverged from the EAT diet, the Veg+ diet was the most similar on a food group level.

Despite the discrepancies, some similarities between the EAT diet and the optimized diets can be found. For example, the optimized vegetarian diet ("Veg+") matched it with respect to Whole grains and Vegetables and the optimized vegan ("Plant") diet was comparable in terms of Legumes and Added sugars. Furthermore, like the EAT-Lancet diet, both Omni models suggest a comparable amount of red meat and poultry to achieve a nutritious and climate-friendly diet. In contrast to the EAT-Lancet diet [5], our diets optimized for similarity may be easier to achieve for adolescents in the Swedish population.

Food-based dietary guidelines (FBDGs) were not considered as constraints in the optimizations. Today, the Nordic Nutrition Recommendations has quantifiable FBDG regarding fruit and vegetables (500 g/day) and fish (2–3 times per week) [62]. Only the "Plant" model met the Swedish FBDGs' recommended intake of 500 g fruits and vegetables (including pulses) per day. The LP algorithm in general did not favor either fruits or vegetables which can be explained by the fact that fruits and certain types of vegetables (such as tomatoes, cucumbers, and onions) may provide smaller amounts of nutrients per gram of CO_2eq compared to other foods such as starchy vegetables and pulses. It thus mirrors research showing that a generous inclusion of fruit and vegetables into the diet can result in higher dietary environmental impacts [6–8]. Another plausible explanation is that our solutions were optimized to be as similar as possible to the baseline diet, where the intake of fruit and vegetables was relatively low. This finding aligns well with findings from the Netherlands, Denmark, and Estonia, where nutritionally adequate diets optimized for acceptability did not meet national FBDG-targets for fruit and vegetables [37,54,63].

One strength of our research is that it highlights the potential of optimized diets, such as those achieved in this study, to be translated into sustainable food-based dietary guidelines. However, for this to happen, other scientific evidence such as the impact on additional environmental factors (blue water usage, land use change, and biodiversity) and other legitimate factors (food safety) must also be considered. Furthermore, additional detailed information may be necessary to be included such as the prioritization of local vs. imported products. Further adaptation towards individual needs may also be necessary before formulating food-based dietary guidelines with support from linear optimization.

Future modeling studies should investigate the feasibility and need for including both DRVs and FBDG in the models as well as aspects on food safety and other environmental aspects such as biodiversity, pollutants, blue water use.

The GHGE values indicated include only the CO_2eq to the factory gate, but not the GHGE associated with transportation to the retailer and to the home or food preparation. Therefore, the final CO_2eq values from different foods might be slightly higher than those calculated in this study.

As the data were recorded in 2016–2017, dietary habits might have changed moderately since then. Furthermore, all optimized diets cover the estimated micronutrient intake of 97.5% of the population. This may be unnecessarily high when using the suggested diets to fulfill average intakes for population groups but guarantees on the other hand the applicability of the optimized diets also for individuals. Another limitation was that no new foods were introduced into the models. There are many new meat and dairy substitutes emerging on the market [64,65]. Including these foods in the optimization of diets could provide certain benefits for the environment without compromising nutritional adequacy [66]. Future studies should further explore the health impacts and environmental effects of also including such foods in the modeling. Since the dietary survey data was averaged, data on the food intake of pescatarians, vegetarians and vegans were not available during optimization. Therefore, the optimization may also be limited for the groups of pescatarians, vegetarians and vegans, as the reported omnivore diet was used as reference. In the case of optimized non-omnivoric diets, the RD represents the deviation after changing to a pescatarian, vegetarian or vegan diet. It is not representative of individuals who already practice these diets.

One of the strengths of this study is that it provides the first guidance for achieving more climate-friendly diets based on the dominating omnivoric dietary pattern of adolescents in Sweden. The results feed into the discussion on how future FBDGs should be shaped. Since comprehensive fiscal measures such as taxes and subsidies to influence on people's food choices are currently not promoted by decision makers in Sweden, information and nudging may be the obvious policy tool available to affect consumer behavior [67]. Therefore, it is critical that messages are simple and clear, yet still sufficiently informative to avoid unintended substitutions and adverse outcomes [18].

5. Conclusions

The results of this study show that an affordable, nutritionally adequate diet with a considerably reduced GHGE can be achieved for omnivorous, pescatarian, vegetarian and vegan Swedish adolescents. Particularly for vegetarians and vegans, this means large deviations from the current reported food pattern. However, even in the climate-friendly diet for omnivores, a considerable reduction in the consumption of red/processed meat (pork and beef), poultry, and solid dairy (cheese) along with an increased intake of potatoes and fish would be needed to meet the desired climate targets. Excluding meat and fish from the diet demands the inclusion of substitutes for meat and dairy, which are fortified with calcium and the vitamins D and B12 to ensure nutritional adequacy. Food fortification is an issue that needs to be discussed in future diet modifications. Our findings can contribute to national recommendations that are simple and clear, yet still sufficiently informative to avoid unintended and adverse outcomes for both human and planetary health. The optimized omnivorous, nutritionally adequate diet in this study differed in several aspects from the EAT-Lancet diet, indicating that there are several ways to define sustainable diets but also that the cultural dietary context is likely to play an important role in characterizing such diets for specific populations. This study provides a basis that can be used in the development of food-based dietary guidelines on affordable, nutritionally adequate diets that are low in GHGE. This methodology can also be applied for other age groups and countries after the basis of the optimization has been adapted to the specific geographical and cultural dietary context.

Supplementary Materials: The following are available online at https://www.mdpi.com/article/10.3390/nu13082507/s1, Table S1: Dietary reference values applied as constraints in the linear programming models, Table S2: Nutrient coverage for baseline intake and optimized solutions, Table S3: Quantities of food groups for a pescatarian diet generating a maximum of 1571 g of CO_2eq per day, based on the "Pesc+" model, Table S4: Quantities of food groups for a pescatarian diet generating a maximum of 1571 g of CO_2eq per day, based on the "Veg+" model, Table S5: Quantities of food groups for a pescatarian diet generating a maximum of 1571 g of CO_2eq per day, based on the "Plant" model.

Author Contributions: Conceptualization, P.E.C., L.S.E., A.K.L. and A.P.; methodology, P.E.C. and A.P.; software, P.E.C. and A.P.; validation, P.E.C. and A.P.; formal analysis, P.E.C. and A.P.; investigation, P.E.C. and A.P.; resources, A.K.L.; data curation, P.E.C.; writing—original draft preparation, P.E.C. and A.P. writing—review and editing P.E.C., L.S.E., A.K.L. and A.P.; visualization, P.E.C.; supervision, A.P.; project administration, P.E.C.; funding acquisition, L.S.E. All authors have read and agreed to the published version of the manuscript.

Funding: This research was funded by Formas, the Swedish Governmental Research Council on sustainable, grant number 2016–00353.

Institutional Review Board Statement: This research used secondary data from the national dietary survey Riksmaten Adolescents 2016–2017. That survey was conducted according to the guidelines of the Declaration of Helsinki and approved by the Regional Ethical Review Board in Uppsala (No. 2015/190).

Informed Consent Statement: Written consent was obtained from all subjects and legal guardians of children younger than 16 years (for pupils providing biological samples, approximately one third of the pupils) participating in the national dietary survey Riksmaten Adolescents 2016–2017. For pupils not involved in biological sampling, verbal opt-out consent was witnessed and formally recorded.

Data Availability Statement: Data used for these analyses can be made available from the Corresponding author upon reasonable request.

Conflicts of Interest: The authors declare no conflict of interest.

References

1. IPCC. *Climate Change and Land: An IPCC Special Report on Climate Change, Desertification, Land Degradation, Sustainable Land Management, Food Security, and Greenhouse Gas Fluxes in Terrestrial Ecosystems*; Intergovernmental Panel on Climate Change: Geneva, Switzerland, 2019.
2. Watts, N.; Amann, M.; Ayeb-Karlsson, S.; Belesova, K.; Bouley, T.; Boykoff, M.; Byass, P.; Cai, W.; Campbell-Lendrum, D.; Chambers, J.; et al. The Lancet Countdown on Health and Climate Change: From 25 Years of Inaction to a Global Transformation for Public Health. *Lancet* **2018**, *391*, 581–630. [CrossRef]
3. Adoption of the Paris Agreement. In Proceedings of the Twenty-First Session Conference of the Parties, Paris, France, 30 November–11 December 2015. Available online: https://unfccc.int/resource/docs/2015/cop21/eng/10.pdf (accessed on 21 July 2020).
4. Government Offices of Sweden. The Climate Policy Framework. Available online: https://www.regeringen.se/artiklar/2016/10/eu-overens-om-har-ratificerat-klimatavtalet-fran-paris/ (accessed on 8 April 2020).
5. Willett, W.; Rockström, J.; Loken, B.; Springmann, M.; Lang, T.; Vermeulen, S.; Garnett, T.; Tilman, D.; DeClerck, F.; Wood, A.; et al. Food in the Anthropocene: The EAT–Lancet Commission on Healthy Diets from Sustainable Food Systems. *Lancet* **2019**, *393*, 447–492. [CrossRef]
6. Ridoutt, B.G.; Hendrie, G.A.; Noakes, M. Dietary Strategies to Reduce Environmental Impact: A Critical Review of the Evidence Base. *Adv. Nutr.* **2017**, *8*, 933–946. [CrossRef]
7. Perignon, M.; Vieux, F.; Soler, L.-G.; Masset, G.; Darmon, N. Improving Diet Sustainability through Evolution of Food Choices: Review of Epidemiological Studies on the Environmental Impact of Diets. *Nutr. Rev.* **2017**, *75*, 2–17. [CrossRef] [PubMed]
8. Aleksandrowicz, L.; Green, R.; Joy, E.J.M.; Smith, P.; Haines, A. The Impacts of Dietary Change on Greenhouse Gas Emissions, Land Use, Water Use, and Health: A Systematic Review. *PLoS ONE* **2016**, *11*, e0165797. [CrossRef] [PubMed]
9. Nelson, M.E.; Hamm, M.W.; Hu, F.B.; Abrams, S.A.; Griffin, T.S. Alignment of Healthy Dietary Patterns and Environmental Sustainability: A Systematic Review. *Adv. Nutr.* **2016**, *7*, 1005–1025. [CrossRef] [PubMed]
10. Hallström, E.; Carlsson-Kanyama, A.; Börjesson, P. Environmental Impact of Dietary Change: A Systematic Review. *J. Clean. Prod.* **2015**, *91*, 1–11. [CrossRef]
11. Allès, B.; Péneau, S.; Kesse-Guyot, E.; Baudry, J.; Hercberg, S.; Méjean, C. Food Choice Motives Including Sustainability during Purchasing Are Associated with a Healthy Dietary Pattern in French Adults. *Nutr. J.* **2017**, *16*, 58. [CrossRef]
12. Vergeer, L.; Vanderlee, L.; White, C.M.; Rynard, V.L.; Hammond, D. Vegetarianism and Other Eating Practices among Youth and Young Adults in Major Canadian Cities. *Public Health Nutr.* **2020**, *23*, 609–619. [CrossRef]
13. Allt Fler Äter Mindre Kött (More People Eat Less Meat). Available online: https://www.ungdomsbarometern.se/allt-fler-ater-mindre-kott/ (accessed on 22 April 2020).
14. Payne, C.L.; Scarborough, P.; Cobiac, L. Do Low-Carbon-Emission Diets Lead to Higher Nutritional Quality and Positive Health Outcomes? A Systematic Review of the Literature. *Public Health Nutr.* **2016**, *19*, 2654–2661. [CrossRef]
15. Sjörs, C.; Hedenus, F.; Sjölander, A.; Tillander, A.; Bälter, K. Adherence to Dietary Recommendations for Swedish Adults across Categories of Greenhouse Gas Emissions from Food. *Public Health Nutr.* **2017**, *20*, 3381–3393. [CrossRef] [PubMed]
16. Warensjö Lemming, E.; Moraeus, L.; Petrelius Sipinen, J.; Anna Karin, L. *Riksmaten Ungdom 2016–2017 Livsmedelskonsumtion Bland Ungdomar i Sverige (Riksmaten Adolescents 2016–2017 Food Intake among Adolescents in Sweden)*; The National Food Agency: Uppsala, Sweden, 2018.
17. Warensjö Lemming, E.; Moraeus, L.; Petrelius Sipinen, J.; Anna Karin, L. *Riksmaten Ungdom 2016–2017 Näringsintag och Näringsstatus Bland Ungdomar i Sverige (Riksmaten Adolescents 2016–2017 Nutrient Intake and Nutrient Status among Adolescents in Sweden)*; The National Food Agency: Uppsala, Sweden, 2018.
18. Irz, X.; Jensen, J.D.; Leroy, P.; Réquillart, V.; Soler, L.-G. Promoting Climate-Friendly Diets: What Should We Tell Consumers in Denmark, Finland and France? *Environ. Sci. Policy* **2019**, *99*, 169–177. [CrossRef]
19. Fresán, U.; Sabaté, J. Vegetarian Diets: Planetary Health and Its Alignment with Human Health. *Adv. Nutr.* **2019**, *10*, S380–S388. [CrossRef]
20. Craig, W.J. Health Effects of Vegan Diets. *Am. J. Clin. Nutr.* **2009**, *89*, 1627S–1633S. [CrossRef]
21. González-García, S.; Esteve-Llorens, X.; Moreira, M.T.; Feijoo, G. Carbon Footprint and Nutritional Quality of Different Human Dietary Choices. *Sci. Total Environ.* **2018**, *644*, 77–94. [CrossRef] [PubMed]
22. Gazan, R.; Brouzes, C.M.C.; Vieux, F.; Maillot, M.; Lluch, A.; Darmon, N. Mathematical Optimization to Explore Tomorrow's Sustainable Diets: A Narrative Review. *Adv. Nutr.* **2018**, *9*, 602–616. [CrossRef] [PubMed]

23. Mertens, E.; Van't Veer, P.; Hiddink, G.J.; Steijns, J.M.; Kuijsten, A. Operationalising the Health Aspects of Sustainable Diets: A Review. *Public Health Nutr.* **2017**, *20*, 739–757. [CrossRef]
24. Perignon, M.; Masset, G.; Ferrari, G.; Barré, T.; Vieux, F.; Maillot, M.; Amiot, M.-J.; Darmon, N. How Low Can Dietary Greenhouse Gas Emissions Be Reduced without Impairing Nutritional Adequacy, Affordability and Acceptability of the Diet? A Modelling Study to Guide Sustainable Food Choices. *Public Health Nutr.* **2016**, *19*, 2662–2674. [CrossRef]
25. Milner, J.; Green, R.; Dangour, A.D.; Haines, A.; Chalabi, Z.; Spadaro, J.; Markandya, A.; Wilkinson, P. Health Effects of Adopting Low Greenhouse Gas Emission Diets in the UK. *BMJ Open* **2015**, *5*, e007364. [CrossRef]
26. Eustachio Colombo, P.; Patterson, E.; Elinder, L.S.; Lindroos, A.K.; Sonesson, U.; Darmon, N.; Parlesak, A. Optimizing School Food Supply: Integrating Environmental, Health, Economic, and Cultural Dimensions of Diet Sustainability with Linear Programming. *Int. J. Environ. Res. Public Health* **2019**, *16*, 3019. [CrossRef]
27. World Wildlife Fund. One Planet Plate 2019—Kriterier Och Bakgrund (One Planet Plate 2019—Criteria and Background). Available online: https://wwwwwfse.cdn.triggerfish.cloud/uploads/2019/04/kriterier-fr-one-planet-plate-rev-2019.pdf (accessed on 30 January 2020).
28. Moraeus, L.; Lemming, E.W.; Hursti, U.-K.K.; Arnemo, M.; Sipinen, J.P.; Lindroos, A.-K. Riksmaten Adolescents 2016–2017: A National Dietary Survey in Sweden-Design, Methods, and Participation. *Food Nutr. Res.* **2018**. [CrossRef] [PubMed]
29. Lindroos, A.K.; Sipinen, J.P.; Axelsson, C.; Nyberg, G.; Landberg, R.; Leanderson, P.; Arnemo, M.; Lemming, E.W. Use of a Web-Based Dietary Assessment Tool (RiksmatenFlex) in Swedish Adolescents: Comparison and Validation Study. *J. Med. Internet Res.* **2019**, *21*, e12572. [CrossRef] [PubMed]
30. Wanselius, J.; Axelsson, C.; Moraeus, L.; Berg, C.; Mattisson, I.; Larsson, C. Procedure to Estimate Added and Free Sugars in Food Items from the Swedish Food Composition Database Used in the National Dietary Survey Riksmaten Adolescents 2016–2017. *Nutrients* **2019**, *11*, 1342. [CrossRef]
31. *Nordic Nutrition Recommendations 2012*, 5th ed.; Nordic Council Of Ministers: Copenhagen, Denmark, 2014.
32. Matpriskollen (Food Price Check). Available online: https://matpriskollen.se/ (accessed on 27 January 2020).
33. Florén, B.; Amani, P.; Davis, J. Climate Database Facilitating Climate Smart Meal Planning for the Public Sector in Sweden. *Int. J. Food Syst. Dyn.* **2017**, *8*, 72–80.
34. International Organization for Standardization. ISO 14040:2006—Environmental Management—Life Cycle Assessment—Principles and Framework. Available online: https://www.iso.org/standard/37456.html (accessed on 9 October 2017).
35. International Organization for Standardization. ISO 14044:2006—Environmental Management—Life Cycle Assessment—Requirements and Guidelines. Available online: https://www.iso.org/standard/38498.html (accessed on 9 October 2017).
36. Parry, M.L.; Intergovernmental Panel on Climate Change. (Eds.) *Climate Change 2007: Impacts, Adaptation and Vulnerability: Contribution of Working Group II to the Fourth Assessment Report of the Intergovernmental Panel on Climate Change*; Cambridge University Press: Cambridge, UK; New York, NY, USA, 2007; ISBN 978-0-521-88010-7.
37. Parlesak, A.; Tetens, I.; Dejgard Jensen, J.; Smed, S.; Gabrijelcic Blenkus, M.; Rayner, M.; Darmon, N.; Robertson, A. Use of Linear Programming to Develop Cost-Minimized Nutritionally Adequate Health Promoting Food Baskets. *PLoS ONE* **2016**, *11*, e0163411. [CrossRef]
38. Dantzig, G.B. Maximization of a linear function of variables subject to linear inequality. In *Activity Analysis of Production and Allocation*; Koopmans, T.C., Ed.; Wiley & Chapman-Hall: New York, NY, USA; London, UK, 1951; pp. 339–347.
39. Nocedal, J.; Wright, S.J. *Numerical Optimization*; Springer: New York, NY, USA, 2006; ISBN 978-0-387-40065-5.
40. Mason, A.J. OpenSolver—An Open Source Add-in to Solve Linear and Integer Progammes in Excel. In *Operations Research Proceedings 2011*; Klatte, D., Lüthi, H.-J., Schmedders, K., Eds.; Springer: Berlin/Heidelberg, Germany, 2012; pp. 401–406. ISBN 978-3-642-29209-5.
41. Institute of Medicine (US) Panel on Micronutrients. *Dietary Reference Intakes for Vitamin A, Vitamin K, Arsenic, Boron, Chromium, Copper, Iodine, Iron, Manganese, Molybdenum, Nickel, Silicon, Vanadium, and Zinc*; National Academies Press (US): Washington, DC, USA, 2001; ISBN 978-0-309-07279-3.
42. Global Warming of 1.5 °C. *An IPCC Special Report on the Impacts of Global Warming of 1.5 °C above Pre-Industrial Levels and Related Global Greenhouse Gas Emission Pathways, in the Context of Strengthening the Global Response to the Threat of Climate Change, Sustainable Development, and Efforts to Eradicate Poverty*; Intergovernmental Panel on Climate Change: Geneva, Switzerland, 2018.
43. Darmon, N.; Ferguson, E.L.; Briend, A. A Cost Constraint Alone Has Adverse Effects on Food Selection and Nutrient Density: An Analysis of Human Diets by Linear Programming. *J. Nutr.* **2002**, *132*, 3764–3771. [CrossRef]
44. Green, R.; Milner, J.; Dangour, A.D.; Haines, A.; Chalabi, Z.; Markandya, A.; Spadaro, J.; Wilkinson, P. The Potential to Reduce Greenhouse Gas Emissions in the UK through Healthy and Realistic Dietary Change. *Clim. Chang.* **2015**, *129*, 253–265. [CrossRef]
45. Mithril, C.; Dragsted, L.O.; Meyer, C.; Blauert, E.; Holt, M.K.; Astrup, A. Guidelines for the New Nordic Diet. *Public Health Nutr.* **2012**, *15*, 1941–1947. [CrossRef]
46. FAO. *The State of World Fisheries and Aquaculture 2020: Sustainability in Action*; FAO: Rome, Italy, 2020; ISBN 978-92-5-132692-3.
47. Fisheries and Aquaculture Dept, Food and Agriculture Organization of the United Nations (FAO). *Review of the State of World Marine Fishery Resources*; FAO Fisheries and Aquaculture Technical Paper; Food and Agriculture Organization of the United Nations: Rome, Italy, 2011; ISBN 978-92-5-107023-9.
48. Clark, M.; Tilman, D. Comparative Analysis of Environmental Impacts of Agricultural Production Systems, Agricultural Input Efficiency, and Food Choice. *Environ. Res. Lett.* **2017**, *12*, 064016. [CrossRef]

49. Berners-Lee, M.; Hoolohan, C.; Cammack, H.; Hewitt, C.N. The Relative Greenhouse Gas Impacts of Realistic Dietary Choices. *Energy Policy* **2012**, *43*, 184–190. [CrossRef]
50. Sáez-Almendros, S.; Obrador, B.; Bach-Faig, A.; Serra-Majem, L. Environmental Footprints of Mediterranean versus Western Dietary Patterns: Beyond the Health Benefits of the Mediterranean Diet. *Environ. Health Glob. Access Sci. Source* **2013**, *12*, 118. [CrossRef]
51. Springmann, M.; Wiebe, K.; Mason-D'Croz, D.; Sulser, T.B.; Rayner, M.; Scarborough, P. Health and Nutritional Aspects of Sustainable Diet Strategies and Their Association with Environmental Impacts: A Global Modelling Analysis with Country-Level Detail. *Lancet Planet. Health* **2018**, *2*, e451–e461. [CrossRef]
52. Hunt, J.R. Bioavailability of Iron, Zinc, and Other Trace Minerals from Vegetarian Diets. *Am. J. Clin. Nutr.* **2003**, *78*, 633S–639S. [CrossRef] [PubMed]
53. Lynch, S.; Pfeiffer, C.M.; Georgieff, M.K.; Brittenham, G.; Fairweather-Tait, S.; Hurrell, R.F.; McArdle, H.J.; Raiten, D.J. Biomarkers of Nutrition for Development (BOND)—Iron Review. *J. Nutr.* **2018**, *148*, 1001S–1067S. [CrossRef]
54. Broekema, R.; Tyszler, M.; van 't Veer, P.; Kok, F.J.; Martin, A.; Lluch, A.; Blonk, H.T.J. Future-Proof and Sustainable Healthy Diets Based on Current Eating Patterns in the Netherlands. *Am. J. Clin. Nutr.* **2020**. [CrossRef]
55. McGivney, A. Almonds Are Out. Dairy Is a Disaster. So What Milk Should We Drink? Available online: http://www.theguardian.com/environment/2020/jan/28/what-plant-milk-should-i-drink-almond-killing-bees-aoe (accessed on 20 August 2020).
56. Baroni, L.; Goggi, S.; Battino, M. Planning Well-Balanced Vegetarian Diets in Infants, Children, and Adolescents: The VegPlate Junior. *J. Acad. Nutr. Diet.* **2019**, *119*, 1067–1074. [CrossRef] [PubMed]
57. Vieux, F.; Darmon, N.; Touazi, D.; Soler, L.G. Greenhouse Gas Emissions of Self-Selected Individual Diets in France: Changing the Diet Structure or Consuming Less? *Ecol. Econ.* **2012**, *75*, 91–101. [CrossRef]
58. Mertens, E.; Kuijsten, A.; Kanellopoulos, A.; Dofková, M.; Mistura, L.; D'Addezio, L.; Turrini, A.; Dubuisson, C.; Havard, S.; Trolle, E.; et al. Improving Health and Carbon Footprints of European Diets Using a Benchmarking Approach. *Public Health Nutr.* **2020**, 1–11. [CrossRef] [PubMed]
59. Perignon, M.; Sinfort, C.; El Ati, J.; Traissac, P.; Drogué, S.; Darmon, N.; Amiot, M.-J.; Amiot, M.J.; Achir, N.; Alouane, L.; et al. How to Meet Nutritional Recommendations and Reduce Diet Environmental Impact in the Mediterranean Region? An Optimization Study to Identify More Sustainable Diets in Tunisia. *Glob. Food Secur.* **2019**, *23*, 227–235. [CrossRef]
60. Vieux, F.; Perignon, M.; Gazan, R.; Darmon, N. Dietary Changes Needed to Improve Diet Sustainability: Are They Similar across Europe? *Eur. J. Clin. Nutr.* **2018**, *72*, 951. [CrossRef]
61. Åkesson, A.; Andersen, L.F.; Kristjánsdóttir, Á.G.; Roos, E.; Trolle, E.; Voutilainen, E.; Wirfält, E. Health Effects Associated with Foods Characteristic of the Nordic Diet: A Systematic Literature Review. *Food Nutr. Res.* **2013**, *57*, 22790. [CrossRef]
62. Brugård Konde, Å.; Bjerselius, R.; Haglund, L.; Jansson, A.; Pearson, M.; Sanner Färnstrand, J.; Lindroos, A.K. *Swedish Dietary Guidelines—Risk and Benefit Management Report*; The Swedish Food Agency: Uppsala, Sweden, 2015.
63. Lauk, J.; Nurk, E.; Robertson, A.; Parlesak, A. Culturally Optimised Nutritionally Adequate Food Baskets for Dietary Guidelines for Minimum Wage Estonian Families. *Nutrients* **2020**, *12*, 2613. [CrossRef]
64. Franco, D.; Martins, A.J.; López-Pedrouso, M.; Purriños, L.; Cerqueira, M.A.; Vicente, A.A.; Pastrana, L.M.; Zapata, C.; Lorenzo, J.M. Strategy towards Replacing Pork Backfat with a Linseed Oleogel in Frankfurter Sausages and Its Evaluation on Physico-chemical, Nutritional, and Sensory Characteristics. *Foods* **2019**, *8*, 366. [CrossRef] [PubMed]
65. Heck, R.T.; Fagundes, M.B.; Cichoski, A.J.; de Menezes, C.R.; Barin, J.S.; Lorenzo, J.M.; Wagner, R.; Campagnol, P.C.B. Volatile Compounds and Sensory Profile of Burgers with 50% Fat Replacement by Microparticles of Chia Oil Enriched with Rosemary. *Meat Sci.* **2019**, *148*, 164–170. [CrossRef] [PubMed]
66. Mertens, E.; Biesbroek, S.; Dofková, M.; Mistura, L.; D'Addezio, L.; Turrini, A.; Dubuisson, C.; Havard, S.; Trolle, E.; Geleijnse, J.M.; et al. Potential Impact of Meat Replacers on Nutrient Quality and Greenhouse Gas Emissions of Diets in Four European Countries. *Sustainability* **2020**, *12*, 6838. [CrossRef]
67. Capacci, S.; Mazzocchi, M.; Shankar, B.; Macias, J.B.; Verbeke, W.; Pérez-Cueto, F.J.A.; Kozioł-Kozakowska, A.; Piórecka, B.; Niedzwiedzka, B.; D'Addesa, D.; et al. Policies to Promote Healthy Eating in Europe: A Structured Review of Policies and Their Effectiveness. *Nutr. Rev.* **2012**, *70*, 188–200. [CrossRef]

nutrients

MDPI

Review

Replacement of Meat with Non-Meat Protein Sources: A Review of the Drivers and Inhibitors in Developed Countries

Marion R. Eckl *, Sander Biesbroek, Pieter van't Veer and Johanna M. Geleijnse

Division of Human Nutrition and Health, Wageningen University, Stippeneng 4, 6708 WE Wageningen, The Netherlands; sander.biesbroek@wur.nl (S.B.); pieter.vantveer@wur.nl (P.v.V.); marianne.geleijnse@wur.nl (J.M.G.)
* Correspondence: marion.eckl@wur.nl

Abstract: The overconsumption of meat has been charged with contributing to poor health and environmental degradation. Replacing meat with non-meat protein sources is one strategy advocated to reduce meat intake. This narrative review aims to identify the drivers and inhibitors underlying replacing meat with non-meat protein sources in omnivores and flexitarians in developed countries. A systematic search was conducted in Scopus and Web of Science until April 2021. In total, twenty-three studies were included in this review examining personal, socio-cultural, and external factors. Factors including female gender, information on health and the environment, and lower price may act as drivers to replacing meat with non-meat protein sources. Factors including male gender, meat attachment, food neophobia, and lower situational appropriateness of consuming non-meat protein sources may act as inhibitors. Research is needed to establish the relevance of socioeconomic status, race, ethnicity, religion, health status, food environment, and cooking skills. Future studies should prioritize standardizing the definitions of meat and non-meat protein replacements and examining factors across different consumer segments and types of non-meat protein sources. Thereby, the factors determining the replacement of meat with non-meat protein sources can be better elucidated, thus, facilitating the transition to a healthier and more sustainable diet.

Keywords: meat replacement; non-meat protein source; environmental sustainability; consumer preference; food choice

Citation: Eckl, M.R.; Biesbroek, S.; van't Veer, P.; Geleijnse, J.M. Replacement of Meat with Non-Meat Protein Sources: A Review of the Drivers and Inhibitors in Developed Countries. *Nutrients* **2021**, *13*, 3602. https://doi.org/10.3390/nu13103602

Academic Editors: Winston Craig and Ujué Fresán

Received: 17 September 2021
Accepted: 12 October 2021
Published: 14 October 2021

1. Introduction

Over the years, there has been an increasing body of research advocating for a reduction in the overconsumption of meat in order to mitigate negative health consequences and environmental burdens [1,2]. Despite being a valuable source of nutrients including protein, vitamin B-12, iron, and zinc [3], red and processed meat, in particular, have been shown to be associated with an increased risk of cardiovascular disease, stroke, cancer, as well as total mortality [3–5]. Plant-based diets have positive health benefits including a reduced risk of type II diabetes mellitus and cardiovascular disease [6–8]. Going further, meat production has been charged with contributing to environmental degradation including increased greenhouse gas emissions, loss of biodiversity, and disturbances in nitrogen-phosphorus soil balance [9–11]. By 2050, the world's population is expected to increase from 8 to 10 billion people [12]. Combined with continued global warming, such population growth will necessitate a further increase in food production, thereby, exacerbating the burden of non-communicable diseases and devastation of the environment [1]. As such, decreasing meat consumption in overconsuming developed countries remains key to abating such disastrous consequences in the coming years.

Overall, strategies to decrease meat consumption exist on a continuum from reduction to elimination [13]. Reduction strategies include decreasing the amount of meat consumed and often increasing the proportion of other non-meat foods at mealtimes (e.g., vegetables) [13,14]. Replacement strategies include either partially substituting meat with non-meat protein sources in traditional meat-based recipes (e.g., replacing a portion of beef with mushrooms in hamburgers) or fully substituting meat with non-meat protein sources (e.g., replacing pork with black beans in tacos) [13,14]. Decreasing the portion size of meat may be more feasible for many consumers as it does not require any alteration of the meal recipe and context or procurement of new cooking skills. In contrast, substituting meat with a non-meat protein source may be more or less feasible depending upon the degree of substitution, type of non-meat protein source, and necessary cooking skills involved to implement the recipe. Besides the sensory pleasure derived from meat, meat continues to maintain a strong symbolic place in many Western cultures often dominating the meal context as the central food item emblematic of higher socioeconomic status and masculinity [15,16]. Consequently, reducing meat consumption regardless of the strategy employed remains a challenge for many consumers.

With this in mind, it is crucial that we elucidate the underlying drivers and inhibitors to reducing meat consumption and particularly replacing meat with non-meat protein sources for public health policy, food industry, and dietitians and other health professionals in order to best facilitate a timely transition to a healthier and more sustainable global food system. To date, many reviews have looked at the factors involved in consumers reducing meat consumption in general but have not specifically examined the factors involved in consumers replacing meat with non-meat protein sources [17–21]. Although important fixtures in transforming the global food system, vegetarians and vegans constitute a small percentage of the population [22–25], and strict elimination of meat may not be realistic or necessary for most consumers particularly as a first step to reducing meat intake [26]. Consumer segments including omnivores, often referred to as meat-eaters, and flexitarians, often referred to as meat-reducers, comprise a much larger percentage of the population in many developed countries [22,23,26]. Therefore, understanding the motivations of omnivores and flexitarians is key to enacting a sizeable and long-term shift towards consuming less meat in developed countries.

As such, the aim of this review is to identify the drivers and inhibitors underlying consumer behavior of replacing meat with non-meat protein sources in omnivores and flexitarians in developed countries. In this way, we can contribute to better elucidating the motivations, attitudes, and behavior of omnivores and flexitarians in order to assist future research investigating the transition to and ultimately acceptance of healthier and more sustainable protein sources in society.

2. Materials and Methods

We conducted a comprehensive literature search in Scopus and Web of Science (Core Collection) in order to identify all studies examining the drivers and inhibitors to replacing meat with non-meat protein sources. The timespan of the search extended from the earliest date available in the databases up to April 2021. The search strings for the respective databases consisted of keywords relating to the replacement of meat with various non-meat protein sources including plant-based and alternative protein sources (Appendices A and B). The initial search was supplemented by a manual search of reference lists of relevant articles to identify studies not retrieved in either Scopus or Web of Science.

Only studies consisting of human consumers that eat meat ((i.e., consumers described as omnivores (or meat-eaters) or flexitarians (or meat-reducers)) from developed countries were included in this review. Studies consisting only of consumers described as pescatarians, vegetarians, or vegans or from developing or transition countries were excluded. Additionally, only studies that utilized a non-meat protein source to replace meat and that examined the drivers and inhibitors relating to the perception, awareness, attitude, motivation, willingness, and behavior to replace meat with a non-meat protein source were included. Figure 1 describes the literature search and provides details for the reasons for excluding studies including: (1) irrelevant topic; (2) irrelevant population; (3) irrelevant exposure; (4) irrelevant outcome; (5) irrelevant study design; (6) no full-text available; and (7) no English translation available.

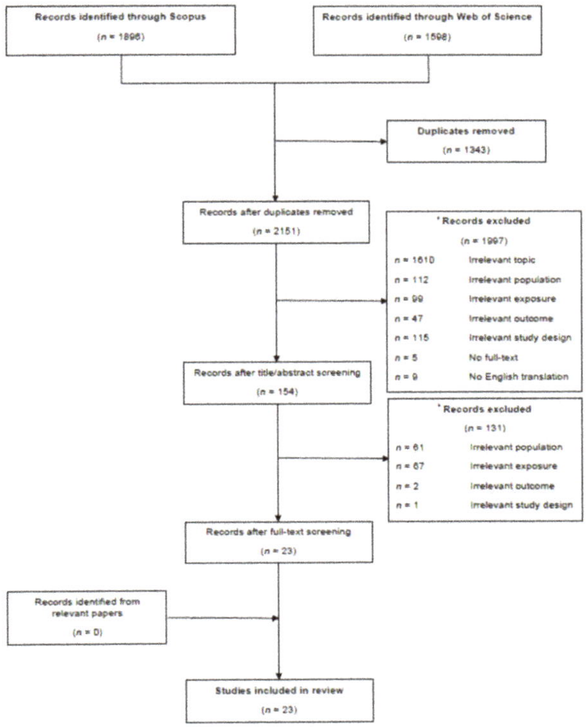

Figure 1. Flowchart of the identification, screening, and inclusion of studies assessing the drivers and inhibitors to replacing meat consumption with non-meat protein sources. * Exclusion criteria for review: (1) irrelevant topic (e.g., food production, animal physiology, or agriculture); (2) irrelevant population (e.g., included vegans, vegetarians, or pescatarians or only specialized populations such as students or armed forces); (3) irrelevant exposure (i.e., lack of replacement of meat with non-meat protein sources or nudging interventions); (4) irrelevant outcome (i.e., not pertaining to the perception, awareness, attitude, intent, willingness, or behavior to replace meat with non-meat protein sources); (5) irrelevant study design (i.e., reviews, protocols, pilot studies, editorials, opinions, or conference proceedings); (6) no full text available; and (7) no English translation available.

The title and abstract of articles were first screened for inclusion and exclusion criteria, and when these criteria were met, the full texts of the articles were retrieved and screened for these criteria. From the selected articles, we extracted data on the authors and year of publication; study location and design; population characteristics; data collection; non-meat protein replacements; explanatory and dependent variables; as well as the outcomes pertaining to the replacement of meat with non-meat protein sources. One researcher was involved in screening articles for inclusion and exclusion criteria and data extraction. A second researcher randomly cross-checked the screening of articles and data extraction and discussed any uncertainties and disagreements with the first researcher.

In this review, we utilized the theoretical framework of factors that influence meat-eating behavior by Stoll-Kleemann and Schmidt (2017) to organize and summarize our findings in the following sections with minor alterations (Figure 2) [17]. This framework was chosen as it provides a comprehensive overview of the personal, socio-cultural, and external factors that influence consumers' meat-eating behavior [17]. This framework considers internal and external incentives related to reducing meat consumption and furthermore the interrelationships among these factors [17]. This framework is based on the pro-environmental model developed by Kollmuss and Agyeman (2010), which also asserts the complexity and synergism of internal and external factors in determining individuals' propensity to partake in pro-environmental behavior that seeks to mitigate the negative impact of an individual's behavior on the environment [27].

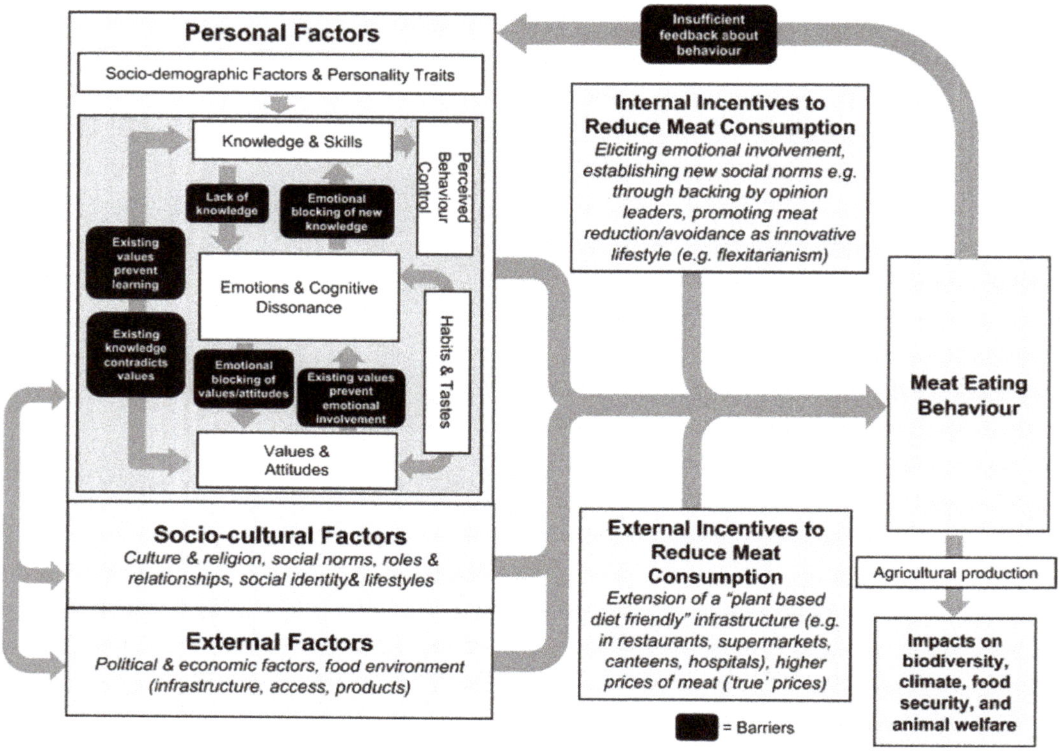

Figure 2. Model of factors that influence meat-eating behaviors. Reprinted from Stoll-Kleemann, S.; Schmidt, U.J. Reducing meat consumption in developed and transition countries to counter climate change and biodiversity loss: a review of influence factors. *Regional Environmental Change* **2017,** *17* (5), 1261–1277. No changes were made to this figure. Creative Common License 4.0 International License available at https://creativecommons.org/licenses/by/4.0/. (accessed on 12 October 2021).

3. Results

3.1. Study Characteristics

In total, twenty-three studies were included in this review. Table 1 provides an overview of the characteristics of these studies. The studies were published from 2011 [28] to 2021 [14,29–36]. Of the twenty-three studies, seventeen were conducted in Europe including [28–34,36–45]: Belgium [34,44], Denmark [29], Finland [29], France [31,36], Germany [29,32–34,36,41,45], Hungary [40], Iceland [29], Italy [39,43], The Netherlands [28,30,42], Romania [29], and the United Kingdom (UK) [36–38]. Five studies were conducted in North America [13,35,46–48] with four studies coming from the United States (US) [13,35,47,48] and one from Canada [46]. One study was conducted in New Zealand [14]. Importantly, three studies were conducted with consumers of multiple countries including: Denmark, Finland, Germany, Iceland, and Romania [29]; Germany, France, and the UK [36]; and Germany and Belgium [34]. Most studies (n = 20) employed quantitative research methods [13,28–31,33–45,47,48] with nine studies utilizing surveys [13,29,33,36,40,41,43–45] and eleven studies an experimental design [28,30,31,34,35,37–39,42,47,48]. One study was considered a qualitative study and utilized semi-structured interviews [14]. Two studies employed a mixed-methods approach of quantitative and qualitative research methods in their study design [32,46]. In terms of replacements of meat, seventeen studies investigated full replacements of meat with non-meat protein sources [14,28–32,36–46], whereas six studies investigated a partial replacement of meat with non-meat protein sources such as mushrooms or legumes [13,33–35,47,48]. Most studies (n = 19) investigated replacing meat with plant-based protein sources, such as Quorn®, tofu, lentils, or legumes [13,14,28–39,41,42,46–48]; and two studies each respectively investigated replacing meat with insects [40,44] and cultured meat [43,45]. Table 2 provides a summary of the findings for each of the personal, socio-cultural, and external factors identified among the included studies.

3.2. Personal Factors

3.2.1. Socio-Demographics

Age

Six studies examined age as a factor influencing the replacement of meat with non-meat protein sources [13,36–38,44,45]. In a discrete choice experiment (DCE) conducted in the UK in 2016, 233 meat-eaters and meat-reducers were segmented into five consumers segments based on their preferences for product-attributes of ground meat and ground meat substitute (i.e., soy, tofu, and Quorn®), which varied by age [37]. Organic (79%), green (45%), and taste-driven (46%) consumers were more likely to be between 18–34 years (yr). Price-conscious consumers were more likely to be between the ages of 35–55 yr (65%), and healthy consumers were more likely to be older than 55 yr (79%) [37]. In another DCE conducted in the UK in 2019 with 400 participants, age also varied among the five consumer segments of meat-eaters and meat-reducers based on their preferences for product attributes of ground meat and ground meat substitute (Quorn®) [38]. Of the meat-eaters, traditional meat-eaters were more likely to be older and showed a greater preference for ground beef compared to price-conscious meat-eaters who were influenced more by, not just the type, but the price of ground meat or ground meat substitute [38]. In an online survey conducted in Belgium in 2015 with 368 participants, every 10-year increase in age was associated with a 27% reduction in the readiness to adopt insects as a meat substitute [44]. Similarly, in a multi-national online survey conducted in Germany, France, and the UK in 2021 with 1734 participants, it was found that older participants were more likely to provide lower ratings for the expected tastiness, healthiness, and environmental friendliness of pea burgers [36]. Nevertheless, it was not found that age was associated with the expectations for taste, health, and environmental friendliness of algae burgers [36].

Furthermore, a 2020 US survey with 602 participants found no association between age and the assessment and acceptance of blending mushrooms into traditionally meat-based foods to reduce meat consumption [13]. Additionally, a 2020 survey in Germany found that age was not shown to moderate the attitudes pertaining to the intention to try, eat, or promote cultured meat to friends among 713 German participants [45].

Table 1. Overview of twenty-three studies included in the review examining the drivers and inhibitors underlying the replacement of meat with non-meat protein sources.

Author, Year	Study Design, Location, and Population	Research Aim	Non-Meat Protein Replacement(s)	Explanatory Variable(s)	Dependent Variables(s)	Main Outcome(s)
Apostolidis & McLeay (2016) [37]	DCE, UK; n = 233; men & women *	Identify the attributes of ground meat (substitute) that influence consumers' choices	Ground meat substitute (i.e., soy, tofu, Quorn®, etc.)	Type of ground meat (substitute) Fat content of ground meat (substitute) Carbon footprint of ground meat (substitute) Method of production of ground meat (substitute) Price of ground meat (substitute) Origin of ground meat (substitute) Brand of ground meat (substitute) Age Gender Income Household Region of residence	Preference for attributes of ground meat (substitute)	Five consumer segments were identified among meat-eaters and meat-reducers: price-conscious (42.5%), green (17%), taste-driven (14.6%), healthy (10.5%), and organic (9.7%) consumers. Strongest influences for price-conscious consumers were type of ground meat (substitute) and origin; for green consumers, carbon footprint and origin; for taste-driven consumers, type of ground meat (substitute) and brand; for healthy consumers, fat content and type of ground meat (substitute); and for organic consumers, fat content and type of ground meat (substitute).
Apostolidis & McLeay (2019) [38]	DCE, UK; n = 400; 61% women *	Compare the importance of sustainability-related labels on consumers' preferences for ground meat (substitute)	Ground meat substitute (Quorn®)	Type of ground meat (substitute) Fat content of ground meat (substitute) Carbon footprint of ground meat (substitute) Method of production of ground meat (substitute) Price of ground meat (substitute) Origin of ground meat (substitute) Brand of ground meat (substitute) Age Gender Income Household Region of residence	Preference for sustainability labels of ground meat (substitute)	Meat-eaters were primarily influenced by the type of ground meat (substitute), origin, and price, whereas the meat-reducers were primarily influenced by the type of ground meat (substitute) fat content, and origin. Meat-eaters were segmented into three consumer segments: price-conscious (63%) primarily influenced by the type of ground meat (substitute) and price; traditional (19%) primarily influenced by the type of ground meat (substitute) and origin; and empowered (18%) primarily influenced by type of ground meat (substitute) and production. Meat-reducers were segmented into two consumer segments: health curtailers (82%) primarily influenced by the fat and origin; and sustainable (18%) primarily influenced by carbon footprint and fat content.

Table 1. *Cont.*

Author, Year	Study Design, Location, and Population	Research Aim	Non-Meat Protein Replacement(s)	Explanatory Variable(s)	Dependent Variables(s)	Main Outcome(s)
Banovic & Sviensdóttir (2021) [29]	Online survey, Denmark, Finland, Germany, Iceland, and Romania; *n* = 1397; 100% women *	Investigate whether general attitudes towards using plant protein in food production and intention to substitute meat protein in the diet are related to the attitude towards rapeseed protein and the attitude and intention to buy meat analogues	Wiener sausages containing rapeseed protein	Attitude towards plant protein in food production Attitude towards rapeseed protein Attitude towards soy protein Attitude towards potato starch Attitude towards gluten Intention to substitute meat protein in the diet Meat consumption frequency Country of origin	Attitude towards and intention to buy meat analogues	Attitude towards using plant protein in food production was shown to influence both the intention to substitute meat and the attitude towards using rapeseed protein as an ingredient in meat analogues in all countries. Attitude towards rapeseed was shown to influence the attitude towards meat analogue.
Castellari et al. (2019) [39]	WTP experiment, Italy; *n* = 119; 47% women *	Evaluate the impact of explanatory messages about health and environment on consumers' WTP for a beef burger and soy burger	Soy burger	Information on health Information on environment	WTP and chosen quantities for beef and soy burgers	Successive rounds of explanatory messaging on health and the environment resulted in a relative decrease of −1.6% in the WTP for beef burgers and relative increase of +3.6% in the WTP soy burgers. Successive rounds of explanatory messaging on health and the environment resulted in a relative decrease of −23.0% in the chosen quantities of the beef burger and a relative increase of +45.6% in the chosen quantities of soy burgers.
Elzerman et al. (2011) [28]	Experiment, The Netherlands; *n* = 93; 77% women §	Obtain insight into the influence of meal context on the acceptance of meat substitutes	Quorn® pieces Quorn® ground Tofu strips Tivall® stir-fry pieces Goodbite® chicken style Vivera® vega stir-fry pieces	Meal context of meat substitute Flavor of meat substitute Texture of meat substitute Form of meat substitute	Overall liking of meat substitute Overall liking of dish Product liking of meat substitute in dish Perceived appropriateness of meat substitute in meal context Intention to use dish with meat substitute	Quorn® pieces were liked more than Quorn® ground when compared separately in the rice and salad dishes, but there was no difference in the overall liking of the meals using Quorn® pieces or Quorn® ground. Shape of meat substitutes appears to influence the appropriateness of the meal more than the flavor and texture of the specific meat substitute.

Table 1. *Cont.*

Author, Year	Study Design, Location, and Population	Research Aim	Non-Meat Protein Replacement(s)	Explanatory Variable(s)	Dependent Variables(s)	Main Outcome(s)
Elzerman et al. (2021) [30]	Experiment, The Netherlands; n = 309; men & women *	Explore the perceived appropriateness of meat products, meat substitutes, and meat alternatives in different usage situations	Vegetarian ground Vegetarian stir-fry pieces Vegetarian hamburger Vegetarian sausages Chickpeas and nuts	Family context Special meal context Vegetarian context Friends context Alone context Children context Flavor context Little time context Health context	Perceived situational appropriateness of meat substitutes in various situations	Situational appropriateness of meat products was higher than meat substitutes and chickpeas and nuts in almost all situations except for vegetarian and health contexts.
Gere et al. (2017) [40]	Online survey, Hungary; n = 400; men & women *	Assess the readiness of Hungarian consumers to adopt insects as a substitute for meat	Insects	Age Gender Education Food neophobia Food technology neophobia Attitude towards health characteristics of food Convenience orientation for food choice Attention to the environmental impact of food Belief that meat is nutritious and healthy Intention to reduce fresh meat intake Familiarities with insects Familiarity with whey Familiarity with algae Familiarity with soy	Readiness to adopt insects as a substitute for meat	Participants intending to reduce their fresh meat intake within the next year had an expected increase of 1.47 in the number of preferred insect types they would be willing to consume as a substitute for meat. Food neophobia was found to be a barrier to the readiness to adopt insects as a substitute for meat.
Gravely & Fraser (2018) [46]	Interviews, Canada; n = 24	Investigate the in-store context for purchasing plant-based protein in major Canadian supermarkets	Plant-based protein	Product availability Product promotions Product location	Extent to which grocery stores are mediating the transition to plant-based protein sources	More space and promotions were allotted to animal-based protein than plant-based protein in grocery stores. Participants found it easier to find animal-based protein compared to plant-based protein in grocery stores.

Table 1. *Cont.*

Author, Year	Study Design, Location, and Population	Research Aim	Non-Meat Protein Replacement(s)	Explanatory Variable(s)	Dependent Variables(s)	Main Outcome(s)
Guinard et al. (2016) [47]	Experiment, USA; n = 147; 58% women *	Test consumer acceptance of meat-based dishes in which meat had been substituted with mushrooms	Low-meat carne asada † Low-meat beef tacos †	Percent substitution with mushrooms Appearance of dish Flavor of dish Texture of dish	Overall liking Liking of appearance Liking of flavor Liking of texture Level of saltiness Level of spiciness Level of moistness	100% beef carne asada was liked more for overall liking, appearance, flavor, and texture compared to the 50% beef carne asada. 100% beef tacos were liked more than mushroom-containing beef tacos for appearance but not more for overall liking, flavor, and texture.
Hartmann & Siegrist (2020) [41]	Online survey, Germany; n = 973; 49% women *	Investigate the impact of unapologetic and apologetic justification strategies on consumers' willingness to substitute meat with meat alternatives	Quorn® Tofu Seitan Soy schnitzel	Unapologetic justification strategy: Pro-meat Denial Hierarchical justification Religion justification Health justification Human destiny Slaughter justification Apologetic justification strategy: Dissociation Dichotomization Avoidance	Willingness to substitute meat for Quorn®, tofu, seitan, or soy schnitzel	Participants who scored higher on the unapologetic justification strategies were less willing to substitute meat with Quorn®, tofu, seitan, or soy schnitzel compared to those who scored lower.
Hoek et al. (2013) [42]	Longitudinal experiment, The Netherlands; n = 89; 78% women ¶	Investigate the hedonic effects of repeated exposure to meat substitutes and meat	Quorn® Tofu	Meal context Type of product Prior consumption of meat substitutes Prior consumption of chicken Repeated exposure to meat or meat substitute Different meals used Hunger Food neophobia Variety-seeking	Desire to eat the product Liking of the product Boredom with the product Amount of eaten product	Liking scores among Quorn®, tofu, and chicken were not different after the repeated exposure period. Most participants who ate tofu showed a mere exposure pattern of increased liking over time, whereas most participants who ate chicken showed a boredom pattern of decreased liking over time. Entire meal was liked better than Quorn®, tofu, or chicken evaluated separately. Food neophobia and variety-seeking did not have an effect on overall product liking over time.

Table 1. *Cont.*

Author, Year	Study Design, Location, and Population	Research Aim	Non-Meat Protein Replacement(s)	Explanatory Variable(s)	Dependent Variables(s)	Main Outcome(s)
Kemper & White (2021) [14]	Semi-structured interviews, New Zealand; *n* = 23; 74% women ¶	Explore young adults' motivations, strategies, and barriers towards flexitarianism	Legumes Lentils Tofu	Cooking skills	Ability to adopt a flexitarian lifestyle	Participants who were more confident and experienced in cooking substituted meat for legumes, lentils, and tofu, whereas participants who were less confident and experienced in cooking preferred substituting meat with meat substitutes like vegetarian patties.
Lang (2020) [13]	Online survey, USA; *n* = 602 *	Explore consumers' response to blending mushrooms into traditional meat-based foods and their lifestyle and motivations influencing the assessment and acceptance of these blended foods	Meat-hybrid products **	Perceived health benefits Perceived cost benefits Perceived taste benefits Perceived culinary benefits Perceived sustainability benefits Assessment of blending Format of blended products Red meat consumption Healthy eating Cooking habits Food innovativeness Food involvement Food knowledge Age Gender Income Education	Acceptance of blending mushrooms into traditionally meat-based foods	Top reasons for consuming blended foods were health benefits followed by price, taste, culinary, and sustainability benefits. Burgers were the preferred format for consuming blended products followed by stir-fry with ground beef, meatloaf, tacos, chili with ground beef, pasta with ground beef, and other. Age, gender, income, and education were not associated with the acceptance of the blending concept, but women assessed blending more positively than men. Participants whose red meat consumption was declining or were contemplating decreasing their consumption were associated with more favorable assessment and acceptance of blending.
Mancini & Antonioli (2019) [43]	Online survey, Italy; *n* = 485; men & women *	Assess the extent to which Italian consumers are willing to accept cultured meat	Cultured meat	Information on positive internalities of cultured meat Information on positive externalities of cultured meat	Perception of cultured meat	Participants showed better agreement with the extrinsic attributes of cultured meat compared to the intrinsic attributes.

Table 1. *Cont.*

Author, Year	Study Design, Location, and Population	Research Aim	Non-Meat Protein Replacement(s)	Explanatory Variable(s)	Dependent Variables(s)	Main Outcome(s)
Martin et al. (2021) [31]	Experiment, France; n = 102; 51% women ¶	Test if information concerning the consequences on health or the environment could be useful in promoting plant-based products	Plant-based sausage	Taste Packaging Information on health Information on environment	Preference to purchase plant-based sausage WTP plant-based sausage	Participants preferred to purchase the pork-based sausage over the plant-based sausage after the blind tasting and tasting with packaging. WTP for the plant-based sausage increased after the second message on health or the environment, and WTP for the pork-based sausage decreased after the second message on the environment.
Michel, Hartmann, et al. (2021) [32]	Online survey, Germany; n = 967; 50% women ¶	Identify barriers that prevent consumers from eating meat alternatives	Meat alternatives Vegetarian nuggets Tofu Vegetarian sausage	Eating alone context Eating with friends context Eating with family on a weekday context Eating with family on Sunday context Invited for dinner in a restaurant context Business meal context Barbecue party context Perceived taste Perceived texture Perceived price Perceived ease of preparation Perceived protein content Perceived fat content Perceived environmental friendliness Perceived masculinity Perceived festivity Perceived healthiness Perceived satiation Perceived naturalness Type of product	Acceptability of eating plant-based meat alternatives	Omnivores and flexitarians rated eating alone as the most appropriate situation to consume meat alternatives. Omnivores rated meat as performing better in regards to taste, texture, price, ease of preparation, protein content, fat content, and environmental friendliness, whereas flexitarians rated meat alternatives as performing better in terms of fat content and environmental friendliness. Participants perceived steak as being the most healthy, protein-rich, filling, natural, festive, masculine, and tasty.

Table 1. *Cont.*

Author, Year	Study Design, Location, and Population	Research Aim	Non-Meat Protein Replacement(s)	Explanatory Variable(s)	Dependent Variables(s)	Main Outcome(s)
Michel, Knaapila, et al. (2021) [36]	Online survey, Germany, France, and UK; *n* = 1734; 48% women ¶	Investigate the taste, healthiness, and environmental expectations of pea and algae burgers as meat alternatives and the factors influencing these expectations	Pea burger Algae burger	Age Sex Country of origin Meat commitment Food neophobia Attitude towards vegans and vegetarians	Expected tastiness Expected healthiness Expected environmental friendliness	Pea and algae burgers were expected to be healthier and more environmentally friendly but less tasty than beef burgers in all countries. Participants who were more committed to meat, food neophobic, and had a negative attitude towards vegans and vegetarians rated the tastiness, healthiness, and environmental friendliness of pea and algae burgers lower. Being older, male, and from France was associated with providing negative ratings for the tastiness, healthiness, and environmental friendliness of pea burgers.
Profeta, Baune, Smetana, Bornkessel, et al. (2021) [33]	Online survey, Germany; *n* = 500; 51% women *	Identify consumer attitudes and preferences for meat-hybrids	Meat-hybrid products ††	Perceived tastiness Perceived healthiness Perceived environmental friendliness Perceived animal welfare Attachment to meat Food neophobia Frequency of purchasing plant-based alternatives	Preference for meat-hybrids	Participants rated meat-hybrids as performing better in terms of perceived healthiness, environmental friendliness, and animal welfare but performing worse in terms of perceived tastiness compared to the 100% meat option. The more attached a participant was to meat and the more food neophobic, the less likely they were to choose the meat-hybrid. The higher the participant rated the meat-hybrid in terms of perceived healthiness, environmental friendliness, and animal welfare, the more likely they were to choose the meat-hybrid with perceived healthiness exerting the largest influence.

Table 1. *Cont.*

Author, Year	Study Design, Location, and Population	Research Aim	Non-Meat Protein Replacement(s)	Explanatory Variable(s)	Dependent Variables(s)	Main Outcome(s)
Profeta, Baune, Smetana, Broucke, et al. (2021) [34]	DCE, Germany and Belgium; $n = 1001$; 51% women *	Identify consumer attitudes and preferences for meat-hybrids	Meat-hybrid meatballs †† Meat-hybrid mortadella †† Meat-hybrid salami †† Meat-hybrid chicken nuggets †† Vegetarian meatballs Vegetarian mortadella Vegetarian salami Vegetarian chicken nuggets	Perceived tastiness Perceived healthiness Perceived environmental friendliness Perceived animal welfare Percent substitution with plant-based protein Organic label Origin label Environmental label Nutritional label Price Attachment to meat Food neophobia Importance of eating healthy	Preference for meat-hybrids	Participants in Germany rated the meat-hybrids as performing better in terms of perceived healthiness, environmental friendliness, and animal welfare, whereas participants in Belgium rated meat-hybrids as performing better in terms of perceived environmental friendliness and animal welfare. Meat was the most preferred option followed by the meat-hybrids with the least preferred option being the 100% vegetarian products.
Spencer et al. (2018) [48]	Experiment, USA; $n = 110$; 58% women §	Test the concept of the Flexitarian Flip™ in a dining venue context by replacing meat with legumes in meat-based recipes	Low-meat pork carnitas arepas † Low-meat chicken tikka masala †	Amount of meat in recipe Flavor Texture Appearance Spiciness	Overall liking of dish Liking of appearance of dish Liking of flavor of dish Liking of texture of dish Liking of spiciness of dish	High-meat arepas were liked more than low-meat arepas and high- and low-meat chicken tikka masala dishes for overall liking. High-meat dishes were liked more than low-meat dishes for overall liking and flavor liking. Spicy versions of the arepas and chicken tikka masala recipes were liked more for flavor and texture than the regular versions across all meat levels.
Spencer et al. (2021) [35]	Experiment, USA; $n = 144$; 65% women; * 58% Caucasian	Investigate implementation of the mixed dish Flexitarian Flip™ strategy in a different geographical area and with a new cuisine	Low-meat East Asian bowls †	Amount of meat in recipe Flavor Texture Appearance Spiciness Satiation Satisfaction Gender	Overall liking of recipe Liking of appearance of recipe Liking of flavor of recipe Liking of texture of recipe Liking of spiciness	No differences in the overall liking, flavor, texture, appearance, satiation, or satisfaction of bowls regardless of the amount of meat. Across all subjects and bowls, not having enough flavor complexity resulted in a decrease in overall liking.

Table 1. *Cont.*

Author, Year	Study Design, Location, and Population	Research Aim	Non-Meat Protein Replacement(s)	Explanatory Variable(s)	Dependent Variables(s)	Main Outcome(s)
Verbeke (2015) [44]	Online survey, Belgium; *n* = 368; 61% women *	Profile consumers who claim to be ready to eat insects as a substitute for meat	Insects	Age Gender Education Familiarity with the idea of eating insects Food neophobia Food technology neophobia Attitude towards health characteristics of food Convenience orientation for food choice Attention to the environmental impact of food Belief that meat is nutritious and healthy Importance of taste when evaluating meat Intention to reduce fresh meat	Readiness to adopt insects as a substitute for meat	Food neophobia was the largest barrier to being ready to adopt insects as a substitute for meat. Male gender, familiarity with the idea of eating insects, convenience orientation for food choice, attention to the environmental impact of food, and planning on reducing fresh meat intake within the next year all increased the likelihood of readiness to adopt insects as a substitute for meat. Increase in age, food neophobia, food technology neophobia, belief that meat is nutritious and healthy, and the importance of taste when evaluating meat all decreased the likelihood of readiness to adopt insects as a substitute for meat.
Weinrich et al. (2020) [45]	Online survey, Germany; *n* = 713; 53% women *	Explore the readiness and intentions of consumers to use cultured meat, their attitudes and their driver strength, and demographic predictors	Cultured meat	Attitude towards cultured meat Age Gender Education Income Region Pre-knowledge Living with children	Intention to try and eat cultured meat and intention to promote cultured meat to friends	Participants' attitudes towards cultured meat were structured into three dimensions: ethics, emotional objections, and global diffusion optimism. Ethics was the strongest predictor for using cultured meat followed by emotional objections and global diffusion. Participants' pre-knowledge of cultured meat was shown to increase the ethical beliefs of cultured meat but did not impact the emotional objections or the global diffusion optimism of cultured meat.

* Study specified investigating gender. † Traditionally meat-based recipes in which a portion of the meat had been substituted with legumes. § Study did not specify whether gender or sex was investigated. ¶ Study specified investigating sex. ** Meat-hybrids consisting of meat-based food products in which a partial portion of the meat had been replaced by mushrooms. †† Meat-hybrids consisting of meat-based food products in which a partial portion of the meat had been replaced by plant-based protein. Abbreviations: DCE, discrete choice experiment; WTP, willingness to pay; UK, United Kingdom; USA, United States of America.

Table 2. Summary of drivers and inhibitors underlying the replacement of meat with non-protein sources examined by the studies included in this review *.

Factors			Summary	Number of Studies Examining the Factor
Personal	Socio-demographics	Age [13,36–38,44,45]	Unclear whether age is a driver or inhibitor.	6
		Gender and sex [13,35–38,44,45]	Female gender may be a driver and male gender may be an inhibitor to replacing meat with non-meat protein sources, but it is unclear if this applies to all alternative protein sources such as insects and cultured meat. Male sex is a possible inhibitor to replacing meat with non-meat protein sources but was only examined by one study.	7
		Socioeconomic status [13,37,38,44,45]	Unclear whether socioeconomic status is a driver or inhibitor.	5
		Ethnicity and race	Factor was not examined by the studies in this review.	0
		Religion	Factor was not examined by the studies in this review.	0
	Sensory and hedonic aspects	Taste, texture, and appearance [13,28,31,32,35,44,47,48]	Taste and texture of meat may be inhibitors to replacing meat with non-meat protein sources depending upon the recipe and meal context, but appearance of a dish appears to exert much less of an influence.	8
	Hunger cues	Hunger and satiety [32,35,42]	Hunger and satiety may be drivers to replacing meat with non-meat protein sources.	3
	Personality traits	Food and food production neophobia [13,33,34,36,40,42,44]	Food neophobia may be an inhibitor to replacing meat with non-meat protein sources, particularly for novel alternative protein sources.	7
		Variety seeking [42]	Variety seeking may be an inhibitor to replacing meat with non-meat protein sources but was only examined by one study.	1

Table 2. *Cont.*

Factors		Summary	Number of Studies Examining the Factor
Knowledge and skills	Information on health and the environment [31,39,43]	Providing information on health and the environment may be drivers to replacing meat with non-meat protein sources.	3
	Cooking skills and food knowledge [13,14,44,45]	Food knowledge of non-meat protein sources may be a driver to replacing meat with non-meat protein sources. Cooking skill is a possible driver to replacing meat with non-meat protein sources but was only examined by one study.	4
Emotions and cognitive dissonance	Justification strategies to consume meat [41]	Unapologetic justification strategies are a possible inhibitor to replacing meat with non-meat protein sources but were only examined by one study.	1
Values and attitudes	Health and the environment [13,32–34,44]	Health may be a driver or inhibitor to replacing meat with non-meat protein sources depending on whether meat or the non-meat protein source is perceived as being healthy by the consumer. Sustainability may be a driver to replacing meat with non-meat protein sources but to a lesser degree than health.	5
	Plant protein sources and food production [13,29,45]	More positive assessments of plant protein sources and production may be drivers to replacing meat with non-meat protein sources.	3
	Vegans and vegetarians [36]	Negative attitude towards vegans and vegetarians is a possible inhibitor to replacing meat with non-meat protein sources but was only examined by one study. The influence of a positive attitude towards vegans and vegetarians was not examined.	1
	Others [13,32]	Perceived naturalness, masculinity, and festivity are possible inhibitors to replacing meat with non-meat protein sources but were only examined by one study. Perceived culinary benefits do not appear to be a driver to replace meat with non-meat protein sources but was only examined by one study.	2

Table 2. *Cont.*

Factors			Summary	Number of Studies Examining the Factor
	Habits and tastes	Healthy eating [13]	Healthy eating is a possible driver to replacing meat with non-meat protein sources but was only examined by one study.	1
		Consumption of meat [29,33,34,36,40,44]	Consuming less meat may be a driver whereas increased commitment or attachment to meat may be an inhibitor to replacing meat with non-meat protein sources.	6
		Consumption of meat substitutes [42]	Consumption of meat substitutes is a possible driver to replacing meat with non-meat protein sources but was only examined by one study.	1
		Cooking habits and food involvement [13,32]	Unclear whether cooking habits or food involvement is a driver or inhibitor.	2
	Perceived behavior control		Factor was not examined by the studies in this review.	0
Socio-cultural	Culture	Country of consumer [29,34,36]	Specific country of participant may be a driver or inhibitor to replacing meat with non-meat protein sources.	3
	Social norms, roles, and relationships	Situational context [30,32]	Perceived lower situational appropriateness of non-meat protein sources may be an inhibitor to replacing meat with non-meat protein sources particularly in formal and social contexts.	2
	Social identity and relationships		Factor was not examined by the studies in this review.	0
External	Political factors		Factor was not examined by the studies in this review.	0
	Economic factors	Price [13,32,34,37,38]	Lower price of non-meat protein sources may be a driver to replacing meat with non-meat protein sources but possibly only for specific consumer segments.	5

Table 2. *Cont.*

Factors		Summary	Number of Studies Examining the Factor
Food environment	External product-attributes [13,31,34,37,38,44]	Packaging information, format, and convenience packaging of the non-meat protein source may be drivers to replacing meat with non-meat protein sources but possibly only for specific consumer segments.	6
	Meal context [28,35,42,48]	Meal context may be a driver to replacing meat with non-meat protein sources and is likely a more pertinent factor in acceptability than evaluating the individual non-meat protein source alone.	4
	Grocery-store context [46]	Grocery store context is a possible inhibitor to replacing meat with non-meat protein sources but was only examined by one study.	1
Animal welfare	Animal welfare [33]	Animal welfare is a possible driver to replacing meat with non-meat protein sources but was only examined by one study.	1

* Grouping of drivers and inhibitors based on the model of factors that influence meat-eating behaviors from Stoll-Kleemann, S.; Schmidt, U.J. Reducing meat consumption in developed and transition countries to counter climate change and biodiversity loss: a review of influence factors. *Regional Environmental Change* **2017**, *17* (5), 1261–1277. Creative Common License 4.0 International License available at https://creativecommons.org/licenses/by/4.0/. (accessed on 12 October 2021).

Gender and Sex

Seven studies investigated the role of gender or sex in the replacement of meat with non-meat protein sources [13,35–38,44,45]. Similar to age, the five consumer segments of meat-eaters and meat-reducers in the 2016 UK DCE varied by gender [37]. Green (62%) and healthy (78%) consumers tended to identify as female, whereas taste-driven (61%) and organic (83%) consumers tended to identify as male. Price-conscious consumers, on the other hand, largely identified equally as male and female [37]. In the 2019 UK DCE, the five consumer segments of meat-eaters and meat-reducers also varied by gender [38]. Of the meat-eaters, empowered consumers, who were influenced more by the type, production, and fat content of ground meat or ground meat substitute, were more likely to identify as female than the traditional and price-conscious consumers. Additionally, meat-reducers (67%) in general were more likely to identify as female than meat-eaters (54%) with the meat-reducer consumer segments of health curtailers and sustainable consumers consisting of 86% and 70% females respectively [38]. In a 2021 US experiment, the results showed that participants identifying as male preferred the high-meat dishes without partial replacement of meat with legumes and vegetables more than the participants identifying as female [35]. In a Belgian survey, participants identifying as male were more than twice as likely to adopt insects as a substitute for meat compared to participants identifying as female [44].

Contrastingly, in a recent US survey, gender was not associated with the acceptance of blending mushrooms into traditional meat-based foods; however, participants identifying as female assessed the blending concept more favorably than the participants identifying as male [13]. Additionally, a 2020 German survey found that gender was not shown to moderate the attitudes pertaining to the intention to try, eat, or promote cultured meat [45].

In a multi-national survey in Germany, France, and the UK investigating the impact of sex, male participants were associated with providing lower ratings for the expected environmental friendliness of pea burgers [36]. Similar to age, however, sex was not shown to influence the expected tastiness, healthiness, and environmental friendliness of algae burgers [36].

Socioeconomic Status

Five studies assessed education and income as factors influencing the replacement of meat with non-meat protein sources [13,37,38,44,45]. Similar to age and gender, the five consumer segments of meat-eaters and meat-reducers varied by income in the 2016 UK DCE [37]. Both price-conscious (66%) and organic (100%) consumers tended to be of lower income [37]. In the 2019 UK DCE, the five consumer segments of meat-eaters and meat-reducers also varied by income [38]. Of the meat-eaters, empowered meat-eaters were found to have a higher proportion of participants earning a higher income [38].

Contrastingly, a recent US survey found that neither income nor education was associated with the assessment or acceptance of blending mushrooms into traditional meat-based foods [13]. In a Belgian survey, education was also not shown to be associated with the readiness to adopt insects as a substitute for meat among participants [44]. Similarly, education was not shown to moderate the attitudes pertaining to the intention to try, eat, or promote cultured meat among participants in a 2020 German survey [45].

3.2.2. Sensory and Hedonic Aspects

Eight studies examined the role of taste, texture, and appearance in the replacement of meat with non-meat protein sources [13,28,31,32,35,44,47,48]. On blind tasting alone, a 2021 experiment in France found that participants preferred to purchase the pork-based sausage over the plant-based sausage [31]. Similarly, a 2021 German survey found that participants perceived all meat products as being tastier than corresponding plant-based meat alternatives (e.g., chicken nuggets versus vegetarian nuggets) [32]. Furthermore, both omnivores and flexitarians in this study rated meat as being expected to perform better than meat alternatives in terms of flavor and texture. In particular, steak was perceived as being the tastiest food product in this survey compared to tofu, chicken nuggets, vegetarian nuggets, wiener sausages, and vegetarian sausages [32]. In a Belgian survey, for every one unit increase in the importance of taste when evaluating meat, there was a 61% decrease in participants' readiness to adopt insects as a substitute [44]. Interestingly, a 2011 experiment conducted in the Netherlands found that the flavor and texture of specific meat substitutes influenced the perceived appropriateness of a meat substitute in a meal context less than the actual shape of the meat substitute itself—whether in pieces or ground [28]. In this experiment, participants liked Quorn® pieces more than Quorn® ground in the rice and salad dishes [28].

In terms of partial replacements of meat, participants in a recent US survey rated the perceived taste benefits as an intermediate reason for blending mushrooms into traditional meat-based foods falling behind perceived health and cost benefits [13]. A US experiment conducted in 2016 examined participants' preferences for beef carne asada and beef taco recipes in which a portion of the beef had been partially replaced by mushrooms [47]. In this experiment, participants liked the 100% beef carne asada recipe more in terms of flavor and texture compared to the carne asada recipe that replaced 50% of the beef with mushrooms. Contrastingly, there were no differences in the flavor and texture among the 100% beef taco recipes and the mushroom-containing beef taco recipes containing either 50% or 80% mushrooms. Although participants rated the appearance of the 100% beef taco recipe higher than the appearance of the mushroom-containing beef taco recipes, a correlation analysis revealed that flavor was the best predictor for the overall liking of the beef taco recipes followed by texture; but, appearance was not related to the overall liking of the beef taco recipes [47].

Another US experiment conducted in 2018 similarly investigated participants' preferences for pork carnitas arepas and chicken tikka masala in which a portion of the meat had been partially replaced by legumes [48]. Participants liked the high-meat arepas recipe more for flavor, texture, and appearance than the low-meat arepas and the high- and low-meat chicken tikka masala recipes. The high-meat versions of the arepas and chicken tikka masala recipes were also liked more than the low-meat versions of these recipes; however, no differences were found in the texture and appearance of the high- and low-meat versions. Notably, however, spicy versions of the arepas and chicken tikka masala recipes were liked more for flavor and texture than the regular versions across all meat levels [48].

In a 2021 US experiment, there were no differences among the East Asian bowls in terms of flavor regardless of whether meat had been replaced by legumes and vegetables or the spiciness of the dish [35]. Nevertheless, it was found that not having enough flavor complexity resulted in a decrease of 10 on a 100 scale in overall liking across all participants and bowls. For participants who felt their bowl was not spicy enough, there was a decrease of 8 in overall liking. For participants who felt the bowl was too spicy, there was an overall mean liking drop of 15 [35].

3.2.3. Hunger Cues

Three studies investigated the role of hunger and satiety in the replacement of meat with non-meat protein sources [32,35,42]. In a longitudinal experiment conducted in the Netherlands, the effect of repeated exposure to Quorn® or tofu on product liking was investigated [42]. In this experiment, it was found that the hungrier the participant, the more likely they were to like Quorn® or tofu [42]. In a 2021 US experiment, however, there were no differences in participants' ratings of satiation, or the feeling of fullness, or satisfaction among East Asian bowls regardless of the meat level [35]. In a 2021 German survey, steak was perceived by participants to being the most filling food product compared to tofu, chicken nuggets, vegetarian nuggets, wiener sausages, and vegetarian sausages [32].

3.2.4. Personality Traits

Food Neophobia and Food Technology Neophobia

Seven studies assessed food neophobia and food technology neophobia as factors influencing replacement of meat with non-meat protein sources [13,33,34,36,40,42,44]. In two online surveys conducted in Hungary and Belgium, food neophobia, or the propensity to avoid consuming new foods, was identified as a barrier to the readiness of participants to adopt insects as a substitute for meat [40,44]. In a Belgian survey, there was an 84% and 55% decrease in the readiness of participants to adopt insects as a substitute for meat respectively for every one unit increase in food neophobia and food technology neophobia, or the propensity to avoid consuming foods produced by new technologies [44]. In the multi-national survey conducted in Germany, France, and the UK, the more food neophobic the participant, the lower the ratings provided for the expected tastiness, healthiness, and environmental friendliness of pea and algae burgers [36]. In a 2021 German survey, it was found that the more food neophobic the participant, the less likely they were to choose the meat-hybrid option consisting of 40% plant-based protein [33]. Similarly, in a recent US survey, food innovativeness, or being open to using new foods or ingredients, was associated with a more positive assessment of blending mushrooms into traditional meat-based foods, which was associated with a greater acceptance of blending [13]. While food innovativeness was associated with a more positive assessment of blending for all participants, it was found to have a greater influence on regular consumers who consumed the same or an increased amount of red meat in contrast to transitional consumers either having reduced or at least considered reducing their red meat consumption [13].

Nevertheless, in a 2021 DCE conducted in Germany and Belgium, food neophobia did not affect the overall product liking of meat-hybrids consisting of 50% or 80% plant-based protein [34]. Additionally, food neophobia did not have an effect on the product liking of Quorn® or tofu in the longitudinal experiment in the Netherlands [42].

Variety-Seeking

One study assessed variety-seeking personality traits as a factor influencing the replacement of meat with non-meat protein sources [42]. In the longitudinal experiment conducted in the Netherlands, variety seeking, or the tendency of consumers to switch between food products to prevent boredom, had an effect on the product liking of Quorn® or tofu only in interaction with product and time [42]. The greater the number of different meals used by the participants, the less likely they were to like Quorn® or tofu [42].

3.2.5. Knowledge and Skills

Information on Health and the Environment

Three studies examined the role of information on health and the environment in the replacement of meat with non-meat protein sources [31,39,43]. In a theoretical willingness to pay (WTP) experiment in Italy in which 119 participants were asked to indicate the amount of money they were willing to pay for beef and soy burgers, successive rounds of explanatory messaging on health and the environment resulted in a relative decrease of −1.6% in the WTP for beef burgers and relative increase of +3.6% in the WTP for soy burgers [39]. Additionally, successive rounds of explanatory messaging on health and the environment resulted in a relative decrease of −23.0% in the chosen quantities of the beef burger and a relative increase of +45.6% in the chosen quantities of soy burgers [39]. In an experiment in France, no difference was found in the willingness to purchase the plant-based sausage after the first message on either health or the environment [31]. After the second message on health or the environment, however, there was an increase in the willingness to purchase the plant-based sausage [31]. In an Italian survey assessing the impact of information on the willingness to try, buy, or purchase cultured meat in Italy, participants showed better agreement with the information provided on the extrinsic attributes of cultured meat, such as its impact on sustainability, security, and animal welfare, in contrast to information provided on the intrinsic attributes of cultured meat, such as the laboratory production and flavor and nutrients [43].

Cooking Skills and Food Knowledge

Four studies investigated the role of cooking skills and food knowledge in the replacement of meat with non-meat protein sources [13,14,44,45]. In a qualitative study conducted in New Zealand with 23 young adults, it was found that participants who described themselves as being more confident and experienced in cooking substituted meat with plant-based proteins such as legumes, lentils, and tofu [14]. Conversely, participants who described themselves as being less confident and less experienced in cooking preferred to substitute meat with more convenience-oriented, plant-based proteins such as vegetarian patties and sausages [14]. In a recent US survey, food knowledge, or the knowledge of foods and cooking, was not associated with a positive assessment of blending mushrooms into traditional meat-based foods and, thereby, was not associated with a greater acceptance of this blending concept by participants [13].

In terms of familiarity with alternative protein sources, a 2020 German survey found that pre-knowledge of or familiarity with cultured meat was shown to increase the ethical beliefs of cultured meat but did not impact the emotional objections of cultured meat being unnatural or disgusting or global diffusion optimism of cultured meat being affordable and capable of solving world nutrition problems [45]. In a Belgian survey, participants who claimed to be more familiar with insects were 2.6 times more likely to be ready to adopt insects as a substitute for meat compared to those who claimed to be unfamiliar with eating insects or did not know what eating insects entailed [44].

3.2.6. Emotions and Cognitive Dissonance

One study assessed cognitive dissonance of meat-eating behavior, or the inconsistency between caring for animals as pets yet consuming animals as meat in the diet, as a factor influencing the replacement of meat with non-meat protein sources [41]. In a 2020 German survey, participants were less willing to substitute meat with meat substitutes such as Quorn®, tofu, seitan, or soy schnitzel when they scored higher for unapologetic justification strategies to consume meat compared to those that scored lower [41]. The unapologetic justification strategies included: pro-meat attitude favoring a taste for meat; denial of animal suffering; hierarchical justification that humans are superior to animals; religious justification; health justification; human destiny that humans are destined to consume animals; and slaughter justification that denies animal suffering in slaughterhouses [41].

3.2.7. Values and Attitudes

Health and Environment

Five studies examined the role of the importance of health and the environment in the replacement of meat with non-meat protein sources [13,32–34,44]. In a 2021 German survey, it was found that the higher a participant rated the meat-hybrid in terms of health, the more likely they were to choose the meat-hybrid consisting of 40% plant-based protein compared to the corresponding meat product consisting of 100% meat [33]. While higher ratings for meat-hybrids in terms of the environment and animal welfare were also associated with an increased likelihood of participants choosing the meat-hybrid compared to the 100% meat product, health was found to exert a larger influence on choosing the meat-hybrid than the environment or animal welfare [33]. Similarly, participants ranked perceived health benefits as the top reason but sustainability benefits as the last reason for consuming blended foods in which mushrooms partially replaced a portion of meat in traditional meat-based foods in a recent US survey [13].

In a DCE conducted in Germany and Belgium, participants rated the meat-hybrid consisting of either 50% or 80% of plant-based protein and the 100% plant-based vegetarian alternative as healthier compared to the corresponding 100% meat product [34]. Nevertheless, the same DCE found that the lower the health consciousness of the participant, the lower their preference for meat [34]. Likewise, a Belgian survey found that for every one unit increase in the belief that meat is nutritious and healthy, there was a 64% reduction in the willingness of participants to adopt insects as a substitute for meat [44]. In terms of the environment, however, this same survey showed that for every one unit increase in the attention participants pay to the environmental impact of food, there was a 71% increase in the readiness to adopt insects as a substitute for meat [44]. In a 2021 Germany survey, omnivores perceived meat as performing better for protein content, fat content, and environmental friendliness compared to meat alternatives [32]. Although flexitarians in this study perceived meat as performing better for protein content, they perceived meat substitutes as performing better for fat content and environmental friendliness. Furthermore, participants perceived steak in particular as being the healthiest and protein-rich food item among tofu, chicken nuggets, vegetarian nuggets, wiener sausages, and vegetarian sausages [32].

Plant Protein Sources and Production

Three studies investigated the attitudes towards specific protein sources and production methods as factors influencing the replacement of meat with non-meat protein sources [13,29,45]. An online survey was conducted with female participants in Denmark, Finland, Iceland, Germany, and Romania to examine attitudes concerning a meat analogue of wiener sausages containing rapeseed protein [29]. In all countries, attitude towards using plant protein in food production was shown to influence both the intention to substitute meat protein in the diet as well as the attitude towards using rapeseed protein as an ingredient in meat analogues. Furthermore, the attitude towards rapeseed was also shown to influence the attitude towards meat analogues [29]. In a 2020 Germany survey, principal component analysis was utilized to identify three attitudinal dimensions of participants in terms of the intention to try, eat, and promote cultured meat: ethical advantage (e.g., ecological, animal welfare), emotional objections (e.g., unnatural, disgusting), and global diffusion optimism (e.g., affordable, possible global solution) [45]. The ethical beliefs were found to be the primary driver in the intention to try, eat, and promote cultured meat in the future followed by emotional objections and finally global diffusion optimism [45]. In a recent US survey, a positive consumer assessment of blending mushrooms into traditional meat-based dishes was associated with a greater acceptance of blending as a means to reduce meat consumption [13].

Vegans and Vegetarians

One study assessed the role of attitudes towards vegans and vegetarians in the replacement of meat with non-meat protein sources [36]. In the multi-national online survey conducted in Germany, France, and the UK, participants who were more negative towards vegan and vegetarian lifestyles provided lower ratings for the expected tastiness, healthiness, and environmental friendliness of pea and algae burgers compared to those who were not negative towards vegan and vegetarian lifestyles [36].

Others

In a 2021 Germany survey, participants perceived steak in particular as being the most natural, masculine, and festive among tofu, chicken nuggets, vegetarian nuggets, wiener sausages, and vegetarian sausages [32]. In a recent US survey, participants rated the perceived culinary benefits as the second to last reason preceding perceived sustainability benefits for blending traditional meat-based dishes with mushrooms [13].

3.2.8. Habits

Healthy Eating

One study examined healthy eating as a factor influencing the replacement of meat with non-meat protein sources [13]. In a recent US survey, healthy eating was associated with a more positive assessment of blending mushrooms into traditional meat-based dishes [13]. While healthy eating was associated with a more positive assessment of blending for all participants, it was found to have a greater influence on transitional consumers either having reduced or at least considered reducing their red meat consumption in contrast to regular consumers who consumed the same or an increased amount of red meat at the time [13].

Consumption of Meat

Six studies investigated the role of consumption of or attachment to meat in the replacement of meat with non-meat protein sources [29,33,34,36,40,44]. In an online survey conducted in Denmark, Finland, Germany, Iceland, and Romania, female participants who consumed less meat were associated with being more likely to purchase meat analogues in Romania [29]. In a Hungarian survey, participants who intended to reduce their meat intake in the next year had an expected increase of 1.47 in the number of preferred insect types that they would eat as a substitute for meat in the next year [40]. Similarly, a Belgian survey found that participants who intended to reduce their meat intake in the next year were 4.5 times more likely to be ready to adopt insects as a substitute for meat in the next year [44]. In the multi-national survey conducted in Germany, France, and the UK, participants who scored higher on the scale assessing commitment to meat provided lower ratings for the expected tastiness, healthiness, and environmental friendliness of pea and algae burgers [36]. Moreover, a 2021 Germany survey found that the higher a participant scored on the questionnaire evaluating attachment to meat, the less likely they were to choose the meat-hybrid option in which a portion of meat was replaced with a plant-based protein [33]. Similarly, in the DCE conducted in Germany and Belgium, the more attached a participant was to meat, the more likely they were to choose the 100% meat option compared to meat-hybrid options [34].

Consumption of Meat Substitutes

One study examined the effects of prior experience with meat substitutes and repeated exposure to meat substitutes on the long-term acceptance of non-meat protein sources as replacements for meat [42]. At the start of this longitudinal experiment in the Netherlands, participants liked Quorn® and tofu less than the reference meat of chicken [42]. Although in general, the liking of Quorn®, tofu, and chicken decreased over the ten-week repeated exposure period, there was no difference in the decrease in liking of Quorn®, tofu, and chicken. Furthermore, the liking scores of Quorn®, tofu, and chicken were notably no longer different from one another after this ten-week repeated exposure period. Additionally, the number of boredom patterns, defined as a decrease in the liking of a product over time, and mere exposure patterns, defined as an increase in the liking of a product over time, differed among the three food products in this experiment. In contrast to the majority of participants who ate chicken and showed a boredom pattern over the repeated exposure period, the majority of participants who ate tofu showed a mere exposure pattern. On the other hand, the participants who ate Quorn® took an intermediate position between chicken and tofu in terms of boredom and mere exposure patterns with the slight majority showing boredom over the repeated exposure period. Prior experience with meat substitutes was also associated with increased product liking of Quorn® and tofu [42].

Cooking Habits and Food Involvement

Two studies explored cooking habits and food involvement as factors influencing the replacement of meat with non-meat protein sources [13,32]. In a recent US survey, cooking habits (i.e., the time and pleasure derived from cooking) were not found to be associated with a more positive assessment of blending mushrooms into traditional meat-based dishes [13]. However, food involvement (i.e., time spent thinking about food and time spent cooking and cleaning up after meals), was found to be associated with a more positive assessment of blending, which was associated with a greater acceptance of blending mushrooms into traditional meat-based dishes [13]. In a 2021 Germany survey, omnivores perceived meat as being easier to prepare than meat alternatives [32].

3.3. Socio-Cultural Factors

3.3.1. Culture

Country of Consumer

Three studies examined the possible role of culture in the replacement of meat with non-meat protein sources by incorporating participants from different countries in their study designs [29,34,36]. In an online survey conducted in Denmark, Finland, Germany, Iceland, and Romania among female participants, the attitude towards using plant protein in food production was shown to influence both the intention to substitute meat in the diet as well as the attitude towards using rapeseed protein as an ingredient in meat analogues in all five countries [29]. Additionally, the attitude towards gluten was associated with a decreased intention to buy meat analogues in all five countries. For Germany only, however, the attitude towards soy protein was associated with an increased intention to buy meat analogues, whereas the attitude towards potato starch was associated with a decreased intention to buy meat analogues. In Romania, those eating less meat were more likely to buy meat analogues [29]. In the multi-national survey conducted in Germany, France, and the UK, it was found that pea and algae burgers were expected to be less tasty but healthier and more environmentally friendly than the beef burger [36]. Nevertheless, being from France was associated with providing lower ratings for the expected tastiness and healthiness of pea burgers compared to being from Germany. Contrastingly, the country of origin was not found to be associated with the ratings for expected tastiness, healthiness, and environmental friendliness of the algae burgers [36]. In a DCE conducted in Germany and Belgium, the majority of participants considered meat to be tastier than meat-hybrids yet considered meat-hybrids to be more environmentally friendly and better for animal welfare compared to meat [34]. Nevertheless, the majority of German participants considered meat-hybrids to be healthier than meat, whereas the majority of Belgian participants considered meat to be healthier than meat-hybrids [34].

3.3.2. Social Norms, Roles, and Relationships

Situational Context

Two studies investigated the situational context of consumption as a factor influencing the replacement of meat with non-meat protein sources [30,32]. In a 2021 experiment in the Netherlands, photographs of meat and plant-based protein products were presented to 309 participants with a question regarding the appropriateness of the meat and plant-based meat substitutes during the hot meal of the day [30]. Overall, the plant-based meat substitutes and chickpeas and nuts were considered less appropriate than the corresponding meat product in almost all situations including: eating alone; with family and friends; cooking for children; to add flavor to a dish; or when there is little time for cooking. Besides the situation of cooking for a vegetarian in which the plant-based meat substitutes and chickpeas and nuts were rated as more appropriate than the corresponding meat product, the vegetarian hamburger was also rated as more appropriate than the hamburger when wanting to eat a healthy meal. Moreover, in situations of wanting to prepare a special meal, the steak was rated highly; but neither the hamburger nor the smoked sausage was rated higher than the corresponding vegetarian burger and vegetarian sausage in this situation [30]. In a 2021 Germany survey, omnivores and flexitarians alike rated the situational appropriateness of consuming a plant-based meat alternative to be most appropriate when eating alone [32]. For omnivores, eating plant-based meat alternatives in more formal settings (i.e., eating a Sunday dinner with family, invited for dinner at a restaurant, at a business meal, or when at a barbecue) was considered as less appropriate than in more casual settings (i.e., eating alone, being invited to eat with friends, or eating dinner with family during the weekday). For flexitarians, eating plant-based meat alternatives when eating alone or when eating with family during the weekday was considered more appropriate than more social or formal settings including: when invited to eat with friends; eating a Sunday dinner with family; invited for a dinner at a restaurant; at a business meal; or when at a barbecue [32].

3.4. External Factors

3.4.1. Economic Factors

Price

Five studies assessed the role of price in the replacement of meat with non-meat protein sources [13,32,34,37,38]. In the 2016 UK DCE, the largest consumer segment was identified as price-conscious consumers making up 43% of meat-eaters and meat-reducers in the study [37]. For price-conscious consumers, price was the third most influential product attribute when determining their preference for ground meat or ground meat substitute preceded only by the type of ground meat or ground meat substitute and the region of production [37]. In the 2019 UK DCE, price was also found to be an influential factor in the preference of ground meat or ground meat substitute but only for some consumer segments of meat-eaters [38]. Among meat-eaters, price-conscious consumers were again the largest consumer segment making up 63% of all meat-eaters in the study. Price-conscious meat-eaters were influenced more by the type and price of ground meat or ground meat substitute, whereas price played a less influential role for traditional and empowered meat-eaters in the study. Among meat-reducers, price was found to be an intermediate factor for the health curtailers and the least influential factor for sustainable meat-reducers in determining their preference for ground meat or ground meat substitute [38]. In a 2021 Germany survey, however, both omnivores and flexitarians alike rated meat as performing better in terms of price compared to meat alternatives [32]. Furthermore, participants perceived steak as being the most expensive among tofu, chicken nuggets, vegetarian nuggets, wiener sausages, and vegetarian sausages [32].

In terms of partially replacing meat-products with plant-based protein, a recent US survey found that price benefits (i.e., reducing the cost of meals and helping with the budget) were rated as one of the top two reasons to consume blended meat products preceded only by health benefits [13]. In the DCE conducted in Germany and Belgium, a decrease in price was found to increase the preferability of the 100% meat and meat-hybrid options except for salami among Belgian participants [34].

3.4.2. Food Environment

Extrinsic Product Attributes

Six studies investigated extrinsic product attributes (i.e., packaging, brand, nutritional information, health claims, and local origin and environmental labels), as factors in the replacement of meat with non-meat protein sources [13,31,34,37,38,44]. In a French experiment, participants preferred to purchase the pork-based sausage rather than the plant-based sausage after tasting and being provided with the packing information for the sausages, which included information on the brand, ingredients, nutrition, preparation, and recycling [31]. Nevertheless, although participants still preferred to purchase the pork-based sausage over the plant-based sausage, participants' preference to purchase the plant-based sausage was higher than it had been during the blind tasting alone without the packaging information provided [31].

When considering the specific information provided on packaging, the 2016 UK DCE found that the brand and type of the ground meat or ground meat substitute primarily influenced taste-driven consumers, whereas brand was found to have only an intermediate or low influence on the other four consumer segments of meat-eaters and meat-reducers [37]. The 2019 UK DCE also found that brand had a low influence on the preference of ground meat or ground meat substitute in meat-eaters and meat-reducers [38]. In terms of health labels, the DCE in Germany and Belgium found that health labels had no effect on the preferability of the meat-hybrid options in which a portion of meat had been replaced by plant-based protein [34]. In the 2016 UK DCE, the fat content of ground meat and ground meat substitutes primarily influenced healthy and organic consumers but only intermediately influenced price-conscious, green, and taste-driven consumers [37]. In the 2019 UK DCE, the fat content of ground meat and ground meat substitutes intermediately influenced consumer segments of meat-eaters but more strongly influenced the health curtailers and sustainable consumers of meat-reducers [38]. For organic and local origin labels, the DCE in Germany and Belgium found that organic and local origin labels had a primarily positive effect on the preferability of the meat-hybrid options [34]. In the 2016 UK DCE, an origin label of the ground meat or ground meat substitute had a strong influence on the preferability of the ground meat or ground meat substitute for the price-conscious, green, taste-driven, and healthy consumers but was the least influential factor for organic consumers [37]. In the 2019 UK DCE, the origin label had an intermediate influence on the preferability of ground meat and ground meat substitute of meat-eaters and meat-reducers [38]. In terms of production labels, both the 2016 and 2019 UK DCEs found that production labels have an intermediate to weak influence on the preferability of ground meat and ground meat substitutes for meat-eaters and meat-reducers [37,38]. For environmental labels, the 2016 and 2019 UK DCEs also found that carbon footprint had an intermediate to low influence on the preferability of ground meat and ground meat substitute for most consumers besides green and sustainable consumers in which it was the primary influence [37,38]. In the DCE in Germany and Belgium, the environmental label had a positive effect on the preference of the meat-hybrid, except for the meat-hybrid options of meatballs and salami in Belgium [34].

Specifically, two studies specifically investigated the impact of the format of the replacement for meat on consumers' preferences [13,44]. In a recent US survey, participants rated burgers as the most preferred format for consuming blended products in which traditional meat-based dishes are partially replaced by mushrooms followed by stir-fry with ground beef, meatloaf, tacos, chili with ground beef, and pasta with ground beef [13]. In a Belgian survey, it was found that every one unit increase in a participant's orientation towards convenience in meal preparation was associated with a 75% increase in the readiness to adopt insects as a substitute for meat [44].

Meal Context

Four studies examined the possible role of meal context on the replacement of meat with non-meat protein sources [28,35,42,48]. In two experiments in the Netherlands, the liking of meat substitutes differed when evaluated individually or within the meal context [28,42]. In a 2011 experiment, participants liked Quorn® pieces more than Quorn® ground when evaluating these meat substitutes individually [28]. Participants also liked Quorn® pieces more than Quorn® ground in the rice and salad dishes; however, there were no differences in the participants' liking of Quorn® pieces and Quorn® ground in the spaghetti and soup dishes. Despite these findings, there were no differences in participants' overall liking of the rice, salad, spaghetti, or soup dishes using either Quorn® pieces or Quorn® mince [28]. Similarly, a longitudinal experiment found that participants liked the entire meals consisting of either Quorn®, tofu, or chicken better than Quorn®, tofu, or reference chicken evaluated individually outside of the meal context [42].

In two experiments conducted in the US, participants' liking of dishes in which meat had been partially replaced by legumes differed depending upon the recipe and thus the meal context [35,48]. A 2018 US experiment found that the high-meat pork carnitas arepas were liked better in terms of overall liking compared to the low-meat arepas and the high- and low-meat chicken tikka masala in which a portion of the meat had been replaced by legumes in the low-meat recipes [48]. Nevertheless, there was no difference in the overall liking among the low-meat arepas and high- and low-meat chicken tikka masala recipes [48]. In a 2021 US experiment, there were no differences in the overall liking of high- and low-meat versions of East Asian bowls with regular and spicy sauces in which a portion of the beef was replaced with legumes and vegetables in the low-meat dishes [35].

Grocery Store Infrastructure

One study investigated the infrastructure of grocery stores as a factor influencing the replacement of meat with non-meat protein sources [46]. In a 2018 Canadian study, participants were interviewed on how well grocery stores in Canada support the transition to plant-based protein [46]. In terms of product availability, it was found that significantly more space was allocated to animal-based protein compared to plant-based protein in grocery stores. Some participants of the study elaborated that meat and dairy sections carry more types of products and brands in comparison to the lack of variety available for plant-based protein products, particularly tofu, grains, and legumes. For product promotion, there were more promotions for animal-based proteins than plant-based proteins in grocery stores. Furthermore, there was a higher percentage of sales and/or descriptive designated for animal-based proteins (32%) compared to plant-based proteins (3%), which was supported by participants noting that meat, seafood, and dairy sections were more "prominent" in grocery stores in contrast to plant-based proteins "hidden" locations throughout the grocery stores' aisles [46]. In regard to product location, participants rated it easier to find animal-based protein sources than plant-based protein sources with one of the most commonly cited obstacles being the inconsistent location of plant-based proteins among different grocery stores [46].

3.4.3. Animal Welfare

One study examined the role of animal welfare in the replacement of meat with non-meat protein sources [33]. In a 2021 German survey, the higher a participant rated meat-hybrids in terms of their animal welfare, the more likely they were to choose the meat-hybrid option consisting of 40% plant-based protein compared to the 100% meat option [33].

4. Discussion

Altogether, this review revealed multiple personal, socio-cultural, and external factors relating to the replacement of meat with non-meat protein sources among omnivores and flexitarians in developed countries. Most importantly, the results indicate that female gender, information on health and the environment, and lower price of non-meat protein sources may act as drivers to replacing meat with non-meat protein sources. Contrastingly, the results show that male gender, food neophobia, attachment to meat, and the perceived lower situational appropriateness of consuming non-meat protein sources in social settings may be inhibitors to replacing meat with non-meat protein sources. Interestingly, although sensory and hedonic attributes of meat such as taste and flavor may act as inhibitors, the recipe and entire meal context appear to be more important than the individual evaluation of non-meat protein sources and thus may act as a driver increasing consumers' acceptability of non-meat protein sources in traditionally meat-based dishes.

Notably, gender, food neophobia, and information on health and the environment are among the factors most researched in literature for their role in reducing meat intake and consuming non-meat protein sources [16,49–55]. Similar to the findings in this review, many studies have shown that gender influences attitudes related to the consumption of meat and non-meat protein sources [16,49–51]. Considering common phrases such as "real men eat meat", studies have shown that men who identified more with traditional beliefs of masculinity that conflate meat consumption and virility were more attached to meat and had a more negative attitude towards a vegetarian diet [50,51]. In accordance with our findings, a recent systematic review incorporating ninety-one articles on consumer acceptance of novel alternative protein sources also highlighted that food neophobia remains a hindrance for consumers to adopt many novel protein sources including insects, seaweed, cultured meat, and plant-based meat substitutes into daily consumption patterns [52]. In terms of information on health and the environment, studies have shown that providing consumers with information on the negative health and environmental consequences of meat consumption increased intentions to reduce meat consumption [53–55], which aligns with our findings.

While this review identified many factors relating to reducing meat consumption, some potentially important factors cannot be substantiated due to mixed results, not being extensively examined, or being entirely missing from the included studies. Age and socioeconomic status have both been cited as influencers of meat consumption in literature [56,57]; yet, findings on these factors were mixed in this review and could not confirm younger age and higher socioeconomic status as being drivers for replacing meat with non-meat protein sources. Going further, only one study each examined sex [36], grocery-store infrastructure [46], and cooking skills [14], making it impossible to definitively draw conclusions on whether these factors act as drivers or inhibitors to replacing meat with non-meat protein sources. Moreover, race, ethnicity, religion, health status, degree of urbanization, and perceived behavior control were not examined by any of the studies included in this review. Since previous studies have shown differences in the consumption of meat and specific meat products among white, Hispanic, and African Americans [58,59], race and ethnicity could be suspected of also influencing the replacement of meat with non-meat protein sources. Additionally, examining perceived behavior control could be instrumental in determining how and to what degree certain knowledge and skills, such as cooking skills, augment consumers' willingness and self-efficacy in being able to institute non-meat protein sources as dietary fixtures [60].

Besides examining the relevance of the aforementioned underresearched factors, studies should prioritize standardizing methods and examining potential drivers and inhibitors across different consumer segments and types of non-meat protein sources in order to foster comparability among studies as well as to identify variations in consumer acceptance and long-term health and environmental consequences. Importantly, studies should standardize their definitions of "meat" namely whether this excludes certain types of meat and includes poultry, fish, and seafood. For replacements of meat, terms such as "plant-based diet", "vegetarian meals", or "meat-less meals" lack adequate description needed for reproducibility in studies and furthermore do not necessitate having sufficient protein content to constitute an actual replacement of meat within a meal context. In regard to study populations, utilizing random versus convenience sampling would provide a more accurate depiction of the population and would avoid volunteer bias [61]. Separate analyses of omnivores, flexitarians, pescatarians, vegetarians, and vegans would also be advantageous in determining subtle distinctions in motivations, willingness, and acceptance that could be employed to build more efficacious public health campaigns to reduce meat [62]. Although most studies included men and women in their analyses, future studies should be explicit in whether they are examining biological differences between the male and female sexes or the psychosocial and cultural differences of male, female, and other genders as implications of such research differ greatly [63]. Beyond different consumer segments, studies should also incorporate various types of non-meat protein sources of different processing levels in order to determine differences in consumer acceptance as well as to forecast the long-term health and environmental consequences of such replacements [64,65]. When possible, it is essential that we amass more experimental evidence in real-life settings to assess if and how these factors truly affect the motivation, willingness, and acceptance of replacing meat with non-meat protein sources.

Ultimately, this review has important implications ranging from public health policy to research collaboration. Firstly, the findings from this review identify relevant drivers and inhibitors that can be used to support more efficacious public health campaigns aiming to reduce meat consumption within developed countries. Next, this research may be relevant for food industries when marketing non-meat protein sources to consumers as replacements for meat during mealtimes. For dietitians and other healthcare professionals, this research could be used as a tool to assist clients and patients in fostering behavior change towards healthier and more sustainable food options. Besides the aforementioned considerations for future research, the findings in this review on the relevance of meal context emphasize the importance of collaboration within the field of nutrition particularly among chefs, dietitians, and scientists to create flavorful dishes with non-meat protein sources in an effort to facilitate and expedite the transition to healthier and more sustainable protein sources among consumers.

Nevertheless, this review has some limitations that should be noted. Although beyond the scope of this review, replacing red and processed meat with white meat, eggs, or dairy products may be a more feasible first step for many consumers attempting to reduce meat consumption [66]. Additionally, this review focused on omnivores and flexitarians given that they comprise a much larger percentage of the population in developed countries compared to pescatarians, vegetarians, and vegans [22,23,26]. However, understanding the motivations, willingness, and behavior of these non-meat-eating subgroups and comparing them with omnivores and flexitarians could be useful in identifying which factors are most decisive in reducing meat. In this review, one researcher screened articles for inclusion and exclusion criteria and extracted relevant data. Although this could have introduced bias in the article section, the inclusion and exclusion criteria were clearly defined, and the search strings were built by two researchers. Furthermore, a second researcher randomly cross-checked article screening and data extraction and discussed any uncertainties or disagreements with the first researcher. Other frameworks, such as the Capability, Opportunity, Motivation behavior model (COM-B) could have been utilized to organize and summarize the findings in this review [67]. However, we chose to use the theoretical framework by Stoll-Kleemann and Schmidt (2017) as it provides a comprehensive overview of the interrelated drivers and inhibitors that may be involved specifically in meat-eating behavior, which relates closely to our aim of identifying the drivers and inhibitors of replacing meat with non-meat protein sources among consumers in developed countries [17]. Furthermore, unlike the COM-B model, this framework is based on a pro-environmental behavior model that analyzes the propensity of individuals to partake in actions that mitigate a negative impact on the environment as well as the dissonance between having environmental awareness and participating in pro-environmental behavior [27]. Yet, this framework does carry some shortcomings [17]. While not covered by the studies in this review or explicitly included in the framework [17], we included race and ethnicity as potentially important factors to consider in the socio-demographics in Table 2.

5. Conclusions

In conclusion, this review revealed multiple personal, socio-cultural, and external factors relating to the replacement of meat with non-meat proteins sources among omnivores and flexitarians in developed countries. The results indicate that female gender, information on health and the environment, and lower price of non-meat protein sources may act as drivers, whereas male gender, food neophobia, attachment to meat, and the lower situational appropriateness of consuming non-meat protein sources act as inhibitors. According to literature, gender, food neophobia, and information on health and the environment are relevant factors in reducing meat and replacing it with non-meat protein sources [16,49–55]. However, more research is needed to establish the relevance of socioeconomic status, race, ethnicity, religion, health status, food environment, and cooking skills. Future research should consider the importance of standardizing methods in order to allow for better comparisons among studies. Additionally, studies should prioritize examining potential drivers and inhibitors across different consumer segments and various non-meat protein sources to determine differences in consumer acceptability and the long-term health and environmental consequences of such replacements. Ultimately, the findings of this review are relevant for supporting more efficacious public health campaigns, product development and marketing in food industry, and behavior change facilitated by healthcare professionals. Given the importance of meal context, this research calls for collaboration particularly among chefs, dietitians, and scientists in research to expedite and facilitate the transition to healthier and more sustainable protein sources.

Author Contributions: The authors' contributions are as follows, M.R.E. conceptualized the review and research aim, developed the search strategy, conducted the literature search, wrote the original draft, and reviewed and edited the manuscript. S.B., P.v.V. and J.M.G. supervised and critically reviewed the manuscript. All authors have read and agreed to the published version of the manuscript.

Funding: This work was supported by a research grant provided by the VLAG Graduate School of Wageningen University, Wageningen, The Netherlands. VLAG Graduate School had no role in this review including its conceptualization, design, execution, interpretation, and writing and revising of the manuscript.

Acknowledgments: The authors would like to acknowledge and thank Jyri Tyrvainen for reviewing and editing the manuscript for English language.

Conflicts of Interest: The authors declare no conflict of interest. The funders had no role in the design of the study; in the collection, analyses, or interpretation of data; in the writing of the manuscript, or in the decision to publish the results.

Appendix A

Search string for Scopus until April 2021 to identify the drivers and inhibitors underlying the replacement of meat with non-meat protein sources.

TITLE-ABS-KEY(((meat W/3 reduc*) OR (meat W/3 substitut*) OR (meat W/3 replace*) OR "less meat" OR flexitarian*) OR ("plant*based meat" OR "plant*based diet" OR "plant*forward diet" OR "alternative protein" AND (pulse OR legume OR bean OR lentil OR seaweed OR alga OR insect OR "cultured meat" OR "in*vitro meat" OR "synthetic meat")) AND (perce* OR aware* OR attitude OR intent* OR willing* OR motiv* OR choice OR prefer* OR accept* OR adopt* OR change OR buy* OR purchas* OR "food choice" OR "behavior* change"))

Appendix B

Search string for Web of Science (Core Collection) until April 2021 to identify the drivers and inhibitors underlying the replacement of meat with non-meat protein sources.

(TS = (((reduc* OR substitut* OR replace*) NEAR/3 meat) OR ("less meat") OR (flexitarian*) OR (("plant*based meat" OR "plant*based diet" OR "plant*forward diet" OR "alternative protein") AND (pulse OR legume OR bean OR lentil OR seaweed OR alga OR insect OR "cultured meat" OR "in*vitro meat" OR "synthetic meat")))) AND TS = ((perce* OR aware* OR attitude OR intent* OR willing* OR motiv* OR choice OR prefer* OR accept* OR adopt* OR change OR buy* OR purchas* OR "food choice" OR "behavior* change"))

References

1. Willett, W.; Rockström, J.; Loken, B.; Springmann, M.; Lang, T.; Vermeulen, S.; Garnett, T.; Tilman, D.; DeClerck, F.; Wood, A.; et al. Food in the Anthropocene: The EAT–Lancet Commission on healthy diets from sustainable food systems. *Lancet* **2019**, *393*, 447–492. [CrossRef]
2. Swinburn, B.A.; Kraak, V.I.; Allender, S.; Atkins, V.J.; Baker, P.I.; Bogard, J.R.; Brinsden, H.; Calvillo, A.; De Schutter, O.; Devarajan, R.; et al. The Global syndemic of obesity, undernutrition, and climate change: The lancet commission report. *Lancet* **2019**, *393*, 791–846. [CrossRef]
3. World Cancer Research fund/American Institute for Cancer Research. Diet, Nutrition, Physical Activity and Cancer: A Global Perspective. *Continuous Update Expert Report 2018*. Available online: Dietandcancerreport.org (accessed on 4 August 2021).
4. Wang, X.; Lin, X.; Ouyang, Y.Y.; Liu, J.; Zhao, G.; Pan, A.; Hu, F.B. Red and processed meat consumption and mortality: Dose-response meta-analysis of prospective cohort studies. *Public Health Nutr.* **2016**, *19*, 893–905. [CrossRef]
5. Yang, C.; Pan, L.; Sun, C.; Xi, Y.; Wang, L.; Li, D. Red meat consumption and the risk of stroke: A dose-response meta-analysis of prospective cohort studies. *J. Stroke Cerebrovasc. Dis.* **2016**, *25*, 1177–1186. [CrossRef] [PubMed]
6. Satija, A.; Bhupathiraju, S.N.; Rimm, E.B.; Spiegelman, D.; Chiuve, S.E.; Borgi, L.; Willett, W.C.; Manson, J.E.; Sun, Q.; Hu, F.B. Plant-based dietary patterns and incidence of type 2 diabetes in US men and women: Results from three prospective cohort studies. *PLoS Med.* **2016**, *13*, e1002039. [CrossRef] [PubMed]
7. Martínez-González, M.A.; Sánchez-Tainta, A.; Corella, D.; Salas-Salvadó, J.; Ros, E.; Arós, F.; Gómez-Gracia, E.; Fiol, M.; Lamuela-Raventós, R.M.; Schröder, H.; et al. A provegetarian food pattern and reduction in total mortality in the Prevención con Dieta Mediterránea (PREDIMED) study. *Am. J. Clin. Nutr.* **2014**, *100*, 320S–328S. [CrossRef] [PubMed]

8. Heidemann, C.; Schulze, M.B.; Franco, O.H.; Van Dam, R.M.; Mantzoros, C.S.; Hu, F.B. Dietary patterns and risk of mortality from cardiovascular disease, cancer, and all causes in a prospective cohort of women. *Circulation* **2008**, *118*, 230–237. [CrossRef]
9. Bouwman, L.; Goldewijk, K.K.; Van Der Hoek, K.W.; Beusen, A.H.W.; Van Vuuren, D.P.; Willems, J.; Rufino, M.C.; Stehfest, E. Exploring global changes in nitrogen and phosphorus cycles in agriculture induced by livestock production over the 1900–2050 period. *Proc. Natl. Acad. Sci. USA* **2013**, *110*, 20882–20887. [CrossRef] [PubMed]
10. Aiking, H.; de Boer, J.; Vereijken, J.M. *Sustainable Protein Production and Consumption: Pigs or Peas?* Springer: Dordrecht, The Netherlands, 2006; Volume 45.
11. Pluhar, E.B. Meat and morality: Alternatives to factory farming. *J. Agric. Environ. Ethics* **2010**, *23*, 455–468.
12. United Nations, Department of Economic and Social Affairs, Population Division. *World Popultaion Prospects 2019: Highlights (ST/ESA/SER.A/423)*; United Nations: New York, NY, USA, 2019.
13. Lang, M. Consumer acceptance of blending plant-based ingredients into traditional meat-based foods: Evidence from the meat-mushroom blend. *Food Qual. Prefer.* **2020**, *79*, 103758. [CrossRef]
14. Kemper, J.A.; White, S.K. Young adults' experiences with flexitarianism: The 4Cs. *Appetite* **2021**, *160*, 105073. [CrossRef]
15. Allen, M.W.; Baines, S. Manipulating the symbolic meaning of meat to encourage greater acceptance of fruits and vegetables and less proclivity for red and white meat. *Appetite* **2002**, *38*, 118–130. [CrossRef]
16. Rothgerber, H. Real men don't eat (vegetable) quiche: Masculinity and the justification of meat consumption. *Psychol. Men Masc.* **2013**, *14*, 363–375. [CrossRef]
17. Stoll-Kleemann, S.; Schmidt, U.J. Reducing meat consumption in developed and transition countries to counter climate change and biodiversity loss: A review of influence factors. *Reg. Environ. Chang.* **2017**, *17*, 1261–1277. [CrossRef]
18. Valli, C.; Rabassa, M.; Johnston, B.C.; Kuijpers, R.; Prokop-Dorner, A.; Zajac, J.; Storman, D.; Storman, M.; Bala, M.M.; Solà, I.; et al. Health-related values and preferences regarding meat consumption a mixed-methods systematic review. *Ann. Intern. Med.* **2019**, *171*, 742–755. [CrossRef]
19. Sanchez-Sabate, R.; Badilla-Briones, Y.; Sabate, J. Understanding attitudes towards reducing meat consumption for environmental reasons. A qualitative synthesis review. *Sustainability* **2019**, *11*, 6295. [CrossRef]
20. Sanchez-Sabate, R.; Sabate, J. Consumer attitudes towards environmental concerns of meat consumption: A systematic review. *Int. J. Environ. Res. Public Health* **2019**, *16*, 1220. [CrossRef] [PubMed]
21. Harguess, J.M.; Crespo, N.C.; Hong, M.Y. Strategies to reduce meat consumption: A systematic literature review of experimental studies. *Appetite* **2020**, *144*, 104478. [CrossRef] [PubMed]
22. Veganz. Veganz Nutrition Study. 2020. Available online: https://veganz.com/blog/veganz-nutrition-study-2020/ (accessed on 4 August 2021).
23. Stahler, C. How Many People are Vegan? How Many Eat Vegan when Eating Out? Asks the Vegetarian Resource Group. Available online: https://www.vrg.org/nutshell/Polls/2019_adults_veg.htm (accessed on 4 August 2021).
24. Paslakis, G.; Richardson, C.; Nöhre, M.; Brähler, E.; Holzapfel, C.; Hilbert, A.; de Zwaan, M. Prevalence and psychopathology of vegetarians and vegans—Results from a representative survey in Germany. *Sci. Rep.* **2020**, *10*, 6840. [CrossRef]
25. Hargreaves, S.M.; Raposo, A.; Saraiva, A.; Zandonadi, R.P. Vegetarian diet: An overview through the perspective of quality of life domains. *Int. J. Environ. Res. Public Health* **2021**, *18*, 4067. [CrossRef] [PubMed]
26. Dagevos, H. Finding flexitarians: Current studies on meat eaters and meat reducers. *Trends Food Sci. Technol.* **2021**, *114*, 530–539. [CrossRef]
27. Kollmuss, A.; Agyeman, J. Mind the gap: Why do people act environmentally and what are the barriers to pro-environmental behavior? *Environ. Educ. Res.* **2002**, *8*, 239–260. [CrossRef]
28. Elzerman, J.E.; Hoek, A.C.; van Boekel, M.; Luning, P.A. Consumer acceptance and appropriateness of meat substitutes in a meal context. *Food Qual. Prefer.* **2011**, *22*, 233–240. [CrossRef]
29. Banovic, M.; Sveinsdottir, K. Importance of being analogue: Female attitudes towards meat analogue containing rapeseed protein. *Food Control.* **2021**, *123*, 107833. [CrossRef]
30. Elzerman, J.E.; Keulemans, L.; Sap, R.; Luning, P.A. Situational appropriateness of meat products, meat substitutes and meat alternatives as perceived by Dutch consumers. *Food Qual. Prefer.* **2021**, *88*, 104108. [CrossRef]
31. Martin, C.; Lange, C.; Marette, S. Importance of additional information, as a complement to information coming from packaging, to promote meat substitutes: A case study on a sausage based on vegetable proteins. *Food Qual. Prefer.* **2021**, *87*, 104058. [CrossRef]
32. Michel, F.; Hartmann, C.; Siegrist, M. Consumers' associations, perceptions and acceptance of meat and plant-based meat alternatives. *Food Qual. Prefer.* **2021**, *87*, 104063. [CrossRef]
33. Profeta, A.; Baune, M.C.; Smetana, S.; Bornkessel, S.; Broucke, K.; Van Royen, G.; Enneking, U.; Weiss, J.; Heinz, V.; Hieke, S.; et al. Preferences of German Consumers for meat products blended with plant-based proteins. *Sustainability* **2021**, *13*, 650. [CrossRef]
34. Profeta, A.; Baune, M.C.; Smetana, S.; Broucke, K.; Van Royen, G.; Weiss, J.; Heinz, V.; Terjung, N. Discrete choice analysis of consumer preferences for meathybrids—findings from Germany and Belgium. *Foods* **2021**, *10*, 71. [CrossRef]
35. Spencer, M.; Rowe, S.; Bonnell, C.; Dalton, P. Consumer acceptance of plant-forward recipes in a natural consumption setting. *Food Qual. Prefer.* **2021**, *88*, 104080. [CrossRef]
36. Michel, F.; Knaapila, A.; Hartmann, C.; Siegrist, M. A multi-national comparison of meat eaters' attitudes and expectations for burgers containing beef, pea or algae protein. *Food Qual. Prefer.* **2021**, *91*, 104195. [CrossRef]

37. Apostolidis, C.; McLeay, F. Should we stop meating like this? Reducing meat consumption through substitution. *Food Policy* **2016**, *65*, 74–89. [CrossRef]
38. Apostolidis, C.; McLeay, F. To meat or not to meat? Comparing empowered meat consumers' and anti-consumers' preferences for sustainability labels. *Food Qual. Prefer.* **2019**, *77*, 109–122. [CrossRef]
39. Castellari, E.; Marette, S.; Moro, D.; Sckokai, P. The impact of information on willingness to pay and quantity choices for meat and meat substitute. *J. Agric. Food Ind. Organ.* **2019**, *17*. [CrossRef]
40. Gere, A.; Szekely, G.; Kovacs, S.; Kokai, Z.; Sipos, L. Readiness to adopt insects in Hungary: A case study. *Food Qual. Prefer.* **2017**, *59*, 81–86. [CrossRef]
41. Hartmann, C.; Siegrist, M. Our daily meat: Justification, moral evaluation and willingness to substitute. *Food Qual. Prefer.* **2020**, *80*, 103799. [CrossRef]
42. Hoek, A.C.; Elzerman, J.E.; Hageman, R.; Kok, F.J.; Luning, P.A.; Graaf, C.D. Are meat substitutes liked better over time? A repeated in-home use test with meat substitutes or meat in meals. *Food Qual. Prefer.* **2013**, *28*, 253–263. [CrossRef]
43. Mancini, M.C.; Antonioli, F. Exploring consumers' attitude towards cultured meat in Italy. *Meat Sci.* **2019**, *150*, 101–110. [CrossRef] [PubMed]
44. Verbeke, W. Profiling consumers who are ready to adopt insects as a meat substitute in a Western society. *Food Qual. Prefer.* **2015**, *39*, 147–155. [CrossRef]
45. Weinrich, R.; Strack, M.; Neugebauer, F. Consumer acceptance of cultured meat in Germany. *Meat Sci.* **2020**, *162*, 107924. [CrossRef]
46. Gravely, E.; Fraser, E. Transitions on the shopping floor: Investigating the role of Canadian supermarkets in alternative protein consumption. *Appetite* **2018**, *130*, 146–156. [CrossRef]
47. Guinard, J.X.; Miller, A.M.; Mills, K.; Wong, T.; Lee, S.M.; Sirimuangmoon, C.; Schaefer, S.E.; Drescher, G. Consumer acceptance of dishes in which beef has been partially substituted with mushrooms and sodium has been reduced. *Appetite* **2016**, *105*, 449–459. [CrossRef] [PubMed]
48. Spencer, M.; Cienfuegos, C.; Guinard, J.X. The Flexitarian Flip (TM) in university dining venues: Student and adult consumer acceptance of mixed dishes in which animal protein has been partially replaced with plant protein. *Food Qual. Prefer.* **2018**, *68*, 50–63. [CrossRef]
49. Schosler, H.; de Boer, J.; Boersema, J.J.; Aiking, H. Meat and masculinity among young Chinese, Turkish and Dutch adults in the Netherlands. *Appetite* **2015**, *89*, 152–159. [CrossRef] [PubMed]
50. Kildal, C.L.; Syse, K.L. Meat and masculinity in the Norwegian Armed Forces. *Appetite* **2017**, *112*, 69–77. [CrossRef]
51. De Backer, C.; Erreygers, S.; De Cort, C.; Vandermoere, F.; Dhoest, A.; Vrinten, J.; Van Bauwel, S. Meat and masculinities. Can differences in masculinity predict meat consumption, intentions to reduce meat and attitudes towards vegetarians? *Appetite* **2020**, *147*, 104559. [CrossRef] [PubMed]
52. Onwezen, M.C.; Bouwman, E.P.; Reinders, M.J.; Dagevos, H. A systematic review on consumer acceptance of alternative proteins: Pulses, algae, insects, plant-based meat alternatives, and cultured meat. *Appetite* **2021**, *159*, 105058. [CrossRef]
53. Cordts, A.; Nitzko, S.; Spiller, A. Consumer response to negative information on meat consumption in Germany. *Int. Food Agribus. Manag. Rev.* **2014**, *17*, 83–106.
54. Graham, T.; Abrahamse, W. Communicating the climate impacts of meat consumption: The effect of values and message framing. *Global Environ. Chang.* **2017**, *44*, 98–108. [CrossRef]
55. Carfora, V.; Catellani, P.; Caso, D.; Conner, M. How to reduce red and processed meat consumption by daily text messages targeting environment or health benefits. *J. Environ. Psychol.* **2019**, *65*, 101319. [CrossRef]
56. Clonan, A.; Roberts, K.E.; Holdsworth, M. Socioeconomic and demographic drivers of red and processed meat consumption: Implications for health and environmental sustainability. *Proc. Nutr. Soc.* **2016**, *75*, 367–373. [CrossRef] [PubMed]
57. Lea, E.; Worsley, A. Influences on meat consumption in Australia. *Appetite* **2001**, *36*, 127–136. [CrossRef] [PubMed]
58. Gossard, M.H.; York, R. Social structural influences on meat consumption. *Hum. Ecol. Rev.* **2003**, *10*, 1–9.
59. Carlos, I.G.; Mishra, A.K. Role of ethnicity in consumption of meat products. *Appl. Econ. Lett.* **2011**, *18*, 665–669.
60. Bertolotti, M.; Carfora, V.; Catellani, P. Different frames to reduce red meat intake: The moderating role of self-efficacy. *Health Commun.* **2020**, *35*, 475–482. [CrossRef] [PubMed]
61. Webb, P.; Bain, C.; Page, A. All that glitters is not gold: The problem of error. In *Essential Epidemiology: An Introduction for Students and Health Professionals*; Cambridge University Press: Cambridge, UK, 2017; pp. 189–190.
62. Cliceri, D.; Spinelli, S.; Dinnella, C.; Prescott, J.; Monteleone, E. The influence of psychological traits, beliefs and taste responsiveness on implicit attitudes toward plant- and animal-based dishes among vegetarians, flexitarians and omnivores. *Food Qual. Prefer.* **2018**, *68*, 276–291. [CrossRef]
63. Del Boca, F.K. Addressing sex and gender inequities in scientific research and publishing. *Addiction* **2016**, *111*, 1323–1325. [CrossRef]
64. Bohrer, B.M. An investigation of the formulation and nutritional composition of modern meat analogue products. *Food Sci. Hum. Wellness* **2019**, *8*, 320–329. [CrossRef]
65. van der Weele, C.; Feindt, P.; Jan van der Goot, A.; van Mierlo, B.; van Boekel, M. Meat alternatives: An integrative comparison. *Trends Food Sci. Technol.* **2019**, *88*, 505–512. [CrossRef]

66. Mertens, E.; Kuijsten, A.; van Zanten, H.H.; Kaptijn, G.; Dofková, M.; Mistura, L.; D'Addezio, L.; Turrini, A.; Dubuisson, C.; Havard, S.; et al. Dietary choices and environmental impact in four European countries. *J. Clean. Prod.* **2019**, *237*, 117827. [CrossRef]
67. Mangurian, C.; Niu, G.C.; Schillinger, D.; Newcomer, J.W.; Dilley, J.; Handley, M.A. Utilization of the behavior change wheel framework to develop a model to improve cardiometabolic screening for people with severe mental illness. *Implement. Sci.* **2017**, *12*, 134. [CrossRef] [PubMed]

nutrients

MDPI

Commentary

Unintended Consequences: Nutritional Impact and Potential Pitfalls of Switching from Animal- to Plant-Based Foods

Rachel Tso [1] and Ciarán G. Forde [1,2,3,*]

1 Clinical Nutrition Research Centre (CNRC), Singapore Institute for Food and Biotechnology Innovation (SIFBI), Agency for Science, Technology and Research (A*STAR), Singapore 117599, Singapore; rachel_tso@sifbi.a-star.edu.sg
2 Department of Physiology, Yong Loo Lin School of Medicine, National University of Singapore, Singapore 117593, Singapore
3 Sensory Science and Eating Behaviour, Division of Human Nutrition and Health, Wageningen University and Research, P.O. Box 17, 6700 AA Wageningen, The Netherlands
* Correspondence: ciaran.forde@wur.nl

Abstract: Consumers are shifting towards plant-based diets, driven by both environmental and health reasons. This has led to the development of new plant-based meat alternatives (PBMAs) that are marketed as being sustainable and good for health. However, it remains unclear whether these novel PBMAs to replace animal foods carry the same established nutritional benefits as traditional plant-based diets based on pulses, legumes and vegetables. We modelled a reference omnivore diet using NHANES 2017–2018 data and compared it to diets that substituted animal products in the reference diet with either traditional or novel plant-based foods to create flexitarian, vegetarian and vegan diets matched for calories and macronutrients. With the exception of the traditional vegan diet, all diets with traditional plant-based substitutes met daily requirements for calcium, potassium, magnesium, phosphorus, zinc, iron and Vitamin B12 and were lower in saturated fat, sodium and sugar than the reference diet. Diets based on novel plant-based substitutes were below daily requirements for calcium, potassium, magnesium, zinc and Vitamin B12 and exceeded the reference diet for saturated fat, sodium and sugar. Much of the recent focus has been on protein quality and quantity, but our case study highlights the risk of unintentionally increasing undesirable nutrients while reducing the overall nutrient density of the diet when less healthy plant-based substitutes are selected. Opportunities exist for PBMA producers to enhance the nutrient profile and diversify the format of future plant-based foods that are marketed as healthy, sustainable alternatives to animal-based products.

Keywords: flexitarian; vegetarian; vegan; plant-based meat alternatives; nutrient intakes

Citation: Tso, R.; Forde, C.G. Unintended Consequences: Nutritional Impact and Potential Pitfalls of Switching from Animal- to Plant-Based Foods. *Nutrients* **2021**, *13*, 2527. https://doi.org/10.3390/nu13082527

Academic Editors: Winston Craig and Ujué Fresán

Received: 30 June 2021
Accepted: 20 July 2021
Published: 23 July 2021

1. Introduction

Global concerns around the consumption of animal products and their adverse effects on health and the environment have led to significant growth in the plant-based protein space, particularly for new products to replace traditional meat and dairy [1–4]. This increased demand for non-animal protein has resulted in more consumers declaring to be "flexitarian" or choosing to reduce meat, dairy and eggs in favour of more plant-based foods to benefit the environment, improve health or both [5]. Consumer market insights suggest that from 2019 to 2020, as many as 5 million consumers in the United States shifted to avoid meat completely, becoming either vegetarians or vegans [6,7]. Non-animal-based proteins include vegetable sources, plant-based meat alternatives (PBMAs) and less commonly consumed sources such as algae, insects and cultured meat [8]. The PBMA market is sizable and growing, valued at USD 4.3 billion in 2020 and projected to reach USD 8.3 billion by 2025 [9]. In addition to improved sustainability, PBMAs are marketed for their nutritional benefits [10,11], driving an increase in their consumption for health reasons [2,3]. However, the health benefits of novel PBMAs—a product category with diverse formulations and nutritional compositions—to replace animal foods long-term remain unclear, due to a lack of longitudinal evidence and randomised controlled trials. There has been the suggestion of a "health halo" around these products, where established health benefits of vegetarian and vegan diets are being conflated with positive messaging around animal welfare, sustainability and the environment for many of the newer PBMA products [8,12]. An additional concern is the current focus on promoting plant-based foods of "unhealthy" product categories and formats such as plant-based burgers, nuggets, meatballs and sausages, which may increase consumption of so-called "junk foods" [8,13].

A concern for many consumers seeking to decrease their meat intake and switch to plant-based diets is potentially compromising their protein intake in terms of quality and quantity [14,15]. To add to consumers' uncertainty, less is known about the wider impact of substituting animal foods with plant-based products—particularly novel ones—on intakes of micronutrients and public health-sensitive nutrients (fat, sugar and sodium). Much more is known about the health benefits of omitting meat in favour of traditional plant sources such as legumes, with extensive evidence demonstrating lowered risks of cardiovascular disease, diabetes, cancer and obesity for vegetarians and vegans [16–19]. Even partially reducing meat while increasing healthful plant-based foods has been shown to be beneficial and is associated with lower risks of disease and mortality [4,20,21]. The 2020–2025 Dietary Guidelines for Americans recommend vegetarian diets as one of three healthful dietary patterns [22], the other two being the Healthy U.S.-Style Dietary Pattern and the Healthy Mediterranean-Style Dietary Pattern [22]. The Mediterranean diet is a well-studied example of a feasible plant-based eating pattern low in red meat and high in unsaturated fat intake [4]. It is associated with numerous health benefits including reduced incidence of cardiovascular disease and greater longevity [23,24] while having a lower environmental impact compared to a typical omnivorous Western diet [25].

Frequent red and processed meat intake is linked to greater risks of cardiovascular disease and cancer [26,27]; however, the World Cancer Research Fund highlights that this does not mean all consumers must completely avoid meat. This is because meat presents a valuable source of protein, iron, zinc and Vitamin B12 [28]. Guidelines recommend a maximum of three portions of red meat weekly, while also limiting or avoiding processed meat [28]. There is also evidence that the relationship between meat intake and risk of stroke differs by type of meat. A systematic review and meta-analysis of total, red, processed and white meat consumption found total stroke incidence to be lowest for consumers of white meat [29]. Whether novel PBMAs and traditional plant-based foods make nutritionally equivalent animal food substitutes remains to be seen.

1.1. Protein and Nutrient Concerns between Animal- and Plant-Based Diets

Animal and plant proteins are both nutritious, and protein quality and quantity are often not compromised when switching to well-designed plant-based diets [30]. This applies whether meat and dairy are reduced or completely excluded in favour of nutritious plant-based alternatives. Animal products provide an important source of nutrients and have been described as "complete" protein sources with "high biological value" since they contain all nine essential amino acids [31]. Many plant foods do not contain all essential amino acids and are termed "incomplete" protein sources [31]. The terms "complete", "incomplete", "high biological value" and "low biological value" have been challenged and termed "misleading" in relation to plant protein because they reflect the quantity and quality of essential amino acids consumed in a single serving, but do not account for plant protein blends or the overall protein quality of the diet to meet requirements [30,32]. Most plant-based foods are consumed in combination and from a variety of sources such that plant-based diets adequately meet requirements for all essential amino acids in a calorie-sufficient diet [33]. Furthermore, evidence is emerging on the equivalence of animal and plant protein sources in sustaining lean muscle mass and strength in extended feeding trials [34]. In this regard, it is possible to achieve equivalent protein quality and quantity through the partial or complete removal of animal products in favour of a plant-based diet. What is less well understood is the impact of such a large dietary shift in consumption patterns on intakes of other dietary components, particularly micronutrients.

On-pack nutritional composition represents the nutrients of the unprepared product and often does not reflect actual nutrients consumed when products require further preparations that influence their final composition. This is especially true for most novel plant-based products. Whereas many traditional vegetarian dishes require minimal oil and salt such as lentil stews and stir-fried tofu, by contrast, most of the current PBMA products are sold in formats such as burgers, sausages or nuggets that require preparation with more oil (i.e., frying) and salt and are often consumed with nutrient-poor sides, condiments and beverages. Regularly consuming these products could potentially lead to higher calorie, fat and salt intakes. Similarly, non-dairy milks targeted towards vegetarians or vegans are often high in added sugar [35], while many vegetarian and vegan spreads, snacks and desserts have high levels of salt and fat. If PBMAs or vegetarian- or vegan-friendly products have poor nutritional profiles but are sold under the pretence of better health or nutritional equivalence to natural plants, consumers who adopt them to support a flexitarian, vegetarian or vegan diet may unintentionally consume nutrient-poor diets that are higher in public health-sensitive nutrients.

1.2. Case-Study: A Comparison of Omnivore, Flexitarian, Vegetarian and Vegan Diets

Similar to all diet and lifestyle changes, switching to plant-based substitutes for meat can be beneficial or not for health. In the absence of clear data on the nutritional impact of such a change, the current case study sought to compare the health impact of substituting animal products (i.e., meat, dairy and eggs) for both traditional and emerging non-animal alternatives. We chose a standard omnivore diet based on NHANES 2017–2018 data (reference diet) and substituted animal products with either traditional plant-based foods or novel non-animal protein alternatives to create flexitarian, vegetarian and vegan diets matched for energy, protein, carbohydrate and fat, to focus on changes in micronutrient intake over a given day. The goal of these comparisons was to establish whether reducing the consumption of animal products using novel PBMA and dairy alternatives results in equivalent nutrient intakes for flexitarian, vegetarian and vegan diets, in comparison to a reference omnivore diet.

2. A Comparison of a Standard Omnivore Western Diet to Plant-Based Alternative Diets

2.1. Selecting a Representative Diet

We compared changes in nutrient intakes when moving from an animal-based diet (omnivore) to plant-based diets (flexitarian, vegetarian and vegan). For the comparison, we selected a reference diet based on the average nutrient intake pattern for an American adult male aged 20–49, based on the NHANES 2017–2018 dietary survey [36] (Table 1). This diet is within recommendations for daily calories and most micronutrients [22,36] but was higher for daily protein, carbohydrate, fat, sugar and sodium and lower in dietary fibre, potassium and magnesium (Table 2). This reference diet is representative of the typical intake of the average consumer since most individuals do not meet recommendations for all nutrients [37].

Table 1. Foods Used for Reference, Flexitarian, Vegetarian and Vegan Diets.

	Reference Diet	Flexitarian-Traditional (Flex-Trad)	Flexitarian-Novel (Flex-New)	Vegetarian-Traditional (Veg-Trad)	Vegetarian-Novel (Veg-New)	Vegan-Traditional (Vegan-Trad)	Vegan-Novel (Vegan-New)
	Example of a Typical Western Diet	• Less Meat • Traditional Plant Sources	• Less Meat • Novel Plant Sources	• No Meat • Traditional Plant Sources	• No Meat • Novel Plant Sources	• No Meat or Dairy • Traditional Plant Sources	• No Meat or Dairy • Novel Plant Sources
Breakfast	White bread (2 slices, 95 g) Butter (13 g) Bacon (29 g) Hardboiled egg (38 g) Low-fat milk (200 mL)	White bread (2 slices, 95 g) Olive-based margarine (14 g) Egg omelette with spinach and low-fat cheese (129 g) Low-fat milk (200 mL)	White bread (2 slices, 95 g) Olive-based margarine (7 g) Egg omelette (85 g) Bacon, meatless (16 g) Ketchup (8 g) Low-fat milk (200 mL)	White bread (2 slices, 95 g) Olive-based margarine (18 g) Egg omelette with spinach and low-fat cheese (119 g) Low-fat milk (200 mL)	White bread (2 slices, 95 g) Olive-based margarine (7 g) Plant-based egg omelette with dairy-free cheese (115 g) Bacon, meatless (18 g) Ketchup (15 g) Low-fat milk (200 mL)	White bread (1 slice, 48 g) Peanut butter (15 g) Plant-based egg omelette with black beans (139 g) Soymilk (300 mL)	White bread (2 slices, 95 g) Coconut spread (14 g) Plant-based egg omelette (85 g) Bacon, meatless (43 g) Ketchup (15 g) Soymilk (300 mL)
Morning Snack	Black coffee (250 mL)	Black coffee (250 mL)	Black coffee (250 mL) Plant-based beef jerky (38 g)	Strawberries (100 g) Black coffee (250 mL)	Black coffee (250 mL) Plant-based beef jerky (35 g)	Strawberries (100 g) Black coffee (250 mL)	Plant-based beef jerky (30 g) Black coffee (250 mL)
Lunch	Beef burger with low-fat cheese (245 g) French fries (117 g) Soda (620 mL)	Bean burrito with low-fat cheese (281 g) French fries (117 g) Soda (620 mL)	Plant-based burger with low-fat cheese (257 g) French fries (31 g) Soda (620 mL)	Bean burrito with low-fat cheese (281 g) French fries (117 g) Soda (620 mL)	Plant-based burger with low-fat cheese (252 g) French fries (71 g) Soda (620 mL)	Bean burrito with hummus and dairy-free cheese (266 g) French fries (31 g) Soda (620 mL)	Plant-based burger with dairy-free cheese (268 g) French fries (31 g) Soda (620 mL)
Afternoon Snack	Low-fat yogurt (140 g) Strawberries (100 g)	Unsalted nuts (35 g) Peach (150 g)	Strawberries (100 g) Coconut yogurt (170 g)	Unsalted nuts (25 g) Peach (150 g)	Strawberries (100 g) Plant-based beef jerky (35 g)	Pumpkin seeds (30 g) Soymilk (300 mL)	Plant-based beef jerky (30 g) Unsalted nuts (8 g)
Evening Meal	Grilled chicken (100 g) Stir-fried spinach (91 g) White rice (125 g)	Grilled chicken (100 g) Stir-fried broccoli (102 g) White rice (85 g)	Plant-based chicken tenders (104 g) Ketchup (8 g) Stir-fried broccoli (93 g) White rice (100 g)	Tofu (127 g) Stir-fried broccoli and spinach (115 g) White rice (75 g)	Plant-based chicken tenders (104 g) Ketchup (15 g) Stir-fried broccoli (93 g) White rice (60 g)	Tofu (157 g) Black beans (81 g) Stir-fried broccoli (94 g) White rice (25 g)	Plant-based chicken tenders (104 g) Ketchup (15 g) Stir-fried broccoli (97 g) White rice (25 g)
Dessert	Ice cream (80 g) Peach (150 g)	Low-fat yogurt (125 g) Banana (50 g)	Vegan coconut ice cream (55 g) Peach (150 g)	Low-fat yogurt (220 g) Unsalted nuts (25 g)	Vegan coconut ice cream (80 g) Peach (150 g)	Coconut yogurt (50 g) Peach (150 g) Pumpkin seeds (30 g)	Strawberries (100 g) Peach (150 g)
Total	2482 kcal, 290 g carbohydrate, 105 g protein, 102 g total fat	2478 kcal, 304 g carbohydrate, 103 g protein, 97 g total fat	2435 kcal, 277 g carbohydrate, 99 g protein, 106 g total fat	2486 kcal, 304 g carbohydrate, 101 g protein, 100 g total fat	2517 kcal, 297 g carbohydrate, 108 g protein, 106 g total fat	2514 kcal, 304 g carbohydrate, 99 g protein, 107 g total fat	2471 kcal, 291 g carbohydrate, 104 g protein, 105 g total fat

Values for mixed foods such as omelettes, burgers, burritos and stir-fried vegetables differ across diets due to varying amounts of individual ingredients such as oil, salt and cheese. Nutrient values for each food item are taken from the United States Department of Agriculture [38]; Health Promotion Board, Singapore [39]; and Internet sources [11,40–42]. These diets used unfortified products and assumed no multivitamins nor supplements were taken.

Table 2. Nutritional Compositions of Reference, Flexitarian, Vegetarian and Vegan Diets.

	Calories (kcal)	Carbs (g)	Protein (g)	Total Fat (g)	Sat Fat (g)	Chol (mg)	Sodium (mg)	Sugar (g)	Fibre (g)	B12 (ug)	Calcium (mg)	Potassium (mg)	Magnesium (mg)	Phosphorus (mg)	Zinc (mg)	Iron (mg)
Reference	2482.1	290.4	104.5	101.9	27.4	383.4	4663.5	130.9	19.5	5.7	1143.0	3122.8	324.5	1589.1	15.1	16.0
Flex-Trad	2478.0	304.0	103.1	96.9	20.1	309.3	3882.3	116.0	23.1	3.5	1685.1	3424.8	400.7	1946.4	11.8	15.1
Flex-New	2434.8	276.6	99.4	106.1	50.4	325.2	6514.0	132.2	21.7	2.3	992.7	2169.8	269.4	1586.3	9.6	20.4
Veg-Trad	2486.1	304.0	100.9	100.3	20.7	235.2	3497.8	115.8	26.1	3.8	1913.4	3808.3	435.5	2091.2	13.4	16.8
Veg-New	2516.7	296.8	107.8	105.8	40.5	23.4	8053.0	131.8	22.1	1.2	853.9	2291.5	256.9	1545.6	8.6	19.3
Vegan-Trad	2514.1	303.9	99.3	107.0	20.7	0	3621.8	136.9	40.7	2.3	552.8	3666.3	799.7	2068.6	13.6	22.2
Vegan-New	2471.2	290.7	104.4	105.4	41.3	0	8166.4	133.4	17.6	1.2	429.6	1987.9	272.1	1308.5	7.0	20.4
Dietary Guidelines	2200–2400	130	56	20–35% (49–93 g)	<27	-	2300	<60	31	2.4	1000	3400	400	700	11	8

Dashes denote unavailable values. Carbs = carbohydrate; Sat Fat = saturated fat; Chol = cholesterol; Fibre = Dietary Fibre; B12 = Vitamin B12. Nutrient values for each food item are taken from the United States Department of Agriculture [38]; Health Promotion Board, Singapore [39]; and Internet sources [11,40–42]. Dietary guidelines are the 2020–2025 Dietary Guidelines for Americans [22].

We compared the reference omnivore diet to traditional and newer versions of flexitarian diets with reduced meat intake (Flex-Trad and Flex-New), traditional and newer versions of a vegetarian diet with no meat (Veg-Trad and Veg-New), and traditional and newer versions of a vegan diet without meat, dairy or eggs (Vegan-Trad and Vegan-New) (Table 1). The Flex-Trad, Veg-Trad and Vegan-Trad diets substituted animal products in the reference diet for traditional plant-based substitutes such as beans, nuts and soy. For the Flex-New, Veg-New and Vegan-New diets, we substituted animal products for novel PBMAs and vegan-friendly packaged products such as coconut- or soy-based dairy alternatives, to represent the recent consumer trend towards these products. To focus comparisons on micronutrient intakes, all diets were matched to within 5% for total calories from carbohydrates, fat and protein (Table 2). Although these diets were hypothetical, they represent practical choices that are available within the current plant-based market and were chosen to be realistic, with nutrient values reflecting each food's composition "as prepared" and ready-to-consume, rather than on-pack composition "as sold". For detailed nutritional information for each individual food item in each diet, please refer to the Supplementary Materials (Tables S1–S7).

2.2. Summary Comparison of Reference Diet to Flexitarian, Vegetarian and Vegan Diets

Flexitarian, vegetarian and vegan diets with less meat and dairy intakes were lower in cholesterol, Vitamin B12 and zinc (Table 2). All diets met daily requirements for phosphorus and iron but fell below fibre requirements (Table 2, Figure 1), with the exception of the traditional vegan diet, which had the greatest quantity of legumes and seeds. Diets with novel plant-based substitutes (Flex-New, Veg-New and Vegan-New) fell below daily requirements for calcium, potassium, magnesium, zinc and Vitamin B12 and exceeded the reference diet for saturated fat, sodium and sugar (Table 2, Figure 1). The shortfall in micronutrients was due to low or no meat and dairy consumption, and the fact that many novel PBMAs had lower micronutrient contents compared to equivalent animal products and many traditional plant-based foods. The increase in public health-sensitive nutrients was due to preparation methods requiring oil and salt and condiments consumed with PBMAs. Many vegan-friendly coconut-based products were low in protein and micronutrients and high in fat. Diets with traditional plant-based substitutes (Flex-Trad and Veg-Trad) apart from the vegan diet (Vegan-Trad) met daily requirements for calcium, potassium, magnesium, phosphorus, zinc, iron and Vitamin B12, and was lower than the reference diet in saturated fat, sodium and sugar (Table 2, Figure 1). This was due to lower meat and dairy intakes and an increase in legumes and seeds. Traditional flexitarian and vegetarian diets were the only two diets to meet all daily micronutrient requirements (Figure 1).

2.3. Comparison of the Reference Omnivore Diet to Two Alternative Flexitarian Diets

The flexitarian diets (Flex-Trad and Flex-New) had a reduced proportion of meat and dairy while increasing plant-based foods, and both were lower in cholesterol, Vitamin B12 and zinc and higher in fibre than the reference diet. The traditional flexitarian diet met daily requirements for all seven micronutrients compared and was lower in saturated fat (27%) and sodium (↓17%) and higher in calcium (↑32%) and magnesium (↑19%) than the reference diet. This was due to a higher proportion of low-fat dairy and plant-based foods such as legumes, which were important protein sources for this diet. The flexitarian diet based on more novel products (Flex-New) only met daily requirements for phosphorus and iron and was the diet highest overall for saturated fat. This was due to a greater proportion of novel plant-based convenience foods such as plant-based snacks, coconut-based products and lower dairy intake. Flex-New had 60% less Vitamin B12, 37% less zinc and 31% less potassium than the reference diet while increasing saturated fat intake by 46% and sodium by 28% (Figure 1, Table 2).

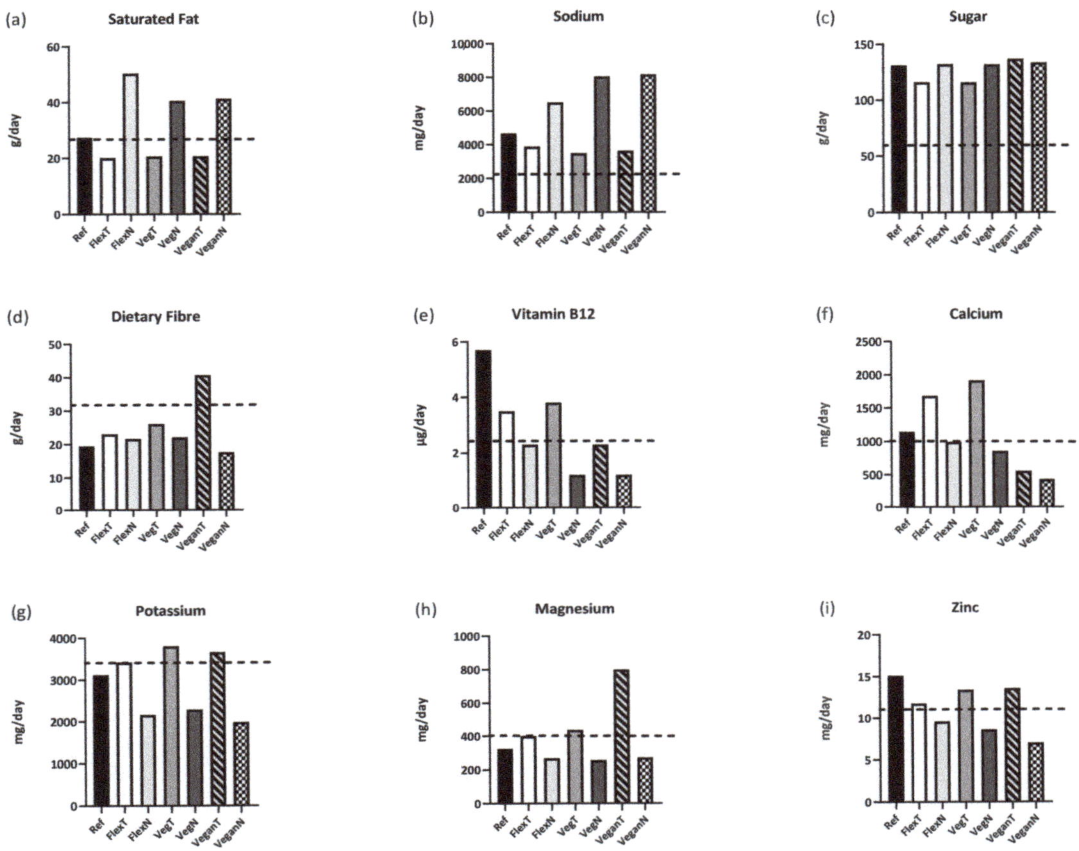

Figure 1. Comparison of nutrient intakes across omnivore, flexitarian (traditional and novel), vegetarian (traditional and novel) and vegan (traditional and novel) diets for (**a**) saturated fat, (**b**) sodium, (**c**) sugar, (**d**) dietary fibre, (**e**) Vitamin B12, (**f**) calcium, (**g**) potassium, (**h**) magnesium and (**i**) zinc. The dotted line in each graph denotes the daily nutritional goal for that nutrient [22]. Ref = reference diet, FlexT = flexitarian-traditional diet, FlexN= flexitarian-novel diet, VegT = vegetarian-traditional diet, VegN = vegetarian-novel diet, VeganT = vegan-traditional diet, VeganN = vegan-novel diet.

2.4. Comparison of the Reference Omnivore Diet to Two Alternative Vegetarian Diets

Removing meat entirely, while retaining dairy and increasing the proportion of energy derived from plant-based foods, resulted in traditional (Veg-Trad) and novel (Veg-New) vegetarian diets, both of which were lower than the reference diet for cholesterol, Vitamin B12 and zinc, and higher in dietary fibre and iron. As with the traditional flexitarian diet, the traditional vegetarian diet met daily requirements for all micronutrients compared and was lower in saturated fat (↓24%) and sodium (↓25%) and higher in calcium (↑40%) and magnesium (↑26%) than the reference diet. This was due to increased intake of dairy and plant-based products such as vegetables, beans, tofu and nuts. The novel vegetarian diet only met daily requirements for phosphorus and iron and not the other micronutrients, due to a greater quantity of novel PBMAs and coconut-based snack products. As a result, the novel vegetarian diet was markedly lower in Vitamin B12 (↓79%), zinc (↓43%), potassium (↓27%) and calcium (↓25%) compared to the reference diet. The novel vegetarian diet also exceeded the reference diet for saturated fat (↑32%) and sodium (↑42%).

2.5. Comparison of the Reference Omnivore Diet to Two Alternative Vegan Diets

Excluding all meat and dairy resulted in vegan diets with no cholesterol and lower Vitamin B12, calcium and zinc content compared to the reference diet. These vegan diets were also higher than the reference diet for iron and sugar. The traditional vegan diet had the highest quantity of legumes and seeds and was the diet highest for iron and fiber, and higher in potassium (↑15%) and magnesium (↑59%) compared to the reference diet. However, the traditional vegan diet also had less Vitamin B12 (↓59%) and calcium (↓52%) than the reference diet. Soymilk was the sole source of Vitamin B12 for both vegan diets. The traditional vegan diet (Vegan-Trad) met all daily micronutrient requirements with the exception of Vitamin B12 and calcium. The novel vegan diet (Vegan-New) provided adequate phosphorus and iron but was the diet lowest for calcium, potassium, phosphorus, zinc and Vitamin B12. Removing all dairy and meat products resulted in daily requirements not being met for Vitamin B12 (79% less than reference diet), calcium (↓62%), potassium (↓36%), magnesium (↓16%) and zinc (↓54%). The novel vegan diet was higher than the reference diet for saturated fat (↑34%) and sodium (↑43%) and had the lowest fibre of all the diets due to the greatest proportion of novel PBMAs and plant-based snack foods. Since many novel PBMAs were rich in iron due to the inclusion of vegetable haem or ferrous sulfate, the novel vegan diet had one of the highest overall iron contents.

3. Trading Places: Feasibility and Nutritional Balance within Specific Diet Comparisons

The traditional flexitarian diet reduced meat using traditional plant-based substitutes and met all daily micronutrient requirements. Retaining dairy and eggs with a smaller portion of meat maintained protein intakes, as well as sustained high micronutrient intakes. This diet did not require the addition of high-protein snacks to meet protein needs, as was needed for vegetarian and vegan diets. An added challenge when making the transition to vegetarian and vegan diets is that unlike meat, many plant-based protein sources tended to be high in carbohydrates, such as dairy and legumes. Similarly, plant sources such as tofu are often lower in protein compared to meats like chicken and need to be consumed in larger portions or supplemented with high-protein snacks such as nuts and yogurt to achieve sufficient intakes, which can increase dietary fat and carbohydrate. The novel vegetarian and vegan diets with PBMAs and dairy alternatives such as coconut ice cream and coconut yogurt were matched for protein intake with the other diets. PBMAs often had lower protein and higher fat, carbohydrate and sodium compared to meat, while dairy alternatives tended to have lower protein yet higher calories, carbohydrate and fat compared to traditional dairy. Non-dairy milk had more added sugar. Accommodating protein from PBMAs and dairy alternatives while matching for energy intake required a reduction in carbohydrate-rich foods such as French fries, rice and bread.

For vegans to go dairy-free, this required the removal of nutrient-rich foods such as cheese and replacing these with dairy-free cheese, which was higher in calories, carbohydrate, fat and sodium but lacked protein. Several foods used to replace animal-source foods in each diet were also low in micronutrients such as coconut yogurt, plant-based egg and dairy-free cheese. To match the reference diet for protein, the novel vegan diet required "supplementation" with protein-rich plant-based snacks such as vegan "beef jerky". Significant protein sources in the vegan diets such as legumes, seeds and PBMAs increased carbohydrate and fat intakes. Whereas vegan diets are known for their low Vitamin B12 content [43,44], this can be overcome with specific vegan food choices such as yeast-fortified dairy-free cheese.

When modelling the various diets to be used in our comparisons, we were mindful of the high variability in nutritional characteristics among products such as plant-based drinks or dairy alternatives [45]. Therefore, to reduce the influence of variability, our case study excluded fortified products and supplements for an equivalent comparison across diets. As a result, the vegan diets had no dietary sources of Vitamin B12 other than soymilk, and neither vegan diet met daily requirements for all seven micronutrients compared, with Vegan-New being the lowest for five of these micronutrients. This suggests that consumption of these diets could result in important micronutrient deficiencies if sustained over time and that vegetarian and vegan diets may require more consideration and dietetic knowledge to implement if they are to meet all daily micronutrient requirements. This includes incorporating a greater proportion of traditional plant-based foods such as beans, nuts and seeds, rather than selecting nutrient-poor PBMAs, and selecting calcium-fortified dairy alternatives alongside Vitamin B12 supplements.

The reference diet against which all other diets were compared reflects a typical daily diet for an American adult male. When calories and % energy from macronutrients were matched, all diets exceeded requirements for macronutrients, sodium and sugar. Only the traditional flexitarian and vegetarian diets met all daily micronutrient requirements, with better intakes than the reference diet. The diets selected for comparison aimed to illustrate in a practical way the possible nutritional implications of adopting flexitarian, vegetarian or vegan diets for the long term. Often, when comparing animal- and plant-based diets, the focus tends to be exclusively on protein quality and quantity. While this is important, the current comparison highlights important implications of selecting plant-based foods believed to be "healthier", and the need to consider non-animal alternatives that are nutrient-dense, and lower in calories, fat, sugar and sodium.

4. Discussion

The current case study demonstrates that traditional plant-based replacements for animal products support the transition to nutritionally adequate flexitarian or vegetarian diets, with flexitarian diets being the most feasible overall. Plant-based diets with larger proportions of novel PBMA substitutes and vegan diets in the absence of nutritional supplements run the risk of being inadequate in a number of important micronutrients.

4.1. Potential Unintended Consequences of Switching to a Plant-Based Diet

A diet that reduces or excludes meat and dairy may have unintended nutritional consequences that arise through selecting foods of lower nutrient density or foods requiring preparation with oil or salt. The current case study shows that diets that increasingly avoid animal-based products result in nutritionally adequate diets if traditional plant-based foods are used as substitutes, with the exception of vegan diets. However, we identified risks when the move to vegetarian and vegan diets was achieved by including novel PBMAs and plant-based dairy, with significantly decreased Vitamin B12, calcium, potassium, magnesium and zinc.

The current comparisons were based on a single day's consumption, which suggests that the continued consumption of diets high in novel PBMAs and plant-based dairy carry the potential risk of increasing intakes of public health-sensitive nutrients while also promoting nutritional deficiencies for a range of micronutrients. For the uninformed consumer, caution should be exercised when making wholesale transitions from diets containing food of animal origin to plant-based diets, particularly those moving to a vegan diet. A diet with sustained low calcium content can increase the risk of low bone mineral density, osteoporosis and fractures. The vegan diets in our case study were the lowest in calcium and the Vegan-New diet was lowest in zinc. These findings are in agreement with a recent cross-sectional comparison that found poorer bone health in vegans compared to omnivores, as well as lower nutritional biomarkers including calcium and zinc [46]. Recent evidence has shown an acceleration of bone turnover when shifting from animal- to plant-based diets [47]. As such, those considering the move to veganism should take steps to ensure adequate consumption of all micronutrients including calcium, zinc and Vitamin B12. As mentioned earlier, this may include calcium-fortified dairy alternatives, Vitamin B12 supplements and traditional plant-based foods. Similarly, consideration should be given to limit specific vegan food choices high in fat, sodium and sugar. This may include several foods catered to vegans, for example falafel and halloumi, which are commonly fried, and non-dairy alternatives such as vegan cheese, yogurt and ice cream, which tend to be lower in protein and micronutrients when compared to animal-based dairy products.

Previous research using dietary data modelling has estimated the impact of switching from animal- to plant-based diets on nutrient intakes. A Canadian study found that increasing the intake of PBMAs by 100% and reducing red and processed meat by 50% improved diet quality as measured by the Nutrient-Rich Foods (NRF) index, but decreased intakes of important nutrients including protein, zinc and Vitamin B12 [48]. Similarly, a French diet simulation study substituted meat, milk and dairy desserts with plant-based substitutes, finding better adequacy for nutrients such as essential fatty acids and fibre but lower adequacies for Vitamin B12, riboflavin, zinc, iron and calcium [49]. There remains a shortage of long-term evidence on the health effects of substituting animal-based foods for newer plant-based meat, dairy and eggs. Two randomized controlled trials to date suggest either a potential long-term negative impact on bone health [47] or beneficial effects including reduced low-density lipoprotein (LDL) cholesterol in an industry-funded trial [50]. However, long-term trials have yet to be completed, and the fluidity of PBMA product formulations and diverse range of products present additional challenges to interpreting the impact of consumption data on health in the future.

Findings from the current case study align with existing literature that flexitarian diets may be the easiest for consumers to transition to in terms of nutritional adequacy and improved health, whereas nutritionally adequate vegetarian and vegan diets require more knowledge and planning. Consuming a mix of proteins from animals and plants while reducing meat and increasing healthful plants is a practical approach for many [51,52]. However, the nutritional profiles of these flexitarian diets depend on the dietary components used as substitutes for animal products [51]. Similarly, when sustaining an omnivorous diet, it is possible to achieve nutrient balance when meat and dairy are low in saturated fat [53] and nutrient-dense foods are prepared with less salt, sugar and oil. However, evidence suggests that reducing meat consumption can have additional health benefits over a balanced omnivore diet. A recent review of Canadian epidemiological studies supports consumption of well-planned vegetarian or vegan diets over omnivore diets to improve nutritional adequacy and reduce risks of chronic conditions and cancer [54]. The current case study suggests consumers will need support and guidance to ensure that when they reduce consumption of animal products, they choose alternatives that avoid unintentional health effects of sustained consumption of nutrient-poor plant-based products.

4.2. *Making the Switch from Animals to Plants: Nutrient Density versus Protein Quantity and Quality*

Much of the commentary around shifting from animal- to plant-based protein diets has centered on the poor protein quantity and quality of plant-based foods [4,30]. However, our comparisons highlight that protein quality or quantity is unlikely to be an issue and many plant proteins and protein blends are capable of meeting daily protein requirements, particularly when consuming a variety of plant foods. Concerns have been raised about consuming adequate protein from vegetarian and vegan diets [14,15], yet today a majority of consumers in developed countries tend to exceed protein requirements [36]. The current comparison suggests the challenge for the emerging plant-based product market will be to enhance nutrient-poor plant sources with adequate micronutrients while reducing consumers' need to add public health-sensitive nutrients such as sugar, salt and fat to enhance the palatability of these products. The protein in new PBMAs is largely based on soy and pea protein and generally sufficient in terms of quantity and quality [50], albeit lower in protein content compared to their animal counterparts. However, the lower nutrient content and requirement for many products to be fried in oil and seasoned with salt could be problematic if adopted as a dietary staple, and the long-term impact of consuming novel PBMAs as a direct replacement for animal-based protein has yet to be tested. In this regard, switching to more frequent PBMA consumption may have the unintended consequence of increasing intakes of nutrient-poor "fast foods", which in turn may not impact protein adequacy but could promote higher intakes of fat, sugar, salt and energy, thus reducing the micronutrient quality of the diet, as highlighted in Table 2 and Figure 1. In recognition of this, recently, PBMA producers such as Beyond Meat have started to fortify their products with B vitamins in order to create burger patties with a micronutrient profile more comparable to that of beef [55].

4.3. Long-Term Impact of Increasing Intakes of Novel PBMAs

Alongside the rapid rise of PBMAs, meat consumption has continued to increase. Global per capita meat consumption has increased approximately 20 kg since 1961 [56], raising the question of whether PBMAs are replacing meat or merely contributing to the increase in overall protein and energy intakes. The long-term health effects of increased PBMA intakes also remain unknown. Excessive haem consumption, the form of iron found in meat, has been associated with increased risks of cancer and type 2 diabetes [57–63]. These associations have not been found for non-haem iron found in plants, which has been used in novel PBMA products to impart the qualities of minced beef. Reducing meat and increasing vegetable intake in an iron-sufficient diet carries health benefits such as promoting the prevention of cancer and chronic diseases [64]. It is unknown whether novel, processed PBMAs carry these same benefits, for instance, soy leghaemoglobin in Impossible Foods. Extensive research over many years has demonstrated the benefits of reducing meat and adopting traditional flexitarian, vegetarian and vegan diets, with associated reductions in cardiovascular disease and cancer risk [17,19]. Claims of the health benefits of novel PBMAs are currently not supported by the same robust long-term evidence from randomised controlled trials, yet these "health halos" are contributing to consumer motivations to consume PBMAs in an effort to improve their health. As highlighted earlier, this may unintentionally support a move to more unbalanced diets and increase intakes of unhealthy dietary components. In the past, unsubstantiated health claims have been shown to drive consumer choice and intake behaviour. For example, when foods are labelled "organic", consumers perceive them to be healthier and to contain fewer calories, despite no evidence to support this [65–67]. Research is now needed to understand the health implications of consuming PBMA products and identify opportunities to enhance their health profile.

Novel PBMAs are marketed to appeal to consumers looking to adopt a "clean" eating dietary regime as part of a global trend to simplify food production and formulations in favour of more natural food products [68]. However, this desire for clean eating conflicts with many novel PBMAs which are highly formulated, processed products that rely on protein isolates, colours, flavours and processing aids to achieve a "meat-like" sensory appeal. A recent study found a greater proportion of "ultra-processed foods" (UPFs) in diets that avoided animal-based foods, with UPFs supplying 33% of daily calories for meat eaters and 39.5% for vegans [69]. Another study saw substitutions of meat, milk and dairy desserts for plant-based substitutes to modify the energy share of UPFs from 29% to 27–40% depending on individual foods used [49]. Thus, many consumers seeking to avoid processed foods when shifting to plant-based diets, especially vegan diets, may find themselves conflicted when faced with messages to consume more of the novel PBMAs. Though processed foods have been associated with increased disease risk [70], not all are unhealthful (i.e., tofu and fortified foods), and it is not clear whether aspects of processing or food formulation are primarily associated with diet-related chronic disease. There is potential for food technology to develop affordable, sustainable and nutritious meat and milk analogues to enhance the nutritional content of diets and benefit consumer health [71]. Our case study comparisons identified a need to enhance the nutrient densities of plant-based proteins, especially novel versions, to create products higher in bioavailable nutrients and lower in public-health-sensitive nutrients. Sugar can be reduced in dairy alternatives while increasing amounts of protein, calcium, Vitamin B12 and other nutrients less often found in non-dairy products. In addition, rather than attempting to construct foods replicating meat or dairy, it may be more nutritious and feasible to encourage consumers to reduce meat and increase consumption of existing nutrient-dense, natural plant sources. Investments and marketing could focus on promoting intakes of fresh vegetables, legumes and seeds rich in protein instead of PBMA products, which are not equivalent nutritionally (as highlighted in Section 2) and are more costly for consumers [9]. Improving the availability of fruits and vegetables in schools, workplaces and communities coupled with choice architecture and strategies to increase palatability of such foods have shown promise [72].

Rather than classifying foods by the degree to which they are processed to determine their healthfulness, measures have been developed that better reflect not only the nutrient density but also the sustainability of foods, which could be used to better inform consumers. Measures such as the plant-based diet index (PDI) may better reflect the nutrient density and impact of dietary changes on health [73]. High-quality plant foods such as wholegrains, fruits, vegetables, nuts and legumes are scored high on the healthful plant-based diet index (hPDI), and lower-quality plant foods such as those high in added sugars and refined grains are high on the unhealthful plant-based diet index (uPDI). In large prospective cohorts, the hPDI was more strongly associated with decreased coronary heart disease (CHD) risk and the uPDI associated with increased CHD risk [73]. Drewnowski and colleagues examined the association between greenhouse gas emissions and energy and nutrient densities of 34 food categories [74]. These approaches could be used to create a standardised diet index that accounts for both the nutritional value and environmental impact of foods selected and consumed. This could be applied to better inform consumers by helping them identify nutrient-dense alternatives to animal products, thus assisting the healthful transition towards sustainable plant-based diets.

This case study's strength was the modelling of seven diets (omnivore, flexitarian, vegetarian and vegan) in a practical manner that follows current nutrient intakes to represent the average consumer. However, it had its limitations. We did not include supplements or fortified products in order to create an equivalent comparison across diets. Therefore, this case study will not fully represent the diets of plant-based consumers who use supplements and fortified products. However, cross-sectional and survey data suggest that supplement use may be lower than expected [43,44] despite an increased risk of poorer micronutrient density in plant-based diets such as vegan diets. Additionally, the single-day dietary pattern, while allowing us to go into detailed comparisons in our data for seven unique diets (i.e., serving sizes, calories and 15 nutrient values of individual foods), forms another limitation of the case study.

5. Conclusions

The current case study highlights that traditional plant-based substitutes can improve nutrient intakes and help consumers to eat more sustainably. However, our findings suggest that it is easy for the uninformed consumer to unintentionally increase public health-sensitive nutrients such as fat, sodium and sugar while also decreasing the nutrient density of the diet. Recent innovation in the plant-based product space has focused more on organoleptic properties (texture, taste and appearance) and formats (nuggets and burgers), rather than developing innovative ways to enhance the nutrient density of plant-derived foods, and ensuring a balanced nutrient profile similar to products of animal origin. Many newer plant-based products are similar to animal products in calories, but lower in protein, calcium, potassium, magnesium, zinc and Vitamin B12 while being higher in sodium and fat after being prepared. If habitually consumed, this could create nutrient shortfalls for consumers motivated to follow healthier, more sustainable diets [8].

Research is needed on how best to guide consumers to choose nutrient-dense plant-based diets that support reduced consumption of public health-sensitive nutrients while sustaining protein quantity and quality. For food producers, there is potential to innovate for the next generation of plant-based foods that provide adequate nutrient intakes alongside protein, and for the development of product formats that do not require the addition of salt, sugar and fat to enhance their sensory appeal.

Supplementary Materials: The following are available online at https://www.mdpi.com/article/10.3390/nu13082527/s1, Table S1: Reference Diet; Table S2: Flexitarian-Traditional (Flex-Trad) Diet; Table S3: Flexitarian-Novel (Flex-New) Diet; Table S4: Vegetarian-Traditional (Veg-Trad) Diet; Table S5: Vegetarian-Novel (Veg-New) Diet; Table S6: Vegan-Traditional (Vegan-Trad) Diet; Table S7: Vegan-Novel (Vegan-New) Diet.

Author Contributions: Conceptualization, C.G.F.; writing—original draft preparation, R.T.; writing—review and editing, C.G.F. and R.T. Both authors have read and agreed to the published version of the manuscript.

Funding: This research was funded by A*STAR-MBIE "Future Food" Catalyst Fund "The Consumer Dimension of Future Foods (Alternative Protein) (2020–23)" (Grant: A20D3b0075) and SFS-2 IAF-PP Future Foods: Alternative Proteins "Understanding functional and nutritional properties of plant-based proteins: application of mixed protein matrices to develop palatable foods for Asian consumers—from lab to table" (Grant: H20H8a0002), both awarded to C.G. Forde (PI).

Data Availability Statement: Data presented are from publicly available online databases: the United States Department of Agriculture's FoodData Central database [36] and Singapore's Health Promotion Board's Energy & Nutrient Composition of Food database [37].

Conflicts of Interest: The authors declare no conflict of interest.

References

1. Bouvard, V.; Loomis, D.; Guyton, K.Z.; Grosse, Y.; Ghissassi, F.E.; Benbrahim-Tallaa, L.; Guha, N.; Mattock, H.; Straif, K. Carcinogenicity of consumption of red and processed meat. *Lancet Oncol.* **2015**, *16*, 1599–1600. [CrossRef]
2. Mintel. Plant-Based Push: UK Sales of Meat-Free Foods Shoot Up 40% between 2014–2019. Available online: https://www.mintel.com/press-centre/food-and-drink/plant-based-push-uk-sales-of-meat-free-foods-shoot-up-40-between-2014-19 (accessed on 1 March 2021).
3. The Business Times. Impossible Foods Launches Plant-Based Sausage Product in HK. Available online: https://www.businesstimes.com.sg/consumer/impossible-foods-launches-plant-based-sausage-product-in-hk (accessed on 3 March 2021).
4. Willett, W.; Rockström, J.; Loken, B.; Springmann, M.; Lang, T.; Vermeulen, S.; Garnett, T.; Tilman, D.; Declerck, F.; Wood, A.; et al. Food in the Anthropocene: The EAT-Lancet Commission on healthy diets from sustainable food systems. *Lancet* **2019**, *393*, 447–492. [CrossRef]
5. Stein, R. How the Rise of "Flexitarians" Is Powering Plant-Based Sales. Available online: https://www.fmi.org/blog/view/fmi-blog/2019/10/22/how-the-rise-of-flexitarians-is-powering-plant-based-sales (accessed on 29 April 2021).
6. Stahler, C. How Many Adults in the U.S Are Vegetarian and Vegan? How Many Adults Eat Vegetarian and Vegan Meals When Eating Out? Available online: https://www.vrg.org/nutshell/Polls/2019_adults_veg.htm (accessed on 9 April 2021).
7. Stahler, C. How Many Adults in the U.S Are Vegetarian and Vegan? How Many Adults Eat Vegetarian and Vegan Meals When Eating Out? Available online: https://www.vrg.org/journal/vj2020issue4/2020_issue4_poll_results.php (accessed on 9 April 2021).
8. Tso, R.; Lim, A.J.; Forde, C.G. A Critical Appraisal of the Evidence Supporting Consumer Motivations for Alternative Proteins. *Foods* **2021**, *10*, 24. [CrossRef]
9. MarketsandMarkets. Plant-based Meat Market by Source (Soy, Wheat, Pea, Quinoa, Oats, Beans, Nuts), Product (Burger Patties, Sausages, Strips & Nuggets, Meatballs), Type (Pork, Beef, Chicken, Fish), Process, and Region—Global Forecast to 2025. Available online: https://www.marketsandmarkets.com/Market-Reports/plant-based-meat-market-44922705.html (accessed on 27 April 2021).
10. Impossible Foods. Health + Nutrition. Available online: https://impossiblefoods.com/health (accessed on 3 March 2021).
11. Beyond Meat. Our Products. Available online: https://www.beyondmeat.com/products/ (accessed on 9 March 2021).
12. Webster, A. Plant-Based Meat Alternatives: Are They Healthy? Available online: https://foodinsight.org/plant-based-meat-alternatives-are-they-healthy/ (accessed on 14 April 2021).
13. ADM. The Plant-Based Protein Market Is Changing What It Takes to Succeed. Available online: https://www.foodnavigator.com/News/Promotional-Features/The-plant-based-protein-market-is-changing-what-it-takes-to-succeed (accessed on 15 April 2021).
14. Sanchez-Sabate, R.; Badilla-Briones, Y.; Sabaté, J. Understanding Attitudes towards Reducing Meat Consumption for Environmental Reasons. A Qualitative Synthesis Review. *Sustainability* **2019**, *11*, 6295. [CrossRef]
15. Fehér, A.; Gazdecki, M.; Véha, M.; Szakály, M.; Szakály, Z. A Comprehensive Review of the Benefits of and the Barriers to the Switch to a Plant-Based Diet. *Sustainability* **2020**, *12*, 4136. [CrossRef]
16. Tonstad, S.; Butler, T.; Yan, R.; Fraser, G.E. Type of vegetarian diet, body weight, and prevalence of type 2 diabetes. *Diabetes Care* **2009**, *32*, 791–796. [CrossRef] [PubMed]
17. Huang, T.; Yang, B.; Zheng, J.; Li, G.; Wahlqvist, M.L.; Li, D. Cardiovascular disease mortality and cancer incidence in vegetarians: A meta-analysis and systematic review. *Ann. Nutr. Metab.* **2012**, *60*, 233–240. [CrossRef] [PubMed]

18. Barnard, N.D.; Katcher, H.I.; Jenkins, D.J.A.; Cohen, J.; Turner-McGrievy, G. Vegetarian and vegan diets in type 2 diabetes management. *Nutr. Rev.* **2009**, *67*, 255–263. [CrossRef]
19. Tantamango-Bartley, Y.; Jaceldo-Siegl, K.; Fan, J.; Fraser, G. Vegetarian diets and the incidence of cancer in a low-risk population. *Cancer Epidemiol. Biomark. Prev.* **2013**, *22*, 286–294. [CrossRef] [PubMed]
20. Zheng, Y.; Li, Y.; Satija, A.; Pan, A.; Sotos-Prieto, M.; Rimm, E.; Willett, W.C.; Hu, F.B. Association of changes in red meat consumption with total and cause specific mortality among US women and men: Two prospective cohort studies. *BMJ* **2019**, *365*, l2110. [CrossRef]
21. Chen, Z.; Drouin-Chartier, J.-P.; Li, Y.; Baden, M.Y.; Manson, J.E.; Willett, W.C.; Voortman, T.; Hu, F.B.; Bhupathiraju, S.N. Changes in Plant-Based Diet Indices and Subsequent Risk of Type 2 Diabetes in Women and Men: Three U.S. Prospective Cohorts. *Diabetes Care* **2021**, *44*, 663–671. [CrossRef] [PubMed]
22. U.S. Department of Agriculture and U.S. Department of Health and Human Services. *Dietary Guidelines for Americans, 2020–2025*, 9th ed.; U.S. Department of Agriculture and U.S. Department of Health and Human Services: Washington, DC, USA, December 2020.
23. Estruch, R.; Ros, E.; Salas-Salvadó, J.; Covas, M.I.; Corella, D.; Arós, F.; Gómez-Gracia, E.; Ruiz-Gutiérrez, V.; Fiol, M.; Lapetra, J.; et al. Primary Prevention of Cardiovascular Disease with a Mediterranean Diet Supplemented with Extra-Virgin Olive Oil or Nuts. *N. Engl. J. Med.* **2018**, *378*, e34. [CrossRef] [PubMed]
24. Kromhout, D.; Keys, A.; Aravanis, C.; Buzina, R.; Fidanza, F.; Giampaoli, S.; Jansen, A.; Menotti, A.; Nedeljkovic, S.; Pekkarinen, M.; et al. Food consumption patterns in the 1960s in seven countries. *Am. J. Clin. Nutr.* **1989**, *49*, 889–894. [CrossRef]
25. Fresán, U.; Martínez-González, M.A.; Sabaté, J.; Bes-Rastrollo, M. Global sustainability (health, environment and monetary costs) of three dietary patterns: Results from a Spanish cohort (the SUN project). *BMJ Open* **2019**, *9*, e021541. [CrossRef] [PubMed]
26. Papier, K.; Fensom, G.K.; Knuppel, A.; Appleby, P.N.; Tong, T.Y.N.; Schmidt, J.A.; Travis, R.C.; Key, T.J.; Perez-Cornago, A. Meat consumption and risk of 25 common conditions: Outcome-wide analyses in 475,000 men and women in the UK Biobank study. *BMC Med.* **2021**, *19*, 53. [CrossRef] [PubMed]
27. Aune, D.; Chan, D.S.; Vieira, A.R.; Navarro Rosenblatt, D.A.; Vieira, R.; Greenwood, D.C.; Kampman, E.; Norat, T. Red and processed meat intake and risk of colorectal adenomas: A systematic review and meta-analysis of epidemiological studies. *Cancer Causes Control* **2013**, *24*, 611–627. [CrossRef]
28. World Cancer Research Fund/American Institute for Cancer Research. Limit Red and Processed Meat. Available online: https://www.wcrf.org/dietandcancer/recommendations/limit-red-processed-meat (accessed on 5 April 2021).
29. Kim, K.; Hyeon, J.; Lee, S.A.; Kwon, S.O.; Lee, H.; Keum, N.; Lee, J.K.; Park, S.M. Role of Total, Red, Processed, and White Meat Consumption in Stroke Incidence and Mortality: A Systematic Review and Meta-Analysis of Prospective Cohort Studies. *J. Am. Heart Assoc.* **2017**, *6*, e005983. [CrossRef]
30. Arens, U.; Convery, L.; Garton, L.; Hood, S.; Mason, P.; Pettinger, C.; Medici, E.; Haffner, T.; Arthur, K.; Lewis, J.; et al. *One Blue Dot Eating Patterns for Health and Environmental Sustainability: A Reference Guide for Dietitians*; British Dietetic Association: Birmingham, UK, 2018; p. 92.
31. Draper, J.; Revilla, M.K.F.; Titchenal, A.; Caacbay, N.A.; Calabrese, A.; Chun, C.N.; Gibby, C.; Hara, S.; Langfelder, G.; Yang, Y.-Y.; et al. Proteins, Diet, and Personal Choices. In *Human Nutrition: 2020 Edition*; University of Hawaii at Mānoa: Honolulu, HI, USA, 2020.
32. Melina, V.; Craig, W.; Levin, S. Position of the Academy of Nutrition and Dietetics: Vegetarian Diets. *J. Acad. Nutr. Diet.* **2016**, *116*, 1970–1980. [CrossRef]
33. Gardner, E. Vegetarianism and Vegan Diets. In *Manual of Dietetic Practice*, 5th ed.; Gandy, J., Ed.; John Wiley & Sons, Ltd.: West Sussex, UK, 2014; pp. 129–140.
34. Hevia-Larraín, V.; Gualano, B.; Longobardi, I.; Gil, S.; Fernandes, A.L.; Costa, L.A.R.; Pereira, R.M.R.; Artioli, G.G.; Phillips, S.M.; Roschel, H. High-Protein Plant-Based Diet Versus a Protein-Matched Omnivorous Diet to Support Resistance Training Adaptations: A Comparison Between Habitual Vegans and Omnivores. *Sports Med.* **2021**, *51*, 1317–1330. [CrossRef]
35. McClements, D.J.; Newman, E.; McClements, I.F. Plant-based Milks: A Review of the Science Underpinning Their Design, Fabrication, and Performance. *Compr. Rev. Food Sci. Food Saf.* **2019**, *18*, 2047–2067. [CrossRef]
36. U.S. Department of Agriculture. *What We Eat in America, NHANES 2017–2018, Individuals 2 Years and Over (Excluding Breast-Fed Children), Day 1*; U.S. Department of Agriculture: Oakland, MD, USA, 2020.
37. Scrafford, C.G.; Bi, X.; Multani, J.K.; Murphy, M.M.; Schmier, J.K.; Barraj, L.M. Health Economic Evaluation Modeling Shows Potential Health Care Cost Savings with Increased Conformance with Healthy Dietary Patterns among Adults in the United States. *J. Acad. Nutr. Diet.* **2019**, *119*, 599–616. [CrossRef] [PubMed]
38. United States Department of Agriculture. FoodData Central. Available online: http://fdc.nal.usda.gov/ (accessed on 10 March 2021).
39. Health Promotion Board. Energy & Nutrient Composition of Food. Available online: https://focos.hpb.gov.sg/eservices/ENCF/ (accessed on 12 March 2021).
40. McDonald's. Nutrition Calculator. Available online: https://www.mcdonalds.com/us/en-us/about-our-food/nutrition-calculator.html (accessed on 13 April 2021).
41. Gardein. Gardein Seven Grain Crispy Tenders. Available online: https://www.gardein.com/chickn-and-turky/classics/seven-grain-crispy-tenders (accessed on 10 March 2021).

42. JUST Egg. JUST Egg | Made from Plants. Available online: https://www.ju.st/en-us/products/consumer/egg/egg (accessed on 15 March 2021).
43. Benham, A.J.; Gallegos, D.; Hanna, K.L.; Hannan-Jones, M.T. Intake of vitamin B12 and other characteristics of women of reproductive age on a vegan diet in Australia. *Public Health Nutr.* **2021**, 1–11. [CrossRef]
44. Gilsing, A.M.J.; Crowe, F.L.; Lloyd-Wright, Z.; Sanders, T.A.B.; Appleby, P.N.; Allen, N.E.; Key, T.J. Serum concentrations of vitamin B12 and folate in British male omnivores, vegetarians and vegans: Results from a cross-sectional analysis of the EPIC-Oxford cohort study. *Eur. J. Clin. Nutr.* **2010**, *64*, 933–939. [CrossRef]
45. Angelino, D.; Rosi, A.; Vici, G.; Dello Russo, M.; Pellegrini, N.; Martini, D.; on behalf of the SINU Young Working Group. Nutritional Quality of Plant-Based Drinks Sold in Italy: The Food Labelling of Italian Products (FLIP) Study. *Foods* **2020**, *9*, 682. [CrossRef]
46. Menzel, J.; Abraham, K.; Stangl, G.I.; Ueland, P.M.; Obeid, R.; Schulze, M.B.; Herter-Aeberli, I.; Schwerdtle, T.; Weikert, C. Vegan Diet and Bone Health—Results from the Cross-Sectional RBVD Study. *Nutrients* **2021**, *13*, 685. [CrossRef]
47. Itkonen, S.T.; Päivärinta, E.; Pellinen, T.; Viitakangas, H.; Risteli, J.; Erkkola, M.; Lamberg-Allardt, C.; Pajari, A.-M. Partial Replacement of Animal Proteins with Plant Proteins for 12 Weeks Accelerates Bone Turnover Among Healthy Adults: A Randomized Clinical Trial. *J. Nutr.* **2020**, *151*, 11–19. [CrossRef] [PubMed]
48. Vatanparast, H.; Islam, N.; Shafiee, M.; Ramdath, D.D. Increasing Plant-Based Meat Alternatives and Decreasing Red and Processed Meat in the Diet Differentially Affect the Diet Quality and Nutrient Intakes of Canadians. *Nutrients* **2020**, *12*, 2034. [CrossRef] [PubMed]
49. Salomé, M.; Huneau, J.-F.; Le Baron, C.; Kesse-Guyot, E.; Fouillet, H.; Mariotti, F. Substituting Meat or Dairy Products with Plant-Based Substitutes Has Small and Heterogeneous Effects on Diet Quality and Nutrient Security: A Simulation Study in French Adults (INCA3). *J. Nutr.* **2021**, nxab146. [CrossRef] [PubMed]
50. Crimarco, A.; Springfield, S.; Petlura, C.; Streaty, T.; Cunanan, K.; Lee, J.; Fielding-Singh, P.; Carter, M.M.; Topf, M.A.; Wastyk, H.C.; et al. A randomized crossover trial on the effect of plant-based compared with animal-based meat on trimethylamine-N-oxide and cardiovascular disease risk factors in generally healthy adults: Study With Appetizing Plantfood—Meat Eating Alternative Trial (SWAP-MEAT). *Am. J. Clin. Nutr.* **2020**, *112*, 1188–1199. [CrossRef]
51. Hemler, E.C.; Hu, F.B. Plant-Based Diets for Cardiovascular Disease Prevention: All Plant Foods Are Not Created Equal. *Curr. Atheroscler. Rep.* **2019**, *21*, 18. [CrossRef] [PubMed]
52. Schmidt, C.V.; Mouritsen, O.G. The Solution to Sustainable Eating Is Not a One-Way Street. *Front. Psychol.* **2020**, *11*, 531. [CrossRef] [PubMed]
53. Richter, C.K.; Skulas-Ray, A.C.; Champagne, C.M.; Kris-Etherton, P.M. Plant protein and animal proteins: Do they differentially affect cardiovascular disease risk? *Adv. Nutr.* **2015**, *6*, 712–728. [CrossRef] [PubMed]
54. Bye, Z.L.; Keshavarz, P.; Lane, G.L.; Vatanparast, H. What Role Do Plant-Based Diets Play in Supporting the Optimal Health and Well-being of Canadians? A Scoping Review. *Adv. Nutr.* **2021**. [CrossRef]
55. Watson, E. Beyond Meat Unveils the Beyond Burger 3.0 with Likeability Scoring 'On-Par with 80/20 Ground Beef Burgers'. Available online: https://www.foodnavigator-usa.com/Article/2021/04/27/Beyond-Meat-unveils-the-Beyond-Burger-3.0-with-likeability-scoring-on-par-with-80-20-ground-beef-burgers (accessed on 1 June 2021).
56. Ritchie, H.; Roser, M. Meat and Dairy Production. Available online: https://ourworldindata.org/meat-production#citation (accessed on 15 April 2021).
57. Bastide, N.M.; Pierre, F.H.F.; Corpet, D.E. Heme iron from meat and risk of colorectal cancer: A meta-analysis and a review of the mechanisms involved. *Cancer Prev. Res.* **2011**, *4*, 177–184. [CrossRef]
58. Chang, V.C.; Cotterchio, M.; Khoo, E. Iron intake, body iron status, and risk of breast cancer: A systematic review and meta-analysis. *BMC Cancer* **2019**, *19*, 543. [CrossRef]
59. Genkinger, J.M.; Friberg, E.; Goldbohm, R.A.; Wolk, A. Long-term dietary heme iron and red meat intake in relation to endometrial cancer risk. *Am. J. Clin. Nutr.* **2012**, *96*, 848–854. [CrossRef] [PubMed]
60. Taunk, P.; Hecht, E.; Stolzenberg-Solomon, R. Are meat and heme iron intake associated with pancreatic cancer? Results from the NIH-AARP diet and health cohort. *Int. J. Cancer* **2016**, *138*, 2172–2189. [CrossRef] [PubMed]
61. World Cancer Research Fund/American Institute for Cancer Research. Meat, Fish and Dairy. Available online: https://www.wcrf.org/dietandcancer/meat-fish-and-dairy/ (accessed on 5 April 2021).
62. Jiang, R.; Ma, J.; Ascherio, A.; Stampfer, M.J.; Willett, W.C.; Hu, F.B. Dietary iron intake and blood donations in relation to risk of type 2 diabetes in men: A prospective cohort study. *Am. J. Clin. Nutr.* **2004**, *79*, 70–75. [CrossRef] [PubMed]
63. Zhao, Z.; Li, S.; Liu, G.; Yan, F.; Ma, X.; Huang, Z.; Tian, H. Body iron stores and heme-iron intake in relation to risk of type 2 diabetes: A systematic review and meta-analysis. *PLoS ONE* **2012**, *7*, e41641. [CrossRef]
64. Clinton, S.K.; Giovannucci, E.L.; Hursting, S.D. The World Cancer Research Fund/American Institute for Cancer Research Third Expert Report on Diet, Nutrition, Physical Activity, and Cancer: Impact and Future Directions. *J. Nutr.* **2020**, *150*, 663–671. [CrossRef]
65. Prada, M.; Garrido, M.V.; Rodrigues, D. Lost in processing? Perceived healthfulness, taste and caloric content of whole and processed organic food. *Appetite* **2017**, *114*, 175–186. [CrossRef]
66. Besson, T.; Bouxom, H.; Jaubert, T. Halo It's Meat! the Effect of the Vegetarian Label on Calorie Perception and Food Choices. *Ecol. Food Nutr.* **2020**, *59*, 3–20. [CrossRef]

67. McCrickerd, K.; Tang, C.S.; Forde, C.G. The independent and combined impact of front-of-pack labelling and sensory quality on calorie estimations and portion selection of commercial food products. *Food Qual. Prefer.* **2020**, *79*, 103766. [CrossRef]
68. Kateman, B. Healthier Plant-Based Meat Is On the Rise. Available online: https://www.forbes.com/sites/briankateman/2021/05/10/healthier-plant-based-meat-is-on-the-rise/?sh=463715a720fd (accessed on 19 May 2021).
69. Gehring, J.; Touvier, M.; Baudry, J.; Julia, C.; Buscail, C.; Srour, B.; Hercberg, S.; Péneau, S.; Kesse-Guyot, E.; Allès, B. Consumption of Ultra-Processed Foods by Pesco-Vegetarians, Vegetarians, and Vegans: Associations with Duration and Age at Diet Initiation. *J. Nutr.* **2020**, *151*, 120–131. [CrossRef]
70. Santos, F.S.D.; Dias, M.D.S.; Mintem, G.C.; Oliveira, I.O.; Gigante, D.P. Food processing and cardiometabolic risk factors: A systematic review. *Rev. Saude Publica* **2020**, *54*, 70. [CrossRef] [PubMed]
71. Gibney, M.J. Food Technology and Plant-Based Diets. *J. Nutr.* **2020**, *151*, 1–2. [CrossRef] [PubMed]
72. Wolfenden, L.; Barnes, C.; Lane, C.; McCrabb, S.; Brown, H.M.; Gerritsen, S.; Barquera, S.; Véjar, L.S.; Munguía, A.; Yoong, S.L. Consolidating evidence on the effectiveness of interventions promoting fruit and vegetable consumption: An umbrella review. *Int J. Behav. Nutr. Phys. Act.* **2021**, *18*, 11. [CrossRef]
73. Satija, A.; Bhupathiraju, S.N.; Spiegelman, D.; Chiuve, S.E.; Manson, J.E.; Willett, W.; Rexrode, K.M.; Rimm, E.B.; Hu, F.B. Healthful and Unhealthful Plant-Based Diets and the Risk of Coronary Heart Disease in U.S. Adults. *J. Am. Coll. Cardiol.* **2017**, *70*, 411–422. [CrossRef]
74. Drewnowski, A.; Rehm, C.D.; Martin, A.; Verger, E.O.; Voinnesson, M.; Imbert, P. Energy and nutrient density of foods in relation to their carbon footprint. *Am. J. Clin. Nutr.* **2015**, *101*, 184–191. [CrossRef] [PubMed]

nutrients

MDPI

Article

Nutritional Content and Health Profile of Non-Dairy Plant-Based Yogurt Alternatives

Winston J. Craig [1,*] and Cecilia J. Brothers [2]

[1] Center for Nutrition, Healthy Lifestyles, and Disease Prevention, School of Public Health, Loma Linda University, Loma Linda, CA 92354, USA
[2] Department of Biology, Walla Walla University, College Place, WA 99324, USA; cecilia.brothers@wallawalla.edu
* Correspondence: wcraig@llu.edu

Abstract: Yogurt is considered a healthy, nutritious food in many cultures. With a significant number of people experiencing dairy intolerance, and support for a more sustainable diet, consumer demand for dairy alternatives has surged. The aim of this study was to conduct a cross-sectional survey of plant-based yogurt alternatives to assess their nutritional content and health profile. A total of 249 non-dairy yogurt alternatives were analyzed from the nutrition label listed on the commercial package. The various yogurt alternatives contained extracts of coconut (n = 79), almonds (n = 62), other nuts or seeds (n = 20), oats (n = 20), legumes (n = 16), and mixed blends (n = 52). At least one-third of the yogurt alternatives had 5 g or more of protein/serving. Only 45% of the yogurt alternatives had calcium levels fortified to at least 10% of daily value (DV), while only about one in five had adequate vitamin D and B12 fortification at the 10% DV level. One-half of the yogurt alternatives had high sugar levels, while 93% were low in sodium. Except for the coconut-based products, the yogurts were not high in fat or saturated fat. The yogurt alternatives were not fortified as frequently or to the same levels as the corresponding non-dairy, plant-based beverages.

Keywords: non-dairy yogurt alternatives; plant-based yogurts; nutrient composition; fortification; calcium; vitamin D; vitamin B12; protein; sugar

Citation: Craig, W.J.; Brothers, C.J. Nutritional Content and Health Profile of Non-Dairy Plant-Based Yogurt Alternatives. *Nutrients* **2021**, *13*, 4069. https://doi.org/10.3390/nu13114069

Academic Editor: Rosa Casas

Received: 10 September 2021
Accepted: 3 November 2021
Published: 14 November 2021

1. Introduction

Yogurt, plain or sweetened, is a very popular food in many cultures. In Europe, the dietary recommendations suggest the consumption of 100–250 g of yogurt/day [1]. Yogurt is considered a tasty, healthy and nutritious food, supplying some important vitamins and minerals. Individuals concerned about following a more sustainable diet, those with a dairy intolerance, or who desire a non-dairy alternative due to dietary preferences, such as vegans, will choose a lactose-free plant–based yogurt alternative. Typically, the comfort level of such individuals, and others who wish to experiment with new foods, is best met when the yogurt alternative has a similar appearance and texture to the dairy product. The growing interest in non-dairy yogurts, combined with the surge of interest in plant-based milk alternatives and meat alternatives has fueled a plant-based food industry valued at USD 5 billion [2] that is re-shaping the future of American cuisine. Over the past 15 years the number of Americans following plant-based diets has surged 300% [2]. The global plant-based yogurt market was valued at USD 1.6 billion in 2019 and is projected to grow at an annual growth rate of nearly 20% from 2020 to 2027 [3]. The US market for plant-based yogurt alternatives was about USD 400 million in 2020 and is expected to be valued at USD 1.3 billion by 2027 [3].

The popularity of yogurt alternatives over the past decade was especially prominent among millennials [3]. A number of manufacturers have made a decided effort to market their plant-based yogurt alternatives to a generation who take sustainability and the health of the planet seriously. Their websites clearly display that emphasis in their stated goals and mission [4–6]. During production of plant-based yogurt alternatives, companies claim they work for the health of the planet, conserve resources, and reduce the environmental impact and the carbon footprint of their activities. Some claim to use fair trade ingredients, and most claim to be vegan, GMO (genetically modified organisms)-free, and organic. The plant-based yogurts currently are based upon almonds, coconut, oats, or a legume (soy or yellow pea). With over 35 flavors available, there are many varieties from which to select. In addition, some plant-based yogurt alternatives possess unique health properties. For example, yogurts based on flax or hemp contain substantial levels of omega-3 fatty acids and fiber.

Consumers who regularly rely upon a non-dairy alternative may place themselves a little short nutritionally. If an appropriate substitution of a dairy product is not made, one could experience a nutritional shortfall in protein, calcium and some micronutrients. For a vegan, and someone who limits their intake of animal products, there are three nutrients of special concern. These are calcium, vitamin D, and vitamin B12 [7]. Calcium and vitamin B12 are supplied by dairy yogurt, while vitamin D is only supplied by those dairy yogurts fortified with vitamin D. On the other hand, these three nutrients are not supplied by plant-based yogurt alternatives unless the products are fortified. When plant-based non-dairy beverages are fortified, they may have similar amounts of calcium, vitamins D and B12 as dairy milk or they may be fortified with only one or two of the three critical nutrients [8,9]. We set out to see how many of the non-dairy yogurt alternatives have adequate protein levels and fortification, and at what level.

People choosing to follow a plant-based diet may need substantial guidance from a health care professional (such as a dietitian) to select a well-fortified non-dairy product. They may not know the nutritional pros and cons of consuming a particular non-dairy beverage versus a particular yogurt alternative. The choice should be influenced by which provides the better level of protein, calcium, vitamin D, and/or vitamin B12. They may prefer to consume a non-dairy beverage long-term rather than a yogurt alternative, or vice-versa, and would be interested to know how their nutritional status would be impacted by their preference. Hence, we set out to see how the nutrient levels in a yogurt alternative compared (on a serving basis) with those from a similar analysis of non-dairy plant-based beverages [8].

Consumers are normally concerned, for health reasons, about the level of sodium, sugar, and fat/saturated fat that exists in the food they purchase. We therefore examined the levels of these nutrients to see how many of the yogurt alternatives had acceptable levels of these nutrients. Furthermore, fermented foods are quite popular due to the perception that they are both nutritious and have health benefits from the bioactive metabolites produced by the fermentation [10,11]. Recently, the consumption of fermented foods containing live microorganisms has been seen as an important dietary approach for improving human health [12]. In addition, water-soluble dietary fibers, such as gums and inulin, are considered probiotic compounds, due to their fermentability by gut microbiota. These fibers provide a variety of health benefits, including an improved immune system defense [13,14]. We also examined the extent to which the plant-based yogurt alternatives contained prebiotics and probiotics.

2. Materials and Methods

The nutritional contents of 249 plant-based yogurt alternatives, representing 34 brands, were analyzed. The yogurts were selected, from March to May 2021, from those available in supermarkets and convenience stores in the western USA. Additional varieties of plant-based, non-dairy yogurt alternatives [15] were analyzed from the nutritional labels given by the manufacturer's website, or from the website of common retailers. The plant-based yogurts with incomplete nutrition data were not included in the analysis.

The nutritional content of each yogurt was recorded from the nutrition label on the commercial package or from the information located on the website of the manufacturer or retailer. The nutrients per serving size, which were available on all packages, included calories, fat, saturated fat, sodium, carbohydrates, dietary fiber, total sugars, protein, and the micronutrients calcium, vitamin D, and vitamin B12. The median values of the nutrients were calculated for each type of yogurt.

A similar analysis was conducted on 326 multi-serve, plant-based, non-dairy beverages available in the USA, so that a comparison could be made between the nutritional value of one serving of the plant-based yogurt alternative with that of one serving of a plant-based beverage. The levels of fortification for calcium, vitamin D, and vitamin B12 were calculated for all yogurts and beverages separately.

The nutritional value of each plant-based yogurt was rated according to the following criterion: calcium, vitamin D and vitamin B12 of at least 10% of daily value (DV)/serving, and at least 5 g of protein/serving (10% of the DV). The health qualities demonstrated by the ingredients were determined by the following criteria: not more than 5 g of total sugars/serving; not more than 1 g of saturated fat/serving; not more than 150 calories/serving; not more than 115 mg sodium/serving; and at least 1.5 g of dietary fiber/serving. In addition, we noted how many yogurts had high levels of sugar, fat, and saturated fat. This was recorded as the number of yogurts having 10 g or more of total sugars/serving (20% DV for sugars), having more than 15.5 g fat/serving (20% DV for fat) and having 4 g or more of saturated fat/serving (20% DV for saturated fat). The US Dietary Guidelines specify, as a general guide, that 5% DV or less of a nutrient/serving is considered low, while 20% DV or more of a nutrient/serving is considered high [16,17]. In the USA, the DV for calcium is 1300 mg, vitamin D is 20 mcg, vitamin B12 is 2.4 mcg, sodium is 2300 mg, protein is 50 g, added sugars is 50 g, saturated fat is 20 g, and dietary fiber is 28 g [17]. For our analyses, we considered a 10% DV as an adequate fortification for calcium, vitamin D and vitamin B12. For sodium and saturated fat we accepted that beverages should not exceed 5% of their DV (a designated low level), namely 115 mg for sodium (5% of 2300 mg) and 1 g of saturated fat (5% of 20 g). We suggest that yogurts have at least 5% of DV for dietary fiber (approx. 1.5 g), at least 10% of the DV/serving for protein (5 g), and no more than 10% DV/serving for sugars. Ten percent was chosen as a mid-stream number between the 5% DV (low value) and the 20% DV (high value). This gave us a minimally acceptable level of 5 g protein/serving, and a level of 5 g sugars/serving.

Statistical Analysis

R software was used to conduct all statistical analyses [18]. Data were tested for normality and homoscedasticity prior to analysis. The median and interquartile range were used for descriptive statistics, as the data were not normally distributed. The nutritional content was compared across the types of non-dairy yogurt bases using a Kruskal-Wallis test for each nutrient, followed by Dunn's post hoc test with Bonferroni adjustment for multiple comparisons. The nutritional content was analyzed between non-dairy yogurt and beverage alternatives using an unpaired Wilcoxon rank sum test for each nutrient. An unpaired Wilcoxon rank sum test was also used to compare the levels of fortified nutrients between non-dairy yogurt and beverage alternatives for each fortified nutrient. The nutritional content was compared across product type (beverage or yogurt) and base type using a two-way analysis of variance for each nutrient, followed by Tukey's honestly significant difference test. Only significant pairwise comparisons between non-dairy yogurt and beverage alternatives for the same base type were reported. A significant p-value of less than 0.05 was used for all analyses.

3. Results

The 249 plant-based yogurt alternatives analyzed were based upon coconut (n = 79), almonds (n = 62), oats (n = 20), cashews (n = 13), soy (n = 11), pea protein (n = 5), hemp (n = 4), pumpkin seed (n = 3), and the following mixtures: coconut with pea protein (n = 23), oats with pea protein and/or fava bean (n = 16), coconut with pili nut (n = 6), almond with fava bean (n = 4), coconut with cashews and watermelon and pumpkin seeds (*n* = 3). Table 1 displays the medians of each nutrient for all of these base types, and differences among the base types are reported.

Two hundred and twelve (85.1%) of the yogurt alternatives were flavored while 24 (9.6%) were unsweetened and 36 were labeled as plain (14.5%). The most common flavors were vanilla (19%), strawberry (14%), and blueberry (10%). Twenty-three (68%) of the brands had no fortification. Only 11 of the 34 brands had calcium fortification. Tricalcium phosphate (TCP) (47%), calcium citrate (17%), and a TCP-calcium citrate mix (18%) were the most commonly used calcium salts. Others included calcium lactate-TCP mix (10%), and calcium carbonate-TCP mix (5%), and calcium carbonate (4%). At least two brands of yogurt alternatives used stevia and/or monk fruit extracts for sweetening. Added fibers included pectin (64% of brands), locust bean gum (45%), agar (12%) and guar (6%). Acacia gum and gellan were rarely used. Six of the 34 brands contained inulin from either chicory root or agave. Each yogurt typically contained 4–6 live cultures. Eight almond-based and three coconut-based yogurts contained mix-ins. All 11 with mix-ins contain 180 calories and above, with a mean of 217 calories/serving. One serving size of yogurt alternative varied according to the size of the container. The most common sizes encountered were 150 g (66.7%), 170 g (14.1%), and 120 g (7.6%).

Table 2 summarizes the data showing the percentage of the yogurt alternatives that contain a) a reasonable level of important and essential nutrients, b) low levels of sugar, sodium and saturated fat, and c) high levels of sugar, fat, saturated fat and calories. In Table 3, the data for the yogurt alternatives that meet or exceed suggested nutrient guidelines are separated out according to the different bases for comparison.

Tables 4–6 compare the nutrient composition of plant-based yogurt alternatives with plant-based beverages. Table 4 summarizes the nutrient composition of the yogurt alternatives and compares the medians for each nutrient with the medians of the nutrients in the non-dairy plant-based beverages. Significant differences were observed between yogurt and beverage alternatives for each nutrient. Table 5 reports the fortification levels of the yogurt alternatives and compares the levels of each of the three nutrients (calcium, vitamins D and B12) in the yogurt alternatives with the corresponding levels in the plant-based beverages. Significant differences were observed between yogurt and beverage alternatives for each fortified nutrient.

The 326 plant-based beverage alternatives analyzed were based upon almonds (*n* = 83), oats (*n* = 62), soy (*n* = 44), coconut (*n* = 24), hemp (*n* = 15), pea protein (*n* = 14), rice (*n* = 11), cashews (*n* = 10), flax (*n* = 6), banana (*n* = 6), macadamia (*n* = 6), hazelnuts (*n* = 4), chia (*n* = 4), quinoa (*n* = 2), pili nut (*n* = 2), walnut (*n* = 1), pistachio (*n* = 1), and the following mixtures: almond and pea (*n* = 7), almond and coconut (*n* = 5), sesame and pea (*n* = 5), flax and pea (*n* = 4), oats and pea (*n* = 2), oats and avocado (*n* = 2), rice and quinoa (*n* = 2), almond and cashew (*n* = 2), coconut, cashew, and oats (*n* = 1), and almond and sesame (*n* = 1; Table 6). For the plant-based beverage alternatives, 121 (37.1%) were unsweetened. While 60% of the beverage alternatives were plain, the most common flavors were vanilla (25%) and chocolate (9.5%). The typical serving size of the beverage was 240 mls.

Table 1. The median (Q1–Q3) values of calories and 10 nutrients of non-dairy yogurts/serving classified according to their bases.

Product Base (n)	Calories	Total Fat (g)	Sat. Fat. (g)	Sodium (mg)	Carbs. (g)	Fiber (g)	Sugar (g)	Protein (g)	Calcium (% DV *)	Vit. D (% DV)	Vit. B12 (% DV)
Almond (62)	150 (140–180) [a]	9.5 (7–11) [ab]	1 (0.5–1) [a]	49 (10–80) [a]	19 (13–23) [a]	3 (2–3) [abc]	12.5 (8–15.8) [a]	4 (3–5) [a]	10 (4–10) [abc]	0 (0–3) [a]	0 (0–0) [ab]
Cashew (13)	140 (140–150) [bcde]	7 (6–7) [cd]	1.5 (1–1.5) [ab]	10 (10–11) [b]	20 (19–20) [abc]	1 (1–1) [de]	12 (12–13) [abc]	3 (3–3) [ab]	2 (2–2) [de]	0 (0–0) [ab]	0 (0–0) [a]
Coconut (79)	160 (127.5–190) [ab]	8 (5.5–12.5) [ac]	7 (5–11.5) [c]	30 (20–55 [a]	17 (10–22) [bc]	2 (0.6–2) [a]	10 (5.5–15) [ab]	1 (0.3–1.5) [c]	6 (0–25) [a]	0 (0–10) [c]	0 (0–25) [b]
Oats (20)	120 (117.5–130) [cf]	3.5 (3–4.8) [f]	2.5 (1.8–2.5) [b]	20 (10–25) [b]	19 (19–20) [ab]	1 (1–2) [abf]	9 (7–9) [cd]	3 (3–3) [b]	10 (2–10) [abf]	0 (0–10) [c]	0 (0–10) [bd]
Pea (5)	160 (160–160) [abde]	6 (6–6) [cdf]	0.5 (0.5–0.5) [ad]	40 (40–40) [ad]	19 (19–20) [abc]	0 (0–0) [e]	15 (15–15) [ae]	6 (6–6) [d]	0 (0–0) [e]	0 (0–0) [ab]	0 (0–0) [ab]
Seeds [1] (7)	130 (130–130) [cef]	4 (4–7) [df]	0.5 (0.5–1) [abd]	25 (0–26) [bc]	14 (14–15) [cd]	3 (3–3) [def]	8 (8–9) [bcd]	8 (5–8) [d]	2 (2–2) [def]	0 (0–0) [ab]	0 (0–0) [a]
Soy (11)	130 (130–145) [cde]	3 (2.5–3.5) [ef]	0.5 (0–0.5) [d]	65 (57.5–87.5) [d]	20 (18.5–23.5) [a]	2 (1–2) [c]	16 (12.5–20) [e]	6 (6–6) [d]	15 (8–15) [c]	10 (0–10) [ab]	0 (0–0) [cd]
Coconut + legume [2] (23)	150 (140–180) [abd]	7 (6.5–9) [acd]	6 (6–7) [c]	25 (15–57.5) [ac]	14 (13–19.5) [bc]	1 (0–1) [bc]	9 (8–9.5) [bcd]	6 (4–8) [d]	20 (5–20) [bc]	10 (5–40) [d]	40 (10–40) [c]
Coconut + seeds/nuts [3] (9)	160 (150–170) [ab]	11 (10–11) [b]	7 (5–7) [c]	65 (10–65) [ad]	11 (9–12) [d]	1 (1–1) [de]	7 (7–7) [d]	2 (2–6) [ab]	2 (2–2) [de]	0 (0–0) [ab]	0 (0–0) [a]
Legume blend [4] (20)	110 (100–122.5) [f]	1.5 (1.5–3.2) [e]	0.5 (0–0.6) [d]	5 (0–25) [b]	18 (18–21) [abc]	2 (1–2) [df]	9.5 (8–12.25) [abc]	6 (5–6) [d]	2 (1.5–4) [df]	0 (0–0) [bc]	0 (0–0) [ab]
Total (249)	150 (120–170)	7 (4–10)	2.5 (1–6)	25 (10–65)	19 (12–21)	2 (1–3)	10 (7–14)	3 (1–5)	10 (2–15)	10 (0–10)	20 (0–40)

[1] hemp ($n = 4$), pumpkin seeds ($n = 3$). [2] coconut with pea protein. [3] coconut with pili nut ($n = 6$), coconut with cashews and watermelon and pumpkin seeds ($n = 3$). [4] oats with pea protein and/or fava bean ($n = 16$), almond with fava bean ($n = 4$). Kruskal–Wallis non-parametric tests for independent samples with multiple pairwise comparisons were used for each nutrient to perform comparisons among base types. Different lowercase letters (a,b,c, etc.) in the same column indicate significant differences among yogurt bases. * DV is the daily value. $P < 0.05$ is considered statistically significant.

Table 2. Percentage of 249 non-dairy yogurt alternatives meeting or exceeding suggested guideline per serving.

At least		
	5 g protein	32.5
	10% DV calcium	45.4
	10% DV vitamin D	17.8
	10% DV vitamin B12	21.7
	1.5 g dietary fiber	57.0
No more than		
	5 g sugar (10% DV)	15.3
	1 g saturated fat (5% DV)	55.0
	115 mg sodium (5% DV)	93.2
High levels		
	More than 150 calories	40.6
	10 g or more of sugar (20% DV)	53.0
	4 g or more saturated fat (20% DV)	39.8
	More than 15.5 g fat (20% DV)	7.6

Table 3. The percentage of non-dairy yogurt alternatives meeting or exceeding a suggested guideline [as in Table 2] separated according to the yogurt base.

Product Base (n)	At least 5 g Protein	At least 10% DV Calcium	At least 10% DV vit D	At least 10% DV vit B12	At least 1.5 g fiber	No more than 115 mg Sodium	No more than 5 g Sugar	No more than 1 g satd. fat	10 g or more Sugar	4 g or more satd. fat	More than 150 Calories	More than 15.5 g fat
Almond (62)	42	56	8	0	95	90	15	81	71	2	48	3
Cashew (13)	0	0	0	0	0	100	15	46	77	0	23	0
Coconut (79)	0	47	27	32	61	86	23	0	53	84	54	22
Oats (20)	0	65	40	40	45	100	10	25	20	0	25	0
Pea (5)	100	0	0	0	0	100	20	100	80	0	80	0
Seeds [1] (7)	100	0	0	0	100	100	0	100	0	0	0	0
Soy (11)	100	64	64	0	64	100	9	100	91	0	0	0
Coconut + legume [2] (23)	52	74	74	74	4	100	9	0	26	100	48	0
Coconut + seeds/nuts [3] (9)	33	0	0	0	0	100	11	0	22	100	56	0
Legume blend [4] (20)	85	20	20	20	55	100	10	95	50	0	0	0

[1] hemp (n = 4), pumpkin seeds (n = 3). [2] coconut with pea protein. [3] coconut with pili nut (n = 6), coconut with cashews and watermelon and pumpkin seeds (n = 3). [4] oats with pea protein and/or fava bean (n = 16), almond with fava bean (n = 4).

Table 4. Medians (Q1–Q3) for calories and nutrients in non-dairy yogurt alternatives versus non-dairy, plant-based multi-serve beverages/serving.

	Yogurts	Beverages
n	249	326
Calories	150 (120–170) [a]	80 (60–120) [b]
Fat (g)	7 (4–10) [a]	4 (2.5–5) [b]
Saturated fat (g)	2.5 (1–6) [a]	0.5 (0–0.9) [b]
Sodium (mg)	25 (10–65) [a]	110 (90–150) [b]
Carbohydrates (g)	19 (12–21) [a]	8 (2–15) [b]
Fiber (g)	2 (1–3) [a]	1 (0–1) [b]
Sugar (g)	10 (7–14) [a]	5 (0.1–8) [b]
Protein (g)	3 (1–5) [a]	2 (1–4) [b]
Calcium (% DV)	10 (2–15) [a]	25 (6–30) [b]
Vitamin D (% DV)	10 (0–10) [a]	15 (0–25) [b]
Vitamin B12 (% DV)	20 (0–40) [a]	0 (0–48.8) [b]

Different lowercase letters in the same row indicate significant differences between the values for the non-dairy yogurt and the plant-based beverages. $P < 0.05$ is considered statistically significant.

Table 5. Medians (Q1–Q3) of fortification levels of calcium, vitamin D and vitamin B12 (expressed as % DV) in fortified non-dairy yogurt alternatives and the fortified non-dairy, plant-based multi-serve beverages.

	Yogurt		Multi–Serve Beverages	
	N (%)	Median	N (%)	Median
Calcium–fortified	117 (47.0%)	15 (10–20) [a]	239 (73.3%)	30 (25–35) [b]
Vitamin D–fortified	77 (30.9%)	10 (10–25) [a]	229 (70.2%)	20 (10–25) [b]
Vitamin B12 fortified	54 (21.7%)	40 (25–47.5) [a]	138 (42.3%)	50 (25–60) [b]

Different lowercase letters in the same row indicate significant differences between the values for the non-dairy yogurt and the plant-based beverages. $P < 0.05$ is considered statistically significant.

Table 6. Comparison of nutrient content of non-dairy yogurts and plant-based non-dairy beverages by base. For each yogurt base, different lowercase letters in the same column indicate significant differences between the plant-based beverages and the non-dairy yogurt alternatives. $P < 0.05$ is considered statistically significant.

Product Base (n)	Calories	Total Fat (g)	Sat. Fat. (g)	Sodium (mg)	Carbs. (g)	Fiber (g)	Sugar (g)	Protein (g)	Calcium (% DV)	Vit. D (% DV)	Vit. B12 (% DV)
Almond											
Yogurt (62)	150 (140–180) [a]	9.5 (7–11) [a]	1 (0.5–1)	49 (10–80) [a]	19 (13–23) [a]	3 (2–3) [a]	12.5 (8–15.8) [a]	4 (3–5) [a]	10 (4–10) [a]	0 (0–3) [a]	0 (0–0)
Beverage (83)	60(35–80) [b]	3 (2.5–3.5) [b]	0 (0–0)	150 (120–170) [b]	3 (1.5–8.5) [b]	1 (0–1) [b]	1 (0–7) [b]	1 (1–1) [b]	30 (4–35) [b]	15 (0–25) [b]	0 (0–0)
Cashew											
Yogurt (13)	140 (140–150) [a]	7 (6–7)	1.5 (1–1.5)	10 (10–11) [a]	20 (19–20) [a]	1 (1–1)	12 (12–13) [a]	3 (3–3)	2 (2–2)	0 (0–0)	0 (0–0)
Beverage (10)	50 (45–80) [b]	4 (2.9–4.4)	0.25 (0–1)	95 (87.5–121.3) [b]	3 (1–7.8) [b]	0 (0–0)	0 (0–1.8) [b]	1 (0.4–1.8)	7 (2.5–10)	0 (0–10)	0 (0–0)
Coconut											
Yogurt (79)	160 (127.5–190) [a]	8 (5.5–12.5) [a]	7 (5–11.5) [a]	30 (20–55)	17 (10–22) [a]	2 (0.6–2) [a]	10 (5.5–15) [a]	1 (0–1.5)	6 (0–25)	0 (0–10) [a]	0 (0–25) [a]
Beverage (24)	60 (50–80) [b]	5 (4.5–5) [b]	4.3 (4–5) [b]	62.5 (30–113.8)	3 (1–7.3) [b]	0 (0–0) [b]	1 (0–7) [b]	0 (0–0.5)	10 (10–30)	10 (10–25) [b]	42.5 (18.8–50) [b]
Oats											
Yogurt (20)	120 (117.5–130)	3.5 (3–4.8)	2.5 (1.8–2.5)	20 (10–25) [a]	19 (19–20)	1 (1–2)	9 (7–9)	3 (3–3)	10 (2–10) [a]	0 (0–10)	0 (0–10)
Beverage (62)	120 (90–130)	4.3 (2–5)	0.5 (0–0.5)	105 (100–120) [b]	16 (14–21)	2 (1–2)	6.5 (4–9)	2 (1–3)	25 (2–25) [b]	10 (0–20)	0 (0–40)
Pea											
Yogurt (5)	160 (160–160)	6 (6–6)	0.5 (0.5–0.5)	40 (40–40) [a]	19 (19–20)	0 (0–0)	15 (15–15)	6 (6–6)	0 (0–0) [a]	0 (0–0) [a]	0 (0–0) [a]
Beverage (14)	105 (90–136.3)	4.5 (4.5–5.8)	0.5 (0.5–0.7)	125 (96.3–160) [b]	7 (6–12.8)	0 (0–1)	5.5 (3–11.8)	8 (4.3–8)	35 (25–35) [b]	30 (25–30) [b]	45 (35–100) [b]
Seeds											
Yogurt (7)	130 (130–130)	4 (4–7)	0.5 (0.5–1)	25 (0–26) [a]	14 (14–15)	3 (3–3) [a]	8 (8–9)	8 (5–8) [a]	2 (2–2) [a]	0 (0–0)	0 (0–0)
Beverage (25)	60 (50–80)	4.5 (2.5–6)	0.1 (0–0.5)	100 (90–125) [b]	4 (1–10)	1 (0–2) [b]	0 (0–7)	2 (2–3) [b]	25 (20–30) [b]	10 (3–10)	0 (0–25)
Soy											
Yogurt (11)	130 (130–145)	3 (2.5–3.5)	0.5 (0–0.5)	65 (57.5–87.5)	20 (18.5–23.5) [a]	2 (1–2)	16 (12.5–20) [a]	6 (6–6) [a]	15 (8–15)	10 (0–10) [a]	0 (0–0) [a]
Beverage (44)	100 (90–130)	4 (3.9–4.6)	0.5 (0.5–0.6)	95 (85–125)	8.5 (4–11.3) [b]	1 (1–2)	6 (2–9) [b]	7 (7–9) [b]	25 (23.8–30)	15 (15–25) [b]	50 (0–75) [b]
Legume blend [1]											
Yogurt (20)	110 (100–122.5)	1.5 (1.5–3.1)	0.5 (0–0.6)	5 (0–25) [a]	18 (18–21) [a]	2 (1–2)	9.5 (8–12.3)	6 (5–6) [a]	2 (1.5–4) [a]	0 (0–0) [a]	0 (0–0)
Beverage (18)	115 (92.5–140)	5 (3.1–6.5)	0.5 (0–0.5)	160 (150–190) [b]	9.5 (2–11.8) [b]	0.3 (0.2–1)	3 (0–10.8)	8 (8–10) [b]	30 (25–30) [b]	12.5 (10–48.8) [b]	0 (0–37.5)

[1] a product with pea protein, soy, or fava bean.

4. Discussion

While non-dairy, plant-based yogurt alternatives are becoming more popular, concerns exist regarding their nutritional value relative to regular dairy yogurts. Of special concern is the level of protein, and the level of fortification of calcium, vitamin D and vitamin B12. The latter three nutrients are especially needed by a vegan and others consuming a plant-based diet that is devoid of adequate fortification. The median values of nutrients in the non-dairy yogurt alternatives, as reported for the overall total of 249 products in Table 1, are similar to those of a leading brand of 2% dairy yogurt (Table S1), except for the higher value of fiber, and lower protein and saturated fat content in the non-dairy alternatives. A recent analysis of dairy yogurts and non-dairy yogurt alternatives revealed that the levels of protein and calcium in the dairy yogurts were significantly higher than the yogurt alternatives (Table S2) [19]. One notes that the nutritional values for a Greek-style yogurt typically show considerably less calcium than a regular yogurt, while its protein content can be as high as 15–20 g/serving [20].

While the median value of protein in our non-dairy plant-based yogurt alternatives was only 3 g, all the legume-containing products (with soy, pea protein or fava bean) had a median of 6 g protein/serving while those based on seeds had a median of 8 g/serving. In fact, one-third (32.5%) of the non-dairy yogurt alternatives had at least 5 g of protein per serving, with 29% of the products containing 5 to 8 g protein/serving, and 4% of the yogurts contained as much as 10–11 g protein/serving. The overall picture for protein (a mean value of only 3 g/serving) was impacted considerably by coconut-based products which accounted for one-third of the yogurt alternatives and had a median value of only 1 g protein/serving. For comparison, the median protein level of 38 non-dairy plant-based yogurt alternatives, in a UK study, was 3.6 g/100 g (equivalent to 5.4 g/150 g serving). The median protein level for flavored dairy yogurts was 17% higher at 4.2 g/100 g (or 6.3 g/150 g serving) [21].

Most of the protein in the non-dairy yogurts comes from legumes (soy and pea protein), and various seeds and nuts (but not coconut). With the exception of soy protein [22], these plant proteins have a biological value lower than that of animal proteins. A Dutch group recently reported the essential amino acid profile for pea protein to be quite similar to that of soy [23]. Recently, a French group pointed out that protein-rich plant foods, such as legumes, nuts and seeds, can achieve protein adequacy in adults consuming balanced plant-based diets [24].

The median level for calcium, in 57 dairy yogurts recently analyzed, was 10% of DV for calcium [19]. In our study, 45% of the yogurt alternatives overall were fortified with calcium to contain at least 10% DV/serving (Table 2). In a UK study, non-dairy plant-based yogurt alternatives had calcium levels of 115 mg/100 g and about 130 mg/100 g for dairy yogurts [21]. This corresponds to 172.5 and 195 mg calcium for a typical 150 g serving of yogurt, or about 15% DV for calcium. Predominantly, the yogurt alternatives based upon almond, coconut, oats, soy and a legume-blend are more likely to have adequate calcium fortification (Table 3). About one-half of the almond-based, coconut-based and legume-blend yogurts were fortified with calcium, while two thirds of the oat-based and 100% of the soy-based yogurt alternatives had calcium fortification. Almond- and oat-based yogurts had a median of 10% DV calcium/serving while soy had 15% DV and coconut/legume blends had a median of 20% DV of calcium/serving. Overall, any kind of fortification is limited to only one in three brands scattered amongst the various bases.

Four different calcium salts were used in the fortification of the yogurt alternatives, with tricalcium phosphate (TCP) and calcium citrate being the most commonly used, followed by calcium lactate and calcium carbonate. Most calcium salts used to fortify foods and beverages exhibit a bioavailability similar to that of milk calcium [25], which has a fractional absorption of about 30%. The absorbability of calcium varies depending upon the food matrix, including the pH and the presence of stabilizers [25,26]. Calcium carbonate absorption is very similar to milk calcium, while calcium lactate and calcium citrate tends to be a little better absorbed; TCP is absorbed less well than milk calcium [25].

Fortification of the non-dairy yogurt alternatives with vitamins D and B12 was not a common feature. It was less commonly observed than calcium fortification. A closer look at the nutritional adequacy of the yogurt alternatives (Table 2) found that only about one in five had vitamin D (17.8%) and vitamin B12 (21.7%) levels of fortification that reached at least at 10% DV level. While dairy yogurts are not commonly fortified with vitamin D, we found 30.9% of the non-dairy yogurts were fortified with vitamin D (Table 5). Most commonly it was the yogurts based on soy, oats, coconut or the legume blends that were vitamin D fortified (Table 3). The median level of vitamin D for our yogurt alternatives was 10% DV/serving (Table 1) which is the typical level of fortification for the dairy yogurts that are fortified. The use of vitamin D-fortified yogurts, compared to plain yogurt, has been found to improve one's vitamin D status along with lower blood lipids and blood glucose levels [27,28]. Non-dairy yogurts are typically fortified with vitamin D2 rather than vitamin D3 since the latter is normally derived from animal sources, and unacceptable to vegans. However, vitamin D_2 is equally as effective as vitamin D_3 in maintaining 25-hydroxyvitamin D levels [29]. Coconut-, and oat-based, and legume-blend yogurt alternatives were the only products fortified with vitamin B12 (Table 3). These products were most commonly fortified at a level of 25–50% of DV/serving. Regular dairy yogurts contain about 1 µg of vitamin B12/serving (about 40% DV). Overall, only one in five non-dairy products (21.7%) had B12 fortification, while less than one-third (30.9%) had vitamin D fortification, and the fortification was brand-specific rather than dependent upon the type of yogurt. Of the 34 brands we analyzed, only 11 brands had calcium fortification while only 6 brands had both vitamin D and B12 fortification. When only the fortified non-dairy yogurt alternatives were analyzed, the calcium, vitamin D and B12 levels were more like those of regular dairy products. In this case, the median values for calcium, vitamin D and B12 were 15% DV, 10% DV, and 40% DV, respectively (Table 5).

4.1. Healthy Profile

The health profile of a product is typically assessed by the levels of salt (sodium), sugar and fat/saturated fat in the product, since these nutrients in excess are known to have negative health effects [30–34]. Consumers, for health reasons, are often concerned about the level of sodium, saturated fat and sugars in their food. A large majority of products (93.2%) had low levels of sodium. Most of the product bases were low in saturated fat except for those containing coconut. Very few products were high fat (only 8% had more than 20% of DV/serving), while more than one-half (53%) had high levels of sugar (at least 10 g sugar/serving). Only about 15% had no more than 5 g of sugar/serving (10% DV). The yogurt alternatives that are based on oats, seeds, or the coconut blends had lower levels of sugar, while those based upon oats, cashews, seeds, and the legume-blend had a lower content of sodium (Table 1). The yogurt alternatives based on oats, seeds, soy and the legume-blend had lower levels of fat and calories/serving. Soy-, pea-, almond-, coconut-, and cashew-based yogurts were the varieties most likely to contain higher levels of sugar and had median sugar contents of 10–16 g/serving. This compares with the 17 g/150 g serving for flavored yogurts analyzed from three different countries [35]. Coconut- and almond-based yogurts showed the highest median levels of fat, while soy-, oat-, seed-based, and legume blend yogurts had the lowest fat levels. The yogurts based on oats, seeds, and a legume blend contained the fewest calories/serving, while those based on almonds, coconut, and pea protein were the highest in calories/serving (Table 1). Furthermore, about one-half of the coconut-based yogurts also contained greater than 150 calories (Table 3). Only 9% of the yogurt alternatives contained over 190 calories/serving and were largely 3 brands. While three-quarters of these products were coconut-based, the balance were mix-ins with special toppings.

Over one-half (57%) of the yogurts studied had at least 1.5 g fiber/serving (Table 2) while the median level was 2 g/serving. The fibers added to the yogurt alternatives were mostly pectin, locust bean gum, and inulin. These water-soluble fibers, used as stabilizers and thickening agents in the yogurt alternatives, have properties that influence glycemia and hypercholesterolemia [36–38]. Inulin, which is added to 13% of the plant-based yogurts, is derived from either chicory root or agave. Inulin-type fructans act as prebiotics, since they have a beneficial effect on the human gut microbiota [39]. The gut bacteria convert inulin and other prebiotics into short-chain fatty acids, which nourish colon cells. Prebiotic fibers are considered a healthy addition since they improve digestive health and may also enhance immune function [36]. Some evidence suggests that prebiotics, as well as probiotics, may be beneficial in the prevention and treatment of colon cancer [40].

Plant-based yogurt alternatives typically contain a variety of live active cultures, similar to those used in dairy yogurts. Some of the products even advertise that they contain probiotic bacteria, as a part of their health messaging. In our study, the yogurt alternatives contained, on average, 4–6 probiotic live cultures. Common cultures that are added to the yogurt alternatives include *Streptococcus thermophilus*, *Lactobacillus acidophilus*, *Bifidobacterium* spp., *L rhamnosus*, L. casei, *Lactobacillus delbrueckii*, and *L.Bulgaricus*. The viable bacteria and bioactive metabolites of fermentation have long been associated with improved gut health. Their impact on the gut microbiome, may also improve overall health of the individual and immune function [41–43]. While the viability of probiotic microorganisms may be more difficult to maintain in a non-dairy matrix [44], the probiotics and bioactive compounds present in fermented non-dairy products have been associated with improved overall intestinal health and immune function by modifying the gut microbiome [44]. This has enabled the manufacturers of plant-based yogurt alternatives containing active cultures to make similar health claims as made for the dairy-based yogurts.

4.2. Comparison to Beverages

When dietary options are available for a vegetarian/vegan to choose between a non-dairy plant-based beverage and a non-dairy yogurt alternative, it is important to know what nutritional advantages may exist for one over the other (Table 4). Both product type and base type had a significant effect on almost all nutritional components (Table S3). The yogurt alternatives have about twice as much fat, sugar and calories/serving and five times more saturated fat than one serving of the non-dairy beverages while the beverages have 4–5 times more sodium than the yogurt alternatives (Table 4). While the non-dairy yogurt alternatives provide more protein/serving, the non-dairy beverages had significantly greater calcium and vitamin D levels.

When the data were divided according to compositional type, marked differences were seen (Table 6). The almond- and seed-based yogurt alternatives had 4 times the protein level of the non-dairy beverage counterparts while the legume-based yogurt alternatives had lower protein levels than the legume-based non-dairy beverages. While all the non-dairy beverages had substantially higher levels of calcium fortification than the yogurt alternatives, the differences were huge in the case of pea-, and seed-based, and the legume-blend products (Table S4). The oat-based and almond-based beverages even had calcium levels 2.5–3 times greater than the similar yogurts. While all yogurt alternatives, except soy-based yogurts, recorded a median of zero level of vitamin D, the plant-based beverages recorded a median vitamin D level of 10–30% DV/serving, except for cashew-based beverages (0% DV). In the same manner, all the non-dairy yogurt alternatives recorded a median of zero vitamin B12, while the soy-, coconut-, pea-based beverages reported a 42.5–50% DV B12/serving. All other beverages reported a median of zero B12/serving. While the non-dairy yogurt alternatives generally had more calories/serving than the non-dairy beverages, the differences were most pronounced for almond-, cashew-, and coconut-based products. Except for oat-based products, all types of non-dairy yogurts had 6–12 g more sugar/serving than the non-dairy beverages. Overall, the yogurt alternatives are more nutrient dense foods and provide more calories and macronutrients than the plant-based beverages/serving. However, fortification is a different story, as seen in Table 5. The beverages overall are twice as often fortified with calcium, vitamins D and B12, and the median level of fortification is substantially greater for the beverages. While many of the non-dairy yogurt alternatives lack fortification (Table 5), the different bases that are fortified with calcium, vitamin D and B12 have similar levels to the corresponding plant-based beverages, or they have significantly less nutrient fortification than the beverages (Table S4).

5. Conclusions

Consumers often choose a non-dairy yogurt alternative as a substitute for dairy yogurt. Sweetened, flavored varieties are the most commonly available types. Non-dairy yogurt alternatives are formulated largely from coconut (32%) and almonds (25%), while a significant number (24%) are either formulated from soy or pea protein or have a blend that includes a legume. At least one-third of the yogurt alternatives have 5 g or more of protein/serving, while a small number (4%) have 10–11 g protein/serving. Those products based upon a legume have the higher protein levels. A majority of the non-dairy yogurt alternatives were not fortified. Only 45% of the plant-based yogurts had calcium levels fortified to at least 10% of DV, while only about one in five had adequate vitamin D and B12 fortification at the 10% DV level. Fortification is very brand-specific, with only one-third of the 34 brands being fortified. The products that are based on oats, soy, or a coconut-legume mix demonstrate the best fortification. In comparison, while most regular dairy yogurts are not fortified with vitamin D, they do naturally contain both calcium (10% DV) and vitamin B12 (40% DV). In addition, the non-dairy, plant-based beverages tended to be better fortified (twice as frequently and at higher levels) than the non-dairy yogurt alternatives. One-half of the yogurt alternatives had high sugar levels, while 93% were low in sodium. Except for the coconut-based products, the yogurts were not high in fat or saturated fat. Less than 10% of the products contained over 190 calories/serving and the majority of these were coconut-based. The median level of dietary fiber in the yogurt alternatives was 2 g per serving. The most common fibers used were pectin, locust bean gum, and inulin. These prebiotic fibers are known to provide a number of health benefits. The presence of active cultures, 4–6 cultures on average, are also known to benefit the gut microbiota and improve overall health.

The observations made from this cross-sectional study represent just a window in time. New products continue to appear on the market, and formulations are not static. Fortification changes do occur with time. In addition, we have noticed that the nutrition information online is not always the same as what is presently seen on the nutrition label on products in the supermarkets, especially with respect to the level of fortification. When the fortification level changes, the company web page may reflect this change before the new products appear on the shelf or older stock has fully moved through the supply line. When a discrepancy was observed, the nutrient values appearing on products from the supermarket were utilized.

Typically, the company will analyze the nutritional composition of their product, and quality control measures will ensure that the final commercial products match the information placed on the nutrition label. However, some products provide the actual mg or gm provided by one serving along with the % DV. In doing the calculations for vitamins and minerals, one will observe that in the USA, numbers for % DV are often rounded to the nearest 5%. Also, for macronutrients, such as sugars, fats, protein, and dietary fiber, numbers are rounded without any decimals present. These procedures stand in contrast to European practices, where rounding is less noticeable and decimals commonly appear on the nutrition label. This gives the opportunity for generating more precise results. However, when we are working in this US project, with large quantities of products, these errors will be largely averaged out or minimized so that the big picture and the trends should still be visible to the researcher.

While consumers may have a taste preference or get a price advantage for a particular non-dairy yogurt alternative, they should be cognizant of the nutritional values of the various choices available to them. Carefully reading nutrition labels will enable the consumer to select a product with a good nutritional profile and a lower content of sugar and saturated fat.

Supplementary Materials: The following are available online at https://www.mdpi.com/article/10.3390/nu13114069/s1, Table S1: Nutrient profile of a leading brand of 2% dairy yogurt; Table S2: Nutritional analysis of dairy and non-dairy plant-based yogurt alternatives; Table S3. The effects of product type (non-dairy yogurts versus non-dairy beverage) and base type; Table S4: Median (Q1–Q3) of the fortification levels of Calcium, Vitamin D and B12 (expressed as % DV) of non-dairy yogurt alternatives and non-dairy, plant-based beverages.

Author Contributions: Conceptualization, W.J.C.; Data curation, W.J.C.; Formal analysis, W.J.C.; Investigation, W.J.C.; Methodology, W.J.C.; Software, C.J.B.; Visualization, W.J.C.; Writing—original draft, W.J.C.; Writing—review & editing, C.J.B. All authors have read and agreed to the published version of the manuscript.

Funding: This research received no external funding.

Conflicts of Interest: The authors declare no conflict of interest.

References

1. Comerford, K.B.; Miller, G.D.; Boileau, A.C.; Masiello Schuette, S.N.; Giddens, J.C.; Brown, K.A. Global review of dairy recommendations in food-based dietary guidelines. *Front. Nutr.* **2021**, *8*, 671999. [CrossRef]
2. FONA International. 2021 Trend Insight: The Opportunity in Plant-Based. Available online: https://www.fona.com/0421plantbased/ (accessed on 15 October 2021).
3. Vegan Yogurt Market Size, Share and Trend Analysis Report by Product (Soy, Almond, Rice) by Distribution Channel (Hypermarket, Supermarket, Convenience Stores, Specialty Stores, Online) by Region, and Segment Forecasts, 2020–2027. Available online: https://www.grandviewresearch.com/industry-analysis/vegan-yogurt-market (accessed on 20 May 2021).
4. Forager Project. Available online: https://www.foragerproject.com/about-us/ (accessed on 14 July 2021).
5. Ripple. Available online: https://www.ripplefoods.com/healthy-environment/ (accessed on 14 July 2021).
6. Sustainability at Silk. Available online: https://silk.com/about-us/sustainability/ (accessed on 14 July 2021).
7. Melina, V.; Craig, W.; Levin, S. Position of the Academy of Nutrition and Dietetics: Vegetarian diets. *J. Acad. Nutr. Diet* **2016**, *116*, 1970–1980. [CrossRef]
8. Craig, W.J.; Fresán, U. International analysis of the nutritional content and a review of health benefits of non-dairy plant-based beverages. *Nutrients* **2021**, *13*, 842. [CrossRef]
9. Turner, H. Vegan Yogurts: A Comparative Analysis. Available online: https://www.todaysdietitian.com/news/exclusive0316.shtml (accessed on 15 May 2021).
10. Rezac, S.; Kok, C.R.; Heermann, M.; Hutkins, R. Fermented Foods as a Dietary Source of Live Organisms. Front Microbiol. *Front. Microbiol.* **2018**, *9*, 1785. [CrossRef] [PubMed]
11. Sanlier, N.; Gokcen, B.B.; Sezgin, A.C. Health benefits of fermented foods. *Crit. Rev. Food Sci. Nutr.* **2019**, *59*, 506–527. [CrossRef] [PubMed]
12. Marco, M.L.; Heeney, D.; Binda, S.; Cifelli, C.J.; Cotter, P.D.; Foligne, B.; Ganzle, M.; Kort, R.; Pasin, G.; Pihlanto, A.; et al. Health benefits of fermented foods: Microbiota and beyond. *Curr. Opin. Biotechnol.* **2017**, *44*, 94–102. [CrossRef]
13. Carlson, J.L.; Erickson, J.M.; Lloyd, B.B.; Slavin, J.L. Health Effects and Sources of Prebiotic Dietary Fiber. *Curr. Dev. Nutr.* **2018**, *2*, nzy005. [CrossRef]
14. Slavin, J. Fiber and Prebiotics: Mechanisms and Health Benefits. *Nutrients* **2013**, *5*, 1417–1435. [CrossRef] [PubMed]
15. Fleming, A. Alphabetical Listing of Dairy-Free Yogurt Reviews and Information. Available online: https://www.godairyfree.org/product-reviews/dairy-free-yogurt-reviews (accessed on 7 May 2021).
16. New Nutrition Facts Label. Available online: https://www.fda.gov/food/new-nutrition-facts-label/lows-and-highs-percent-daily-value-new-nutrition-facts-label (accessed on 3 June 2021).
17. Daily Value and Percent Daily Value: Changes on the New Nutrition and Supplements Facts Labels. Available online: https://www.fda.gov/media/135301/download (accessed on 3 June 2021).
18. R Core Team. *R: A Language and Environment for Statistical Computing*; R Foundation for Statistical Computing: Vienna, Austria, 2019. Available online: https://www.R-project.org/ (accessed on 17 August 2021).
19. Moyer, L.; Mitchell, K. Get some culture. What's new in the yogurt aisle. *Nutr. Action Health Lett.* **2021**, *48*, 13–15.
20. What Is Greek Yogurt? Available online: https://www.usdairy.com/news-articles/what-is-greek-yogurt?ref=www.nationaldairycouncil.org (accessed on 9 August 2021).
21. Moore, J.B.; Horti, A.; Fielding, B.A. Evaluation of the nutrient content of yogurts: A comprehensive survey of yogurt products in the major UK supermarkets. *BMJ Open* **2018**, *8*, e021387. [CrossRef]

22. Young, V.R.; Puig, M.; Queiroz, E.; Scrimshaw, N.S.; Rand, W.M. Evaluation of the protein quality of an isolated soy protein in young men: Relative nitrogen requirements and effect of methionine supplementation. *Am. J. Clin. Nutr.* **1984**, *39*, 16–24. [CrossRef]
23. Gorissen, S.H.M.; Crombag, J.J.R.; Senden, J.M.G.; Waterval, W.A.H.; Bierau, J.; Verdijk, L.B.; van Loon, L.J.C. Protein content and amino acid composition of commercially available plant-based protein isolates. *Amino Acids* **2018**, *50*, 1685–1695. [CrossRef] [PubMed]
24. Mariotti, F.; Gardner, C.D. Dietary Protein and Amino Acids in Vegetarian Diets—A Review. *Nutrients* **2019**, *11*, 2661. [CrossRef]
25. Rafferty, K.; Walters, G.; Heaney, R.P. Calcium fortificants: Overview and strategies for improving calcium nutriture of the U.S. population. *J. Food Sci.* **2007**, *72*, R152–R158. [CrossRef] [PubMed]
26. Palacios, C.; Cormick, G.; Hofmeyr, G.J.; Garcia-Casal, M.N.; Peña-Rosas, J.P.; Betrán, A.P. Calcium-fortified foods in public health programs: Considerations for implementation. *Ann. N.Y. Acad. Sci.* **2021**, *1485*, 3–21. [CrossRef]
27. Beauchet, O.; Launay, C.P.; Galery, K.; Vilcocq, C.; Dontot-Payen, F.; Rousseau, B.; Benoit, V.; Allali, G. Effects of vitamin D and calcium fortified yogurts on gait, cognitive performances, and serum 25-hydroxyvitamin D Concentrations in older community-dwelling females: Results from the gait, memory, dietary and vitamin D (GAME-D2) randomized controlled trial. *Nutrients* **2019**, *11*, 2880. [CrossRef] [PubMed]
28. Gasparri, C.; Perna, S.; Spadaccini, D.; Alalwan, T.; Girometta, C.; Infantino, V.; Rondanelli, M. Is vitamin D-fortified yogurt a value-added strategy for improving human health? A systematic review and meta-analysis of randomized trials. *J. Dairy Sci.* **2019**, *102*, 8587–8603. [CrossRef]
29. Holick, M.F.; Biancuzzo, R.M.; Chen, T.C.; Klein, E.K.; Young, A.; Bibuld, D.; Reitz, R.; Salameh, W.; Ameri, A.; Tannenbaum, A.D. Vitamin D2 is as effective as vitamin D3 in maintaining circulating concentrations of 25-hydroxyvitamin D. *J. Clin. Endocrinol. Metab.* **2008**, *93*, 677–681. [CrossRef]
30. He, F.J.; MacGregor, G.A. Salt and sugar: Their effects on blood pressure. *Pflügers Archiv Eur. J. Physiol.* **2015**, *467*, 577–586. [CrossRef]
31. Clifton, P.M.; Keogh, J.B. A systematic review of the effect of dietary saturated and polyunsaturated fat on heart disease. *Nutr. Metab. Cardiovasc. Dis.* **2017**, *27*, 1060–1080. [CrossRef]
32. DiNicolantonio, J.J.; Lucan, S.C.; O'Keefe, J.H. The Evidence for Saturated Fat and for Sugar Related to Coronary Heart Disease. *Prog. Cardiovasc. Dis.* **2016**, *58*, 464–472. [CrossRef]
33. Wang, L.; Folsom, A.R.; Zheng, Z.-J.; Pankow, J.S.; Eckfeldt, J.H.; ARIC Study Investigators. Plasma fatty acid composition and incidence of diabetes in middle-aged adults: The Atherosclerosis Risk in Communities (ARIC) Study. *Am. J. Clin. Nutr.* **2003**, *78*, 91–98. [CrossRef]
34. Morenga, L.T.; Mallard, S.; Mann, J. Dietary sugars and body weight: Systematic review and meta-analyses of randomized controlled trials and cohort studies. *BMJ* **2012**, *346*, e7492. [CrossRef] [PubMed]
35. Coyle, D.H.; Ndanuko, R.; Singh, S.; Huang, P.; Wu, J.H. Variations in sugar content of flavored milks and yogurts: A cross-sectional study across 3 countries. *Curr. Dev. Nutr.* **2019**, *3*, nzz060. [CrossRef]
36. Anderson, J.W.; Baird, P.; Davis, R.H.J.; Ferreri, S.; Knudtson, M.; Koraym, A.; Waters, V.; Williams, C.L. Health benefits of dietary fiber. *Nutr. Rev.* **2009**, *67*, 188–205. [CrossRef] [PubMed]
37. Lattimer, J.M.; Haub, M.D. Effects of dietary fiber and its components on metabolic health. *Nutrients* **2010**, *2*, 1266–1289. [CrossRef] [PubMed]
38. Jensen, C.D.; Haskell, W.; Whittam, J.H. Long-term effects of water-soluble dietary fiber in the management of hypercholesterolemia in healthy men and women. *Am. J. Cardiol.* **1997**, *79*, 34–37. [CrossRef]
39. Vandeputte, D.; Falony, G.; Vieira-Silva, S.; Wang, J.; Sailer, M.; Theis, S.; Verbeke, K.; Raes, J. Prebiotic inulin-type fructans induce specific changes in the human gut microbiota. *Gut* **2017**, *66*, 1968–1974. [CrossRef]
40. Geier, M.S.; Butler, R.N.; Howarth, G.S. Probiotics, prebiotics and synbiotics: A role in chemoprevention for colorectal cancer? *Cancer Biol. Ther.* **2006**, *5*, 1265–1269. [CrossRef]
41. Markowiak, P.; Slizewska, K. Effects of probiotics, prebiotics, and synbiotics on human health. *Nutrients* **2017**, *9*, 1021. [CrossRef]
42. Shiby, V.K.; Mishra, H.N. Fermented milks and milk products as functional foods—A review. *Crit. Rev. Food Sci. Nutr.* **2013**, *53*, 482–496. [CrossRef] [PubMed]
43. La Fata, G.; Weber, P.; Mohajeri, M.H. Probiotics and the gut immune system: Indirect regulation. *Probiotics Antimicrob Proteins* **2018**, *10*, 11–21. [CrossRef] [PubMed]
44. Valero-Cases, E.; Cerda-Bernad, D.; Pastor, J.-J.; Frutos, M.-J. Non-dairy fermented beverages as potential carriers to ensure probiotics, prebiotics, and bioactive compounds arrival to the gut and their health benefits. *Nutrients* **2020**, *12*, 1666. [CrossRef] [PubMed]

nutrients

Article

Nutritional Quality of Plant-Based Cheese Available in Spanish Supermarkets: How Do They Compare to Dairy Cheese?

Ujué Fresán [1,*] and Holly Rippin [2]

1 eHealth Group, Instituto de Salud Global Barcelona (ISGlobal), 08036 Barcelona, Spain
2 Independent Nutritionist, Cumbria LA22, UK; rippinholly@gmail.com
* Correspondence: ujuefresan@gmail.com

Abstract: Plant-based cheese is one of the most increasingly consumed dairy alternatives. Evidence is lacking on their nutritional quality. We aimed to evaluate the nutritional composition of the plant-based cheese options available in Spanish supermarkets, and how they compare with dairy cheese. An audit of plant-based cheese alternatives has been conducted in seven of the most common supermarkets. For each product, the nutritional content per 100 g and ingredients were collected. Data on generic dairy cheese were retrieved from the BEDCA website. Descriptive statistics (median, minimum and maximum) were used to characterize the plant-based cheese products, for both all the products and grouped by main ingredients (i.e., coconut oil, cashew nuts and tofu). Mann–Whitney U tests were used for comparisons between dairy and different types of plant-based cheese. The coconut oil-based products (the large majority of plant-based cheese products, n = 34) could not be considered as healthy foods. Their major ingredients were refined coconut oil and starches and were high in saturated fats and salt. The other smaller groups, cashew nut- (n = 4) and tofu-based (n = 2), showed a healthier nutritional profile. Replacing dairy cheese with these groups could be nutritionally beneficial. Future investigations should address the health effects of substituting dairy cheese with these products.

Keywords: dairy alternative; dairy substitute; cheese analogues; vegan cheese; vegetarian cheese; plant-based alternatives

Citation: Fresán, U.; Rippin, H. Nutritional Quality of Plant-Based Cheese Available in Spanish Supermarkets: How Do They Compare to Dairy Cheese? *Nutrients* **2021**, *13*, 3291. https://doi.org/10.3390/nu13093291

Academic Editor: Stephen Ives

Received: 19 July 2021
Accepted: 16 September 2021
Published: 21 September 2021

1. Introduction

An increasing number of people in developed countries, including in Europe, are moving away from animal-rich diets to adopt plant-based dietary patterns [1]. Different reasons are behind this transition, such as animal welfare, environmental degradation and health, among others [2–4]. The general adoption of plant-based diets has been highlighted as one of the most effective options for mitigating climate change [5], and a necessary shift if we are to feed the 10 billion people expected by 2050 within planetary boundaries [6]. At the same time, the general adoption of plant-based diets could benefit the health of populations globally [7,8]. While the consumption of animal-sourced foods such as red and processed meat has been linked to detrimental health outcomes such as colorectal cancer, cardiovascular disease and diabetes [9], the health benefits of whole plant-based foods, such as legumes, whole grains, nuts, vegetables and fruits, have also been reported in the scientific literature [9]. Indeed, replacing animal products with whole plant-based options has been linked to health benefits [10,11].

With the aim of facilitating this dietary transition, more and more plant-based products imitating animal-sourced equivalents are reaching the market, such as meat and dairy analogs [12,13]. Moreover, forecasts suggest that their consumption will continue growing [12,13]. This is especially relevant in Europe; four out of five of the top countries in the world with the highest share of global vegan new product launches in food and drink in 2018 were European, namely Germany, the United Kingdom, France and Spain [1]. The emergence and increased popularity of these products is concerning public health and nutrition professionals. Although some studies have aimed to assess how nutritionally sound and healthy these options are, they are relatively few, and the World Health Organization (WHO) has called for more studies on the topic [14]. The investigations carried out to date have been focused on the assessment of plant-based meat and milk alternatives [15–18], while less attention has been paid to other food products, such as plant-based cheese.

A recent Spanish report assessing those eating a plant-based diet found that 43% of interviewees had eaten plant-based cheese during the last three months. Indeed, plant-based cheese is one of the most consumed plant-based alternatives, after milk, meat and yogurts [19]. Evidence is lacking on the nutritional adequacy of plant-based cheese alternatives. The aim of this study is, therefore, to evaluate the nutritional and ingredient composition of plant-based cheese options available in Spanish supermarkets and compare the nutrient composition with that of dairy cheese.

2. Materials and Methods

2.1. Plant-Based Cheese Alternatives

An audit of plant-based cheese alternatives was conducted in seven of the most common supermarkets in Spain, namely Carrefour, Mercadona, Alcampo, Día, Lidl, Eroski and El Corte Inglés, to reflect choices available to the majority of Spanish consumers. The websites of those supermarkets were assessed during April and May 2021, using the keywords "queso vegano" and "queso vegetal" ("vegan cheese" and "plant-based cheese", in Spanish, respectively) to ensure all available products were captured. Supermarkets were visited in person to capture any additional products that were sold in situ but not online. For each product, several data were collected: name, commercial brand, supermarkets in which the product was available, selling format (i.e., block, slices, grated, spreads), nutritional content and ingredients. Nutritional data (i.e., calories, total fat, saturated fat, carbohydrates, sugars, fiber, protein and salt per 100 g of product) and ingredient lists were obtained from the nutritional label on the commercial package or from the information located on the website of the retailer. Retail data for all products were double checked against data from the website of the producer of each product. Supplementary Table S1 shows the main ingredient, selling format, calories and nutritional content per 100 g of the assessed plant-based cheese alternatives.

2.2. Dairy Cheese Data

Data on the most commonly consumed generic types of dairy cheese in Spain were retrieved from the BEDCA (Spanish food composition database) website [20]. The format in which each cheese types are regularly sold in stores is not specifically reported for all products. In cases where this was missing, the lead author assigned the most common format according to their experience. One of the following three formats was selected: block/slices, grated and spreads. The BEDCA does not report nutrition information about sugars content in foods. Dairy cheese products formatted as block/slices have a content of total carbohydrates of 0 g/100 g of cheese; only two products are reported as having 0.3 and 0.5 g of carbohydrates per 100 g. We, therefore, assigned a sugars value of 0 g for all block/slices products. However, the two examples of dairy spreadable cheese had 2.3 and 3.1 g of carbohydrates per 100 g of cheese. The researcher in charge of the database confirmed that all carbohydrates in those spreadable products were sugars (personal communication); thus, the sugars content in those products was considered as 2.3 and 3.1 g per 100 g, respectively. As salt content is reported as mg of sodium per 100 g, sodium values were multiplied by 2.5 and divided by 1000 in order to obtain g of salt per 100 g of product. Specific types of dairy cheese were considered; their selling format (i.e., block/slices, grated, spreads) and their caloric and nutritional values are reported in Supplementary Table S2.

2.3. Statistical Analysis

Normality of data distribution was tested through the Shapiro–Wilk test and rejected. Descriptive statistics (median and range) of energy and selected nutrients (total fats, saturated fats, carbohydrates, sugars, fiber, protein and salt) per 100 g were stated to describe plant-based cheese products. This characterization was performed for all plant-based cheese products as a whole, and also by groups. Groups were assigned according to their main ingredients (i.e., coconut oil, cashew nuts and tofu) and within each of these groups, by selling format (i.e., block/slices, grated and spreads). Comparisons between dairy and plant-based cheese were carried out using the Mann–Whitney U test for independent samples. We compared all dairy products to all plant-based alternatives as a whole, and also by group (by main ingredient and selling format). The statistical analysis was performed through the statistical software jamovi (version 1.6) [21], with the significance level set at $p < 0.05$.

3. Results

3.1. Plant-Based Cheese Alternatives

In total, 40 different plant-based cheese products were detected. One supermarket had by far the largest product offer, selling 38 out of the 40 assessed products; the other supermarkets had a much lower number of products available (n = 5, 3, 1, 1, 0 and 0). The most common selling formats were blocks (n = 15), slices (n = 10), grated (n = 8) and spreads (n = 7). The products fell into three different categories regarding their ingredients' composition: the largest group (82.5%, n = 34) was mainly composed of refined coconut oil and starches; two smaller groups had cashew nuts (n = 4) and tofu (n = 2) as the major ingredients. Not all the coconut oil-based products reported on the percentage of coconut oil they contain. In those reporting that value, the coconut oil content ranged between 20–29%. In the case of cashew nut-based products, around 50% of their weight were cashews, followed by water and lemon juice. In the case of tofu-based products, 98.5% of the product weight was soy milk. The coconut oil-based products contained many food additives including thickeners, preservatives, flavorings and colorings. In the case of the cashew nut- and tofu-based products, there were fewer additives, primarily natural flavorings, and agar-agar (a gelling agent) in cashew nut-based options. Six commercial brands manufactured coconut oil-based products, while all the cashew nut-based and tofu-based products were produced by a single company each. Supplementary Table S1 shows the main ingredient, selling format, energy and nutritional content per 100 g of the assessed plant-based cheese alternatives.

Overall, the plant-based cheese alternatives did not have a good nutritional profile (Table 1) [22–24]. They were high in calories (median: 288 kcal/100 g), fats (median: 23 g/100 g), saturated fats (median: 20 g/100 g) and salt (median: 1.5 g/100 g), while they contained a low amount of protein (median: 0.5/100 g) and fiber (median: 0 g/100 g). Assessing the products grouped by major ingredient, the coconut oil-based cheese (Table 1) contained a median of 287 kcal/100 g. They were high in fats (median: 23 g/100 g), being mainly saturated fats (median: 21 g/100 g) and salt (median: 1.6 g/100 g). Their median carbohydrate content was 20 g/100 g, with negligible sugars. They contained a very low amount of protein (median: 0.4 g/100 g) and no fiber. Assessing the nutritional composition of the coconut oil-based products by selling format (i.e., blocks/slices, grated and spreads), the same general pattern was observed in the three groups, except for the spreadable products, which had a lower amount of carbohydrates (Supplementary Table S3). The cashew nut-based cheese products (Table 1) were energy-dense, with a median of 323 kcal/100 g. They were high in fats (median: 25 g/100 g), and their saturated fat content was 5.7 g/100 g. They were a good source of protein, with 11 g/100 g, 2.7 g/100 g of (natural) sugars, and 2.6 g/100 g of fiber. Their salt content was moderate (0.6 g/100 g). All the cashew nut-based products were sold as blocks; thus, no analysis by selling format was carried out. The tofu-based cheese products (Table 1) provided a median of 185 kcal, 11 g of total fats and 1.7 g of saturated fats per 100 g of product. They were a good source of protein (median: 18 g/100 g) and fiber (median: 6.2 g/100 g). Their median amount of salt was 1.0 g/100 g. As in the case of the cashew nut-based options, all the tofu-based cheese were sold as blocks; therefore, no analysis by selling format was carried out.

Table 1. Median (minimum-maximum) values of calories and nutritional content in plant-based cheese alternatives per 100 g, by ingredients' composition.

	All Products $n = 40$	Coconut Oil-Based $n = 34$	Cashew Nut-Based $n = 4$	Tofu-Based $n = 2$
	Median (Min–Max)	Median (Min–Max)	Median (Min–Max)	Median (Min–Max)
Calories (kcal)	288 (185–328)	287 (200–327)	328 (306–328)	185 (185–185)
Total fat (g)	23.0 (11.0–29.0)	23.0 (16.7–29.0)	25.0 (21.0–25.9)	11.0 (11.0–11.0)
Saturated fat (g)	20.0 (1.7–26.0)	21.0 (8.3–26.0)	5.7 (4.4–6.3)	1.7 (1.7–1.7)
Carbohydrate (g)	19.9 (0.5–30.0)	20.0 (1.3–30.0)	13.3 (11.9–17.1)	0.5 (0.5–0.5)
Sugars (g)	0.2 (0.0–7.10)	0.1 (0.0–7.1)	2.7 (2.7–3.5)	0.0 (0.0–0.0)
Fibre (g)	0.0 (0.0–6.2)	0.0 (0.0–5.9)	2.6 (2.2–2.7)	6.2 (6.2–6.2)
Protein (g)	0.5 (0.0–18.0)	0.4 (0.0–6.0)	11.0 (10.6–11.0)	18.0 (18.0–18.0)
Salt (g)	1.5 (0.5–3.5)	1.6 (1.0–3.5)	0.6 (0.5–0.6)	1.0 (1.0–1.0)

3.2. *Plant-Based Cheese Alternatives vs. Dairy Cheese*

The dairy cheese products were higher in calories (median: 364 vs. 288 kcal ($p < 0.001$)), total fats (median: 31 vs. 23 g ($p < 0.001$)) and proteins (median: 23 vs. 0.5 g ($p < 0.001$)), per 100 g, than the plant-based cheese alternatives. Dairy cheese was lower in carbohydrates (median: 0 vs. 19.9 g ($p < 0.001$)), sugars ($p = 0.002$) and fiber ($p < 0.001$). There were no significant differences in the amount of saturated fats and salt between the dairy cheese and the plant-based alternatives ($p > 0.05$) (Table 2).

Table 2. Median (minimum–maximum) values of calories and nutritional content in dairy cheese and plant-based cheese alternatives per 100 g.

	Dairy Cheese $n = 22$	Plant-Based Cheese $n = 40$	p Value
	Median (Min–Max)	**Median (Min–Max)**	
Calories (kcal)	364 (201–467)	288 (185–328)	<0.001
Total fat (g)	31 (11.2–40.50)	23.0 (11.0–29.0)	<0.001
Saturated fat (g)	18.9 (7.0–25.4)	20.0 (1.7–26.0)	0.729
Carbohydrate (g)	0.0 (0.0–3.1)	19.9 (0.5–30.0)	<0.001
Sugars (g)	0.0 (0.0–3.1)	0.2 (0.0–7.10)	0.002
Fibre (g)	0.0 (0.0–0.0)	0.0 (0.0–6.2)	<0.001
Protein (g)	23.0 (6.5–32.30)	0.5 (0.0–18.0)	<0.001
Salt (g)	1.7 (0.1–3.8)	1.5 (0.5–3.5)	0.492

Mann–Whitney U non-parametric test for independent samples was used to perform comparisons among dairy and plant-based cheese. $p < 0.05$ is considered statistically significant.

When assessing the differences by type of products according to the major ingredient, the nutritional composition of the coconut oil-based products was quite similar to that of dairy products, aside from being significantly lower in protein ($p < 0.05$). The products sold as blocks or slices were slightly lower in calories (367 vs. 289 kcal/100 g ($p < 0.001$)) and total fats (31.3 vs. 23 g/100 g ($p < 0.001$)), but not in saturated fatty acids ($p = 0.392$) (Table 3). As shown in Table 4, compared to the dairy cheese, both the cashew nut-based and tofu-based products were the less caloric options (cashew nut-based: 367 vs. 328 kcal/100 g ($p = 0.007$); tofu-based: 367 vs. 185 kcal/100 g ($p = 0.024$)) and lower in fats, both total (cashew nut-based: 31.3 vs. 25 g/100 g ($p = 0.009$); tofu-based: 31.3 vs. 11 g/100 g ($p = 0.024$)) and especially saturated fats (cashew nut-based: 19.4 vs. 5.7 g/100 g ($p = 0.002$); tofu-based: 19.4 vs. 1.7 g/100 g ($p = 0.023$)). Their salt content was also lower (cashew nut-based: 1.7 vs. 0.6 g/100 g ($p = 0.002$); tofu-based: 1.7 vs. 1.0 g/100 g ($p = 0.045$)). As expected, their fiber content was higher ($p < 0.05$). The cashew nut-based products had more sugars (0 vs. 2.7 g/100 g ($p < 0.001$)), and less protein (23.4 vs. 11.0 g/100 g ($p = 0.002$)) than dairy cheese. No difference was observed among the amount of protein between the tofu-based and milk-based cheese options ($p = 0.143$).

Table 3. Median (minimum-maximum) values of calories and nutritional content in dairy cheese and coconut oil-based cheese alternatives per 100 g, by format.

	Block/Slices			Grated			Spreads		
	Dairy Cheese $n = 17$	Coconut Oil-Based $n = 19$	p Value	Dairy Cheese $n = 3$	Coconut Oil-Based $n = 8$	p Value	Dairy Cheese $n = 2$	Coconut Oil-Based $n = 7$	p Value
	Median (Min–Max)	Median (Min–Max)		Median (Min–Max)	Median (Min–Max)		Median (Min–Max)	Median (Min–Max)	
Calories (kcal)	367 (201–467)	290 (273–327)	<0.001	367 (223–395)	289 (240–313)	0.473	306 (251–361)	267 (200–290)	0.500
Total fat (g)	31.3 (11.2–40.5)	23.0 (17.0–29.0)	<0.001	26.5 (16.1–32.1)	21.8 (16.7–25.9)	0.473	28.2 (23.9–32.4)	25.0 (16.7–28.8)	0.500
Saturated fat (g)	19.4 (7.0–25.4)	20.9 (16.0–26.0)	0.392	16.7 (8.7–17.1)	19.5 (8.3–22.0)	0.357	17.0 (14.3–19.6)	23.0 (15.0–24.2)	0.184
Carbohydrate (g)	0.0 (0.0–0.5)	20.0 (6.1–30.0)	<0.001	0.0 (0.0–0.0)	21.4 (18.3–26.7)	0.017	2.7 (2.3–3.1)	6.0 (1.3–13.30)	0.186
Sugars (g)	0.0 (0.0–0.0)	0.1 (0.0–7.1)	<0.001	0.0 (0.0–0.0)	0.0 (0.0–0.2)	0.292	2.7 (2.3–3.1)	0.2 (0.0–1.1)	0.053
Fibre (g)	0.0 (0.0–0.0)	0.0 (0.0–5.9)	0.014	0.0 (0.0–0.0)	0.0 (0.0–4.0)	0.449	0.0 (0.0–0.0)	0.0 (0.0–2.2)	0.384
Protein (g)	23.4 (13.8–29.0)	0.5 (0.0–2.2)	<0.001	26.9 (19.5–32.3)	0.0 (0.0–1.6)	0.014	11.1 (6.5–15.6)	0.2 (0.0–6.0)	0.053
Salt (g)	1.7 (0.7–3.8)	1.8 (1.0–3.5)	0.514	1.5 (0.9–2.1)	1.9 (1.0–2.3)	0.474	1.2 (0.1–2.3)	1.2 (1.0–1.6)	1.000

Mann–Whitney U non-parametric test for independent samples was used to perform comparisons among dairy and plant-based cheese. $p < 0.05$ is considered statistically significant.

Table 4. Median (minimum-maximum) values of calories and nutritional content of those dairy cheese and cashew nut- and tofu-based cheese alternatives formatted as block and slices per 100 g.

	Dairy Cheese $n = 17$	Cashew Nut-Based $n = 4$	p Value	Dairy Cheese $n = 17$	Tofu-Based $n = 2$	p Value
	Median (Min–Max)	Median (Min–Max)		Median (Min–Max)	Median (Min–Max)	
Calories (kcal)	367 (201–467)	328 (306–328)	0.008	367 (201–467)	185 (185–185)	0.028
Total fat (g)	31.3 (11.2–40.5)	25 (21–25.9)	0.011	31.3 (11.2–40.5)	11.0 (11.0–11.0)	0.028
Saturated fat (g)	19.4 (7.0–25.4)	5.7 (4.4–6.3)	0.003	19.4 (7.0–25.4)	1.7 (1.7–1.7)	0.028
Carbohydrate (g)	0.0 (0.0–0.5)	13.3 (11.9–17.1)	<0.001	0.0 (0.0–0.5)	0.5 (0.5–0.5)	0.008
Sugars (g)	0.0 (0.0–0.0)	2.7 (2.7–3.5)	<0.001	0.0 (0.0–0.0)	0.0 (0.0–0.0)	NaN
Fibre (g)	0.0 (0.0–0.0)	2.5 (2.2–2.7)	<0.001	0.0 (0.0–0.0)	6.2 (6.2–6.2)	<0.001
Protein (g)	23.4 (13.8–29.0)	11.0 (10.6–11.0)	0.003	23.4 (13.8–29.0)	18.0 (18.0–18.0)	0.163
Salt (g)	1.7 (0.7–3.8)	0.6 (0.5–0.6)	0.003	1.7 (0.7–3.8)	1.0 (1.0–1.0)	0.053

NaN: not a number. Mann–Whitney U non-parametric test for independent samples was used to perform comparisons among dairy and plant-based cheese. $p < 0.05$ is considered statistically significant.

4. Discussion

The present study describes the ingredients and nutritional profile of the plant-based cheese alternatives currently available in Spanish supermarkets and how they compare nutritionally to dairy cheese. Our findings indicate that the availability of plant-based cheese products is relatively low; only one supermarket had a considerable number of options (n = 38). However, the rest of the supermarkets have at the disposal of customers few products, if any. This is in accordance with the fact that Spanish plant-based dieters demand for a greater offer of these products [19]. Our findings also support their concern for the low nutritional quality of available products [19]. The majority of these products were high in calories, fats, especially saturated fat and salt, while they were low in fiber and proteins. Therefore, these products would not be considered healthy plant-based products [22–24]. However, not all the products had the same ingredients' composition or nutritional profile.

Coconut oil-based cheese (the largest group of available products) had a no healthy nutritional profile. Those plant-based products were mainly composed of refined coconut oil with some starches. They were, therefore, high in saturated fats. No significant difference was observed between the saturated fat content of dairy cheese and the coconut oil-based alternatives. The evidence suggests that limiting the consumption of foods high in saturated fats is necessary for health reasons [25]. Specifically, the cardiovascular health benefits of consuming virgin coconut oil, which is high in saturated fats, have been disproved [26,27]. In addition, the coconut-oil based products were made of refined fats, of which consumption should be limited [28]. However, coconut oil is preferred by the plant-based cheese industry because its high amount of saturated fatty acids enables a creamy texture and also gives firmness to the product at refrigerated temperatures. Additional ingredients such as starches, carrageenan or agar-agar are also used in coconut oil-based cheese to mimic the density and texture of real cheese [29,30]. Refined coconut oil is preferred by the industry over its virgin counterpart because it is less intense in flavor and could be easily masked by the use of flavorings [30,31]. This may explain the presence of several food additives in those products. The coconut oil-based products were also high in salt, with over 1.5 g salt per 100 g [22,23]. High salt intake has been recognized as one of the main diet-related risk factors for global mortality and for loss of quality-adjusted life years [32]. These alternatives also contained negligible amounts of protein; therefore, they cannot be considered as a dietary protein source, as dairy cheese is.

The ingredient composition of cashew nut-based and tofu-based products seem nutritionally healthier. In the case of cashew nut-based products, around 50% of their weight were cashews, being the major ingredient, followed by water and lemon juice. The health benefits of whole nuts consumption is well-known [33,34], and it has been reported that other nut-based processed products, such as peanut butter without added sugars, could provide some health benefits, reducing the risk of type 2 diabetes incidence [35]. In the case of tofu-based products, 98.5% of the products' weight is soy milk. Similarly, some studies suggest that tofu has health benefits for type 2 diabetes prevention [36] and soy milk consumption for dyslipidemia management [37]. Further studies could investigate whether cashew nut-based and tofu-based cheese can also improve people's health.

The difference in the calorie content between dairy cheese and cashew nut-based options, besides being statistically different, is minimal. However, the replacement of dairy cheese by tofu-based alternatives could help reduce energy intake. In addition, the replacement of dairy cheese with cashew nut-based and particularly tofu-based options may be helpful in reducing the intake of total fats, and would contribute to replacing the intake of saturated fats by unsaturated options, which could provide health benefits [25]. However, unsaturated fatty acids are sensitive to processing, being easily oxidized. Lipid oxidation lowers the nutritional quality of lipid-containing foods [38,39]. Future analysis should determine to what extent the lipid profile of the primary ingredients of these food products is affected by processing. The cashew nut-based and tofu-based products were good sources of protein (median: 11 g and 18 g/100 g for cashew nut-based and tofu-based, respectively) [24]. Indeed, there is no significant difference in the protein level of dairy cheese and tofu-based alternatives. In addition, soy protein could also be considered as a complete protein, containing all the essential amino acids. On the other hand, the substitution of dairy cheese with cashew nut-based options would moderate the intake of protein. Considering that a large proportion of Spaniards have an excessive intake of proteins [40], along with the fact that the distribution of macronutrients in diets of Spanish vegetarians and vegans corresponds well to that proposed by the Spanish standards for dietary reference intake, including the protein content [41], the replacement of dairy cheese by cashew nut-based alternatives would not necessarily be a problem in terms of dietary protein intake if integrated in a nutritionally balanced diet. They could also provide some fiber, especially tofu-based products. Fiber consumption has been linked to several health benefits, such as a reduced risk of all-cause and cardiovascular-related mortality, and incidence of coronary heart disease, stroke incidence and mortality, type 2 diabetes and colorectal cancer, among others [42]. Both types of products contained lower levels of salt compared to dairy cheese, although the salt content in tofu-based cheese was relatively high [23].

Altogether, the cashew nut-based and tofu-based products currently available in Spanish supermarkets seem to provide an overall nutritionally healthier profile and could be interesting substitutes for dairy cheese. However, they are not a common option; only four cashew nut-based and two tofu-based cheese products were available in supermarkets at the time of the study, and just one out of the seven stores visited sold them. Our findings, along with the fact that Spanish people eating plant-based diets demand more and healthier plant-based cheese products in supermarkets [19], indicate that there is room for the food industry to commercialize new plant-based cheese options with a better nutritional profile. Currently, just one brand manufactures each of those healthier cashew and tofu-based options for Spanish supermarkets. New nut- and tofu-based products to be marketed should avoid the addition of unhealthy ingredients, such as refined oils, starches and excess salt that would offset the benefits of using those healthy main ingredients. However, the food industry may face some barriers in reformulating foods, such as higher costs or lower consumer acceptability if the organoleptic properties do not sufficiently mimic dairy cheese. More efforts into mitigating this is needed [43,44].

It should be noted, however, that not all plant-based cheese consumers opt for these alternatives for health reasons, but for other reasons such as environmental or animal welfare motives. The evidence suggests that plant-based meat and milk analogs have a lower environmental impact than the animal-based products they are intended to replace [17,45,46]. To the authors' knowledge, the ecological footprint of plant-based cheese has not been addressed in the scientific literature. A Danish database reports that the carbon footprint of producing 1 kg of vegan cheese would be 1 kg of carbon dioxide equivalents (CO_2 e) [47], a value significantly lower than that assigned to dairy cheese, ranging from 5.33 to 16.35 kg CO_2 e per kg of product [48]. Without concrete data, it can only be assumed that—as long as the main ingredients (i.e., coconut oil, cashew and soy) are grown using low environmental impact agricultural techniques—these plant-based products would be environmentally friendlier alternatives to dairy cheese, which is one of the food products with the highest impact on the environment [46].

A major strength of our study is that, for the first time, the nutritional profile and ingredients' composition of the plant-based cheese alternatives available in Spanish supermarkets has been carried out. In addition, their nutritional composition has been compared with that of dairy cheese. Our results debunk the common assumption that all plant-based products are healthy options, highlighting the necessity of assessing other plant-based products on the market. Some limitations should be also considered. Only products available at supermarkets were targeted, not considering those products exclusively sold in specialist vegan stores; this study's objective was to assess the products available for consumers on a mass scale. This would be a good area for future research. Further research could also investigate to what extent the protein and lipid profile of ingredients have been affected by processing [38,39,49]. Other data gaps, such as the micronutrient content, particularly those that are characteristic of dairy cheese, such as calcium, also deserve further attention.

5. Conclusions

This study shows that the majority of plant-based cheese alternatives available in Spanish supermarkets do not have a good nutritional profile. Nevertheless, although relatively few products are available, healthier options could be found, such as those goods composed mainly of cashew nuts and tofu. The replacement of dairy cheese by cashew nut- and tofu-based plant-based alternatives could reduce intakes of salt and total fats, while replacing the intake of saturated with unsaturated fats. Future investigations should address the health effects of substituting dairy cheese with plant-based cheese products. The assessment of the environmental impact of plant-based cheese also deserves further attention.

Supplementary Materials: The following are available online at https://www.mdpi.com/article/10.3390/nu13093291/s1, Table S1: Main ingredient, format, calories and nutritional content per 100 g of plant-based cheese alternatives, Table S2: Format, calories and nutritional content per 100 g of different types of dairy cheese, Table S3: Median (minimum-maximum) values of calories and nutritional content in coconut oil-based cheese alternatives per 100 g, by selling format.

Author Contributions: Conceptualization, U.F.; methodology, U.F.; formal analysis, U.F.; investigation, U.F.; writing—original draft preparation, U.F.; writing—review and editing, H.R.; visualization, U.F. Both authors have read and agreed to the published version of the manuscript.

Funding: U.F. acknowledges support from the Spanish Ministry of Science and Innovation and State Research Agency through the "Centro de Excelencia Severo Ochoa 2019–2023" Program (CEX2018-000806-S), and support from the Generalitat de Catalunya through the CERCA Program.

Institutional Review Board Statement: Not applicable.

Informed Consent Statement: Not applicable.

Data Availability Statement: Data is contained within the article or supplementary material.

Conflicts of Interest: The authors declare no conflict of interest.

References

1. European Data Journalism Network. Europe Is Going Veg 2019. Available online: https://www.europeandatajournalism.eu/eng/News/Data-news/Europe-is-going-veg (accessed on 8 June 2021).
2. Hopwood, C.J.; Bleidorn, W.; Schwaba, T.; Chen, S. Health, environmental, and animal rights motives for vegetarian eating. *PLoS ONE* **2020**, *15*, e0230609. [CrossRef] [PubMed]
3. De Backer, C.J.; Hudders, L. From meatless Mondays to meatless Sundays: Motivations for meat reduction among vegetarians and semi-vegetarians who mildly or significantly reduce their meat intake. *Ecol. Food Nutr.* **2014**, *53*, 639–657. [CrossRef]
4. Fresán, U.; Errendal, S.; Craig, W.J. Influence of the Socio-Cultural Environment and External Factors in Following Plant-Based Diets. *Sustainability* **2020**, *12*, 9093. [CrossRef]
5. Wynes, S.; Nicholas, K.A. The climate mitigation gap: Education and government recommendations miss the most effective individual actions. *Environ. Res. Lett.* **2017**, *12*, 074024. [CrossRef]
6. Springmann, M.; Clark, M.; Mason-D'Croz, D.; Wiebe, K.; Bodirsky, B.L.; Lassaletta, L.; de Vries, W.; Vermeulen, S.J.; Herrero, M.; Carlson, K.M.; et al. Options for keeping the food system within environmental limits. *Nature* **2018**, *562*, 519–525. [CrossRef] [PubMed]
7. Springmann, M.; Godfray, H.C.J.; Rayner, M.; Scarborough, P. Analysis and valuation of the health and climate change cobenefits of dietary change. *Proc. Natl. Acad. Sci. USA* **2016**, *113*, 4146–4151. [CrossRef]
8. Springmann, M.; Wiebe, K.; Mason-D'Croz, D.; Sulser, T.B.; Rayner, M.; Scarborough, P. Health and nutritional aspects of sustainable diet strategies and their association with environmental impacts: A global modelling analysis with country-level detail. *Lancet Planet. Health* **2018**, *2*, e451–e461. [CrossRef]
9. Clark, M.A.; Springmann, M.; Hill, J.; Tilman, D. Multiple health and environmental impacts of foods. *Proc. Natl. Acad. Sci. USA* **2019**, *116*, 23357–23362. [CrossRef]
10. Fresán, U.; Sabate, J. Vegetarian Diets: Planetary Health and Its Alignment with Human Health. *Adv. Nutr.* **2019**, *10*, S380–S388. [CrossRef]

11. Aleksandrowicz, L.; Green, R.; Joy, E.J.M.; Smith, P.; Haines, A. The Impacts of Dietary Change on Greenhouse Gas Emissions, Land Use, Water Use, and Health: A Systematic Review. *PLoS ONE* **2016**, *11*, e0165797. [CrossRef]
12. Markets and Markets. Plant-Based Meat Market by Source (Soy, Wheat, Pea, & Other Sources), Product (Burger Patties, Strips & Nuggets, Sausages, Meatballs, & Other Products), Type (Beef, Chicken, Pork, Fish, & Other Types), Process, and Region-Global Forecast to 2025. 2020. Available online: https://www.marketsandmarkets.com/Market-Reports/plant-based-meat-market-44922705.html?gclid=EAIaIQobChMIos2j9dCF5gIVye$\neg$J3Ch1Z5gKsEAAYASAAEgJedfD_BwE (accessed on 8 June 2021).
13. Markets and Markets. Dairy Alternatives Market by Source (Soy, Almond, Coconut, Oats, Rice, Hemp), Application (Milk, Yogurt, Ice creams, Cheese, Creamers), Distribution Channel (Supermarkets, Health Food Stores, Pharmacies), Formulation, and Region-Global Forecast to 2026. 2021. Available online: https://www.marketsandmarkets.com/Market-Reports/dairy-alternative-plant-milk-beverages-market-677.html (accessed on 8 June 2020).
14. Wickramasinghe, K.; Breda, J.; Berdzuli, N.; Rippin, H.; Farrand, C.; Halloran, A. The shift to plant-based diets: Are we missing the point? *Glob. Food Secur.* **2021**, *29*, 100530. [CrossRef]
15. Angelino, D.; Rosi, A.; Vici, G.; Dello Russo, M.; Pellegrini, N.; Martini, D.; On Behalf Of The Sinu Young Working Group. Nutritional Quality of Plant-Based Drinks Sold in Italy: The Food Labelling of Italian Products (FLIP) Study. *Foods* **2020**, *9*, 682. [CrossRef]
16. Curtain, F.; Grafenauer, S. Plant-Based Meat Substitutes in the Flexitarian Age: An Audit of Products on Supermarket Shelves. *Nutrients* **2019**, *11*, 2603. [CrossRef]
17. Santo, R.E.; Kim, B.F.; Goldman, S.E.; Dutkiewicz, J.; Biehl, E.M.B.; Bloem, M.W.; Neff, R.A.; Nachman, K.E. Considering Plant-Based Meat Substitutes and Cell-Based Meats: A Public Health and Food Systems Perspective. *Front. Sustain. Food Syst.* **2020**, *4*, 134. [CrossRef]
18. Craig, W.J.; Fresán, U. International Analysis of the Nutritional Content and a Review of Health Benefits of Non-Dairy Plant-Based Beverages. *Nutrients* **2021**, *13*, 842. [CrossRef]
19. ProVeg International. Encuesta de Consumidores de Alternativas Vegetales en España-Descripción del Panorama de los Productos Vegetales y de las Prioridades para el Desarrollo y la Mejora de los Mismos (Survey to Plant-Based Alternatives Consumers in Spain—An Overview of the Plant-Based Products and the Priorities for Their Development and Improvement) Madrid. 2020. Available online: https://proveg.com/es/wp-content/uploads/sites/2/2021/02/PVES_Encuesta_de_Consumidores_de_Alternativas_Vegetales_En_Espan%CC%83a.pdf (accessed on 8 June 2020).
20. BEDCA (Spanish Food Composition Database). Available online: https://www.bedca.net/bdpub/index.php (accessed on 8 June 2021).
21. The Jamovi Project. Jamovi (Version 1.6) [Computer Software]. 2021. Available online: https://www.jamovi.org (accessed on 25 April 2021).
22. NHS. Food Labels. Available online: https://www.nhs.uk/live-well/eat-well/how-to-read-food-labels/ (accessed on 30 June 2021).
23. Ministerio de Sanidad, Servicios Sociales e Igualdad. Agencia Española de Seguridad Alimentaria y Nutrición. *Plan Cuidate más.* Available online: http://www.plancuidatemas.aesan.msssi.gob.es/ (accessed on 25 August 2021).
24. Ministerio de Sanidad, Consumo y Bienestar Social. Agencia Española de Seguridad Alimentaria y Nutrición. Declaraciones Nutricionales Autorizadas en el Anexo del Reglamento (CE) Nº 1924/2006. Available online: https://www.aesan.gob.es/AECOSAN/docs/documentos/seguridad_alimentaria/gestion_riesgos/Tabla_declaraciones_NUTRICIONALES_autorizadas.pdf (accessed on 25 August 2021).
25. Dietary Guidelines Advisory Committee. *Scientific Report of the 2020 Dietary Guidelines Advisory Committee: Advisory Report to the Secretary of Agriculture and the Secretary of Health and Human Services*; U.S. Department of Agriculture, Agricultural Research Service: Washington, DC, USA, 2020. Available online: https://www.dietaryguidelines.gov/sites/default/files/2020-07/ScientificReport_of_the_2020DietaryGuidelinesAdvisoryCommittee_first-print.pdf (accessed on 1 June 2021).
26. Jayawardena, R.; Swarnamali, H.; Ranasinghe, P.; Misra, A. Health effects of coconut oil: Summary of evidence from systematic reviews and meta-analysis of interventional studies. *Diabetes Metab Syndr* **2021**, *15*, 549–555. [CrossRef] [PubMed]
27. Neelakantan, N.; Seah, J.Y.H.; van Dam, R.M. The Effect of Coconut Oil Consumption on Cardiovascular Risk Factors: A Systematic Review and Meta-Analysis of Clinical Trials. *Circulation* **2020**, *141*, 803–814. [CrossRef] [PubMed]
28. Liu, R.; Guo, X.; Cheng, M.; Zheng, L.; Gong, M.; Chang, M.; Jin, Q.; Wang, X. Effects of chemical refinement on the quality of coconut oil. *J. Food Sci. Technol.* **2019**, *56*, 3109–3116. [CrossRef]
29. Food Revolution Network. What's So Great about Nut Cheese–and How to Make Your Own. Available online: https://foodrevolution.org/blog/nut-cheese/ (accessed on 24 August 2021).
30. Hill, A. A Guide to Vegan Cheese: What's the Best Dairy-Free Option? Healthline, 2018. Available online: https://www.healthline.com/nutrition/vegan-cheese (accessed on 24 August 2021).
31. Marcus, J.B. (Ed.) Chapter 6—Flavor Enhancement Ingredients. In *Aging, Nutrition and Taste*; Academic Press: Cambridge, MA, USA, 2019; pp. 173–206. [CrossRef]
32. Afshin, A.; Sur, P.J.; Fay, K.A.; Cornaby, L.; Ferrara, G.; Salama, J.S.; Mullany, E.C.; Abate, K.H.; Abbafati, C.; Abebe, Z.; et al. Health effects of dietary risks in 195 countries, 1990–2017: A systematic analysis for the Global Burden of Disease Study 2017. *Lancet* **2019**, *393*, 1958–1972. [CrossRef]
33. Becerra-Tomás, N.; Paz-Graniel, I.; Kendall, C.W.C.; Kahleova, H.; Rahelić, D.; Sievenpiper, J.L.; Salas-Salvadó, J. Nut consumption and incidence of cardiovascular diseases and cardiovascular disease mortality: A meta-analysis of prospective cohort studies. *Nutr. Rev.* **2019**, *77*, 691–709. [CrossRef] [PubMed]

34. Zhang, D.; Dai, C.; Zhou, L.; Li, Y.; Liu, K.; Deng, Y.-J.; Li, N.; Zheng, Y.; Hao, Q.; Yang, S.; et al. Meta-analysis of the association between nut consumption and the risks of cancer incidence and cancer-specific mortality. *Aging* **2020**, *12*, 10772–10794. [CrossRef] [PubMed]
35. Becerra-Tomás, N.; Paz-Graniel, I.; Hernández-Alonso, P.; Jenkins, D.J.A.; Kendall, C.W.C.; Sievenpiper, J.L.; Salas-Salvadó, J. Nut consumption and type 2 diabetes risk: A systematic review and meta-analysis of observational studies. *Am. J. Clin. Nutr.* **2021**, *113*, 960–971. [CrossRef]
36. Tang, J.; Wan, Y.; Zhao, M.; Zhong, H.; Zheng, J.-S.; Feng, F. Legume and soy intake and risk of type 2 diabetes: A systematic review and meta-analysis of prospective cohort studies. *Am. J. Clin. Nutr.* **2020**, *111*, 677–688. [CrossRef]
37. Eslami, O.; Shidfar, F. Soy milk: A functional beverage with hypocholesterolemic effects? A systematic review of randomized controlled trials. *Complement. Ther. Med.* **2019**, *42*, 82–88. [CrossRef] [PubMed]
38. German, J.B. Food processing and lipid oxidation. *Adv. Exp. Med. Biol.* **1999**, *459*, 23–50. [CrossRef]
39. Akoh, C.C. (Ed.) Polyunsaturated Lipid Oxidation in Aqueous Systems. In *Food Lipids: Chemistry, Nutrition, and Biotechnology*, 4th ed.; CRC Press: Boca Raton, FL, USA, 2017. [CrossRef]
40. Ruiz, E.; Ávila, J.M.; Valero, T.; Del Pozo, S.; Rodriguez, P.; Aranceta-Bartrina, J.; Gil, Á.; González-Gross, M.; Ortega, R.M.; Serra-Majem, L.; et al. Macronutrient Distribution and Dietary Sources in the Spanish Population: Findings from the ANIBES Study. *Nutrients* **2016**, *8*, 177. [CrossRef] [PubMed]
41. Menal-Puey, S.; Morán del Ruste, M.; Marques-Lopes, I. Food and nutrient intake in Spanish vegetarians and vegans. *Prog. Nutr.* **2018**, *20*, 189–196. [CrossRef]
42. Reynolds, A.; Mann, J.; Cummings, J.; Winter, N.; Mete, E.; Te Morenga, L. Carbohydrate quality and human health: A series of systematic reviews and meta-analyses. *Lancet* **2019**, *393*, 434–445. [CrossRef]
43. Short, E.C.; Kinchla, A.J.; Nolden, A.A. Plant-Based Cheeses: A Systematic Review of Sensory Evaluation Studies and Strategies to Increase Consumer Acceptance. *Foods* **2021**, *10*, 725. [CrossRef]
44. McClements, D.J.; Grossmann, L. A brief review of the science behind the design of healthy and sustainable plant-based foods. *NPJ Sci. Food.* **2021**, *5*, 17. [CrossRef]
45. Henderson, A.; Unnasch, S. Life Cycle Assessment of Ripple Non-Dairy Milk. Life Cycle Associates Report LCA.6121.159.2017, Prepared for Ripple Foods, Inc. 2017. Available online: https://www.ripplefoods.com/pdf/LIfe%20Cycle%20Assessment%20of%20Non-Dairy%20Milk.pdf (accessed on 15 June 2021).
46. Poore, J.; Nemecek, T. Reducing food's environmental impacts through producers and consumers. *Science* **2018**, *360*, 987–992. [CrossRef] [PubMed]
47. CONCITO, 2.-0 LCA Consultants and the Salling Foundations. The Large Climate Database. Available online: https://denstoreklimadatabase.dk/ (accessed on 1 June 2021).
48. Clune, S.; Crossin, E.; Verghese, K. Systematic review of greenhouse gas emissions for different fresh food categories. *J. Clean. Prod.* **2017**, *140*, 766–783. [CrossRef]
49. Meade, S.J.; Reid, E.A.; Gerrard, J.A. The Impact of Processing on the Nutritional Quality of Food Proteins. *J. AOAC Int.* **2019**, *88*, 904–922. [CrossRef]

nutrients

MDPI

Article

Acceptability of Plant-Based Diets for People with Chronic Kidney Disease: Perspectives of Renal Dietitians

Jordan Stanford [1,2,*], Mikaela Zuck [1], Anita Stefoska-Needham [1,2], Karen Charlton [1,2] and Kelly Lambert [1,2]

1 School of Medicine, Faculty of Science, Medicine and Health, University of Wollongong, Northfields Ave, Wollongong, NSW 2522, Australia; mz849@uowmail.edu.au (M.Z.); anitasn@uow.edu.au (A.S.-N.); karenc@uow.edu.au (K.C.); klambert@uow.edu.au (K.L.)

2 Illawarra Health and Medical Research Institute, Wollongong, NSW 2522, Australia

* Correspondence: js096@uowmail.edu.au

Abstract: The purpose of this study was to explore the perspectives of renal dietitians regarding plant-based diets for chronic kidney disease (CKD) management and evaluate the acceptability of a hypothetical plant-based dietary prescription aiming for the consumption of 30 unique plant foods per week. This study used an exploratory mixed methods design. Forty-six renal dietitians participated in either an online survey (n = 35) or an in-depth interview (n = 11). Dietitians perceived that plant-based diets could address multiple clinical concerns relevant to CKD. Forty percent of survey respondents reported the hypothetical dietary prescription was realistic for people with CKD, 34.3% were unsure, and 25.7% perceived it as unrealistic. Strengths of the hypothetical prescription included shifting the focus to whole foods and using practical resources like recipes. Limited staffing, time, and follow-up opportunities with patients, as well as differing nutrition philosophies were the most commonly reported challenges to implementation; while a supportive multidisciplinary team was identified as an important enabler. To increase patient acceptance of plant-based dietary approaches, education about plant food benefits was recommended, as was implementing small, incremental dietary changes. Successful implementation of plant-based diets is perceived to require frequent patient contact and ongoing education and support by a dietitian. Buy-in from the multidisciplinary team was also considered imperative.

Keywords: plant-based diets; chronic kidney disease; implementation; barriers; enablers; cross-sectional survey; qualitative research

Citation: Stanford, J.; Zuck, M.; Stefoska-Needham, A.; Charlton, K.; Lambert, K. Acceptability of Plant-Based Diets for People with Chronic Kidney Disease: Perspectives of Renal Dietitians. *Nutrients* **2022**, *14*, 216. https://doi.org/10.3390/nu14010216

Academic Editor: Peter Pribis

Received: 17 December 2021
Accepted: 1 January 2022
Published: 4 January 2022

1. Introduction

Diet plays a central role in the management of chronic kidney disease (CKD) [1]. However, dietary prescriptions are often confusing and divergent from standard healthy eating guidelines and may limit healthy plant foods such as fruits, vegetables, and whole-grain products due to their potassium content [2]. Such restrictive dietary guidance has broader implications for people living with CKD by resulting in limited intakes of health-protective food components such as dietary fibre and phytochemicals, as well as potentially contributing to poor dietary adherence overall. Concerns about the contribution of potassium from plant-based foods may be outdated given the emerging evidence that suggests the bioavailability of potassium in whole non-processed fruits and vegetables are lower than initially estimated [3], at around 50–60% [4]. A recent study confirmed that dietary potassium intake was not associated with hyperkalaemia or death in patients receiving haemodialysis treatment [5]. Research indicates that habitual dietary patterns rich in plant-based sources are protective against disease progression and risk of mortality in people with CKD, even at advanced disease stages [6,7]. Additionally, there is growing recognition of the role of plant-based diets in modulating the composition and metabolic activity of the human gut microbiome, which in turn may lead to improved health outcomes relevant to individuals with CKD [8–11].

In light of this evidence, a less didactic and more liberal educational approach to the renal diet may be possible for patients with CKD. However, successful implementation of novel diet therapies in clinical practice requires acceptance from practitioners before making them available to their patients. Dietitians provide extensive education to patients, caregivers, and their families to facilitate appropriate food choices and improve long-term dietary adherence, which may alleviate disease progression [2]. Additionally, dietitians have unique first-hand insights into the challenges faced by patients regarding the prescription of a complex therapeutic diet [12]. Therefore, the objectives of this study were to explore renal dietitians' perspectives regarding plant-based diets for CKD management and evaluate the acceptability of a hypothetical plant-based dietary prescription and accompanying print resources. The hypothetical dietary prescription, which aimed to increase the amount and variety of plant foods in diets for people with CKD, was used to stimulate discussion and facilitate recommendations for implementing plant-based diets in future clinical trials and current practice.

2. Materials and Methods

This exploratory mixed-methods study was approved by the joint Illawarra Shoalhaven Local Health District/University of Wollongong Human Research Ethics Committee (2019/ETH00397). The Checklist for Reporting Results of Internet E-Surveys [13] was used as a guide to create the online survey. The consolidated criteria for reporting qualitative research (COREQ) checklist (Item S1, Supplementary Materials) was used to facilitate a detailed and comprehensive reporting of the qualitative component of the study [14].

2.1. Study Sample and Recruitment

Accredited Practising Dietitians (APDs) or those eligible for APD status actively working in primary, secondary, or tertiary care employed to provide dietary advice to people with CKD in Australia were eligible to take part in this study. A convenience sample of renal dietitians was recruited using two approaches to provide the greatest possible coverage and to maximise the participation of the specialist target group [15]. Recruitment took place between April and August 2019, where eligible dietitians were contacted via: (i) a professional e-mail distribution network for renal dietitians in Australia; and (ii) attendance at the 2019 World Congress of Nephrology Renal Nutrition, Nursing, and Allied Health Professionals Symposium. Conference attendees were identified with the assistance of conference organisers. Eligible dietitians were invited to participate by e-mail or in person and given the option to partake in either an in-depth, semi-structured interview, or complete a short online survey. Advertising materials included an overview of the study, a participant information sheet with investigators' contact details, interview questions, and a direct URL link to the online survey. Participants wanting to participate in an interview rather than the survey were advised to contact one of the investigators to schedule a meeting. No incentives were offered, and participation was voluntary.

2.2. Hypothetical Plant-Based Dietary Prescription

Survey and interview participants were informed that, in the context of this study, a 'plant-based dietary prescription' referred to a diet dominated by a variety of vegetables, fruits, legumes, whole grains, nuts, seeds, herbs, and spices. It did not infer that the eating pattern was exclusively vegetarian or vegan and could include small to moderate amounts of animal-based products such as dairy, meats, poultry, and fish. The hypothetical plant-based dietary prescription designed by members of the research team aimed to increase the amount and variety of plant foods in the diets of people with CKD. The philosophy to achieve a plant-based diet introduced to all participants in this study was simplified by encouraging patients to consume 30 or more unique plant foods over a seven-day period. This concept was informed by the findings of the largest observational study to date investigating the human gut microbiome [16], whereby individuals consuming a higher plant-based diet (defined as consuming more than 30 different varieties of plant foods per week) had increased microbial diversity and lower antibiotic-resistant microbial genes [16].

Further details about how the hypothetical prescription could be implemented with patients in practice were explored with interview participants exclusively. For instance, to supplement the target of consuming 30 or more unique plant foods per week, specific advice was proposed about the number of daily food servings patients would need to consume to more closely align to the Australian Dietary Guidelines (i.e., five servings of vegetable, two servings of fruits, etc.) [17], while still adhering to the evidence-based guidelines for nutrition in kidney disease [18,19]. Five ancillary print resources were also developed to accompany the hypothetical plant-based dietary prescription. The print resources included a recipe book, a seven-day template for participants to fill out and plan meals, an A-Z food guide to build meals, a food swap list, and an instruction manual on how these resources could be used.

2.3. Data Collection and Analysis

The anonymous online survey consisted of eight questions (Item S1, Supplementary Materials), including close-ended questions (multiple-choice) and open-ended questions, which were pilot-tested with three dietitians to assess face validity. Multiple responses were accepted for questions 2, 4, 5, and 6. This online survey was self-administered using SurveyMonkey (San Mateo, CA, USA) and was open for 10 weeks from 26 June 2019, to 4 September 2019. Tacit consent was implied by online survey completion. Participants were only able to complete the survey once and could review and change their answers on any survey page until they submitted the survey. Analysis of quantitative data was facilitated using Microsoft Excel (2010). Data presented in figures were produced using R version 3.5.0 [20]. Open-ended responses were analysed using deductive content analysis [21]. Only participants with complete survey data were included.

Semi-structured interviews, conducted by two research members, lasted between 30–45 min and were undertaken either face-to-face, via Skype® or telephone. The semi-structured interview guide covered four key topic areas (Item S3, Supplementary Materials, Item S3). For those interviewed by phone, the print resources were emailed before the interview once written consent had been obtained. Demographic details (age, gender, length of practice as a dietitian, length of practice as a renal dietitian, current full-rime equivalent (FTE) load, practice setting) were collected. Interviews were recorded using a digital recorder and transcribed verbatim. Dedoose software was used to manage and code the data [22]. Transcripts were analysed inductively using Braun and Clarke's six phases of thematic analysis [23]. Specifically, two researchers read and reread the transcribed verbatim independently for immersion in the data. Quotes relevant to the research question were highlighted, and codes were systematically applied to identify elements of interest. Codes were collated into potential subthemes. The inter-reliability of codes was examined in a subset of transcripts by a third member of the research team to ensure the credibility of the coding analysis. Researchers worked collaboratively to reach a consensus on the key themes that emerged in the assigned codes and subthemes. Ongoing analysis took place to refine each key theme to ensure that they were reflective of the coded extracts and the entire data set. Compelling extracts were selected for final analysis, relating back to the research question and literature. Participants did not provide feedback on the key themes.

3. Results

Approximately 120 renal dietitians were invited to take part in this study, and 46 (response rate: 38.3%) either completed the online survey (n = 35) or participated in an in-depth interview (n = 11).

3.1. Online Survey

The median duration of time the 35 survey respondents currently worked as renal dietitians was seven years (Table 1). The case mix of patients seen by the survey participants was dominated by those receiving haemodialysis (n = 31, 88.6%) and patients in the pre-dialysis stage (n = 23, 65.7%).

Table 1. Demographic characteristics of the study populations.

Characteristics	Interview (n = 11)	Survey (n = 35)
Gender (female, %)	11 (100%)	-
Age (range)	25–64	-
Years actively working as a renal dietitian (median-IQR)	9.17 (3.67–27.50)	7 (3–12.25)
Current employment status (Full time equivalent: median, IQR)	0.6 (0.3–1.0)	-
Practice setting		
Community settings	1 (9%)	-
Private practice	1 (9%)	-
Public health/hospitals	10 (91%)	-
Area of practice		
Early CKD	2 (18.2%)	11 (31.4%)
Pre-dialysis	7 (63.6%)	23 (65.7%)
Haemodialysis	8 (72.7%)	31 (88.6%)
Peritoneal dialysis	5 (45.5%)	20 (57.1%)
Renal transplant	4 (36.4%)	12 (34.3%)
Renal supportive care/palliative care	5 (45.5%)	13 (37.1%)
Other	0 (0%)	3 (8.6%)

'-' data not available for survey respondents. IQR, Interquartile range.

Forty percent of survey participants reported that the proposed hypothetical dietary prescription recommending consumption of 30 unique plant-based foods per week was realistic for people with CKD (Figure 1A). The following quotes echo the most common justifications provided: "this amount reflects the healthy eating guidelines"; "I have already had success in implementing this in my role as a renal dietitian", and "when breaking it down into 4–5 different plant foods per day from a variety of sources, it [the hypothetical plant-based dietary prescription] doesn't seem excessive". However, support was not unanimous, and several participants expressed that they were unsure (n = 12, 34.3%) or felt the hypothetical plan was unrealistic (n = 9, 25.7%). For example, "misinformation provided by other health care professionals", "financial burdens", and "limited accessibility to fresh foods" were concerns impacting implementation.

Lack of cooking and preparation skills were also considered substantial barriers to implementation, in addition to personal food preferences (Figure 1B). Other barriers identified were "managing patients' fear around potassium control" (n = 17, 48.57%). Furthermore, when implementing plant-based diets for patients with CKD, renal dietitians consistently reported they would be cautious about prescribing more dried fruit, followed by nuts and seeds (Figure 1C). The main reason for using a cautious approach to these items was fear of inducing hyperkalaemia and/or hyperglycaemia. Suggestions to enhance implementation included educating patients about the health benefits of plant-based eating and providing recipes (Figure 1D). Other recommendations included "gaining support from the multidisciplinary team", "education with motivated patients at earlier stages of CKD", and "availability of supplementary educational materials such as food checklists, meal plans, and pictorial resources".

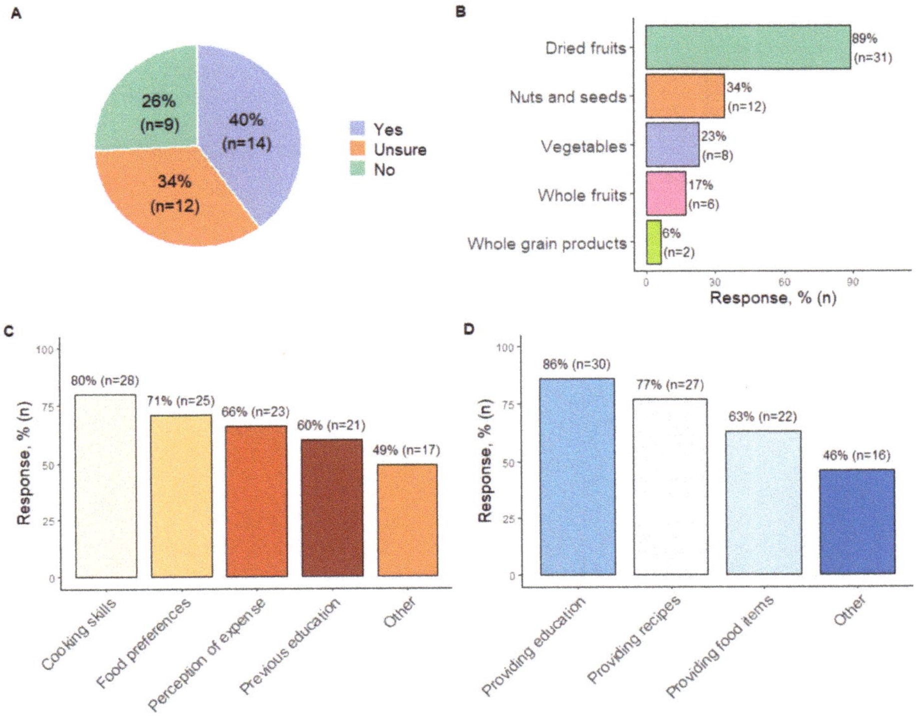

Figure 1. Summary of responses to multiple-choice survey questions. (**A**) Answers to whether the target of consuming 30 unique plant-based foods over seven days is realistic for patients with CKD. (**B**) Responses to the question about which plant foods dietitians felt cautious about prescribing to people with CKD. (**C**) Potential challenges to implementing plant-based diets for people with CKD. (**D**) Potential enablers to implementing plant-based diets for people with CKD. Multiple responses were accepted for the questions presented in figures (**B–D**).

3.2. In-Depth Interviews

Twelve renal dietitians expressed interest, of which eleven participated in an in-depth interview. Non-participation was due to scheduling conflicts. Data saturation was reached by the 11th interview, with no new themes subsequently identified. All interview participants were female (age range 25–64 years). Participants worked in various geographic locations, with diverse CKD populations and varying levels of renal dietetic experience (Table 1). Two overarching themes from the interviews were: (i) the value of plant-based diets and strengths of the hypothetical dietary prescription; and (ii) existing barriers and enablers to successful implementation. A further eight sub-themes were apparent.

3.2.1. Value of Plant-Based Diets and Strengths of the Hypothetical Dietary Prescription Addresses Multiple Clinical Concerns

All interviewees acknowledged that a plant-based diet could prevent or manage various risk factors for disease progression, including comorbidities relevant to individuals with CKD.

> "... so if they are eating this way ... glycemic control will be better, hypertension control will be better, proteinuria is likely to be better, ... all of the things that you would be worried about.. a plant-based diet is going to help with that"
>
> (Dietitian 11)

Shifts Focus from Nutrients to Whole Foods

Participants repeatedly indicated that traditional approaches to dietary counselling for CKD, which tend to focus on individual nutrients, can lead to patients feeling confused. Respondents discussed that a focus on nutrients does not consider overall diet quality and often results in whole foods, like fruits and vegetables, being erroneously restricted or removed by patients because they are rich sources of potassium.

> "We're very focused on guidelines and millimoles of potassium, but I get concerned ... when I provide my education ... that people are not able to put that into practice or they get the message wrong and then end up cutting out things unnecessarily because they think it's bad for them ... It [the hypothetical dietary prescription] focuses on making healthy food choices because I think people are misunderstanding the information about potassium and phosphate and see it as cutting out fruit and vegetables ... and that is a very big concern ... When in fact you want to maintain a healthy diet overall and variety ... it [the hypothetical dietary prescription] allows for flexibility, and it allows people to transition from receiving some abstract information about food into this is what I'm going to eat today."
>
> (Dietitian 10)

Dietitians in this study perceived that using a simple target such as '30 unique plant foods per week' might help alleviate patient confusion. Respondents also felt that the increased focus on foods rather than nutrients in the hypothetical dietary prescription would likely improve overall diet quality. Positive dietary messaging used in the hypothetical prescription was preferred (i.e., include these foods) rather than negative messaging (i.e., limit consumption of these foods) as a strategy to reduce patient anxiety and encourage adherence to dietary recommendations.

A Need for Practical Complementary Resources

Additional strengths of the hypothetical dietary prescription were the inclusion of recipes and the food swap list. These resources were perceived to be valuable tools to help demonstrate how various plant foods could be incorporated into meals, aid in building customisable meal plans, and accommodate individual food preferences.

> "The recipes are a great suggestion because there is no point giving this if people don't have the skills or the knowledge of how to incorporate those foods to make it into a meal ... demonstrating it is achievable."
>
> (Dietitian 5)
>
> "I like the variety ... they have this list of foods that they can swap, like each meal, each fruit serve out with, so that really helps."
>
> (Dietitian 5)

3.2.2. Barriers and Enablers to Implementation

Organisational Norms and System Inadequacies Are Barriers

Dietitian participants reported that tensions heightened by larger and more significant organisational norms and system inadequacies presented deterring barriers. For example, limited time compounded by limited staffing resources may often result in the inadequate follow-up of patients, which were expressed as common barriers in existing practice and suspected to be problematic, particularly for implementing plant-based diets in practice.

> "I think it would be crucial about our follow up ... in our pre-dialysis clinic, we do not see patients for like another two to three months, just because of our wait times ... "
>
> (Dietitian 7)

> "Generally, dietitians do not have the time to do these sorts of detailed meal plans."
>
> (Dietitian 3)

> "...you would need to spend a fair bit of time making sure that the patients understand and checking it."
>
> (Dietitian 2)

The dietitians stressed the importance of having supportive systems (such as adequate staffing and time) in place that are conducive to ongoing dietetic review and patient monitoring. Respondents suggested this as essential to ensure patient safety, monitor the overall nutritional adequacy of the diet (avoiding any nutrient deficiencies), and support patients to implement dietary recommendations as intended.

Differing Nutrition Philosophies and Perceptions about Diet Are Barriers

Similar to the survey respondents, there was noted to be differences in philosophies by some medical and nursing colleagues about the 'renal diet'. This was identified to be a major obstacle to implementation. Dietitians reported that some other members of the multidisciplinary team (MDT) may provide overly restrictive dietary advice and do not consider the quality of overall dietary patterns, nutrient bioavailability, nor acknowledge other non-dietary causes of electrolyte abnormalities relevant to people with CKD.

> "On this sort of diet [high plant-based diet] ... that perception, that they have to restrict them because of the potassium, and doctors or people, other people who might also have that perception and not, who do not sort of understand the difference in bioavailability."
>
> (Dietitian 8)

> "Depending on the medication they're on, they can be more susceptible to hyperkalaemia regardless of what they're eating..."
>
> (Dietitian 11)

The importance of shifting philosophies and perspectives regarding the renal diet was not limited to the MDT but also the patients themselves. Conflicting sources of information, in addition to previous dietary advice, were perceived to impact patient willingness to adopt new dietary recommendations.

> "You have a patient who is very compliant with previous dietary recommendations; they often do not like to go against that."
>
> (Dietitian 4)

> "Trying to move towards [a plant-based diet] ... which will undoubtedly include some of the foods that a lot of them just will not touch. It has been entrenched in them ... I cannot eat that sort of food ... "
>
> (Dietitian 7)

Supportive Multidisciplinary Networks Facilitates Implementation

Collaboration with MDT members to ensure consistency of messaging about plant-based diets was considered to be essential for translating plant-based diets into clinical practice. Interviewees suggested that regular in-services, team meetings, or journal clubs with the MDT may be helpful to disseminate and discuss up-to-date nutritional literature.

> "[Implementation] would [need] convincing other health staff about the research because many others other than dietitians provide dietary education...you have got doctors ... or nurses. So making sure they are aware of the evidence ... because ... if a patient hears something from their doctor, they [the patient] will

listen to them over what we [dietitians] recommend so, I think making sure that the message is consistent."

(Dietitian 4)

Timing of Implementation

Dietitians suggested that careful consideration should be given to the patient's overall treatment plan when judging an appropriate time to implement a plant-based dietary approach. Like survey respondents, many interviewees agreed that a plant-based diet would most benefit individuals in earlier stages of CKD in terms of clinical outcomes, safety, and patients might be more motivated by preventative measures to delay the initiation of renal replacement therapies.

"[if] they are preparing for dialysis ... patients can be quite overwhelmed ... , so I think [managing] the complexity of the diet you are trying to prescribe, but also managing their cognitive and emotional states as well can be quite challenging."

(Dietitian 11)

" ... they're unlikely to have seen a dietitian at CKD stage 3, so you know they don't have all those restrictions placed on them ... they typically are more motivated as they don't want dialysis and are more likely to benefit from using nutrition as a preventative to delay disease progression ... I think if you can get them early, then I think that's wonderful."

(Dietitian 1)

'Marketing' the Plant-Based Approach

A recommended strategy to enhance implementation was to "market" the plant-based diet to patients. It was suggested that the benefits of the diet should be explained clearly in the first session to motivate patients and encourage adherence, especially for those patients who are asymptomatic at the time of education.

"I think if we are very clear on the outcomes we are looking at, we could sell this to patients ... what outcomes can we expect because if people know that it might actually maintain their kidney function for another two years, or three years, or five years, that could be a very big motivator for them. But if we are just talking about general health, they already feel okay, especially when they are in the earlier stages."

(Dietitian 11)

Furthermore, a graduated goal system to encourage patients to achieve the target of 30 different plant foods over time rather than immediately was also recommended to improve implementation. Improvements to the presentation of dietary targets (30 unique plant foods) in the accompanying print resources of the hypothetical prescription were suggested.

" ... a starting point of ... I am having x amount of plant-based foods ... over the next two weeks can I increase it by an extra 5 or 10 foods a week ... so it can be a little bit more of a step guided approach ... that could be another strategy to help them with goal setting.".

(Dietitian 11)

Several suggested improvements to the supplementary print resources were also offered. Respondents felt that the format of the seven-day template (Item S4, Supplementary Materials) should provide greater detail by including headings to break down into individual meals and snacks for each day. Providing a separate column for 'oils and spreads' instead of grouping these products in with the 'extras' category, and using standardised serving sizes for each food group listed on the swap list (Item S5, Supplementary Materials) were other recommendations. Participants also emphasised that the resources provided need to be attentive to the health literacy levels of patients and should consider including more pictures.

> "Having it . . . mapped out . . . how many meals they like to have over the day and being able to fit it in that way . . . some people might find it a little bit easier to . . . actually to see how it fits into . . . breakfast, lunch and dinner, or how many meals a day."
>
> (Dietitian 11)
>
> "I think it's good how you have under free vegetables you say one serve is one cup or half a cup cooked. I think its nice to have a standard serve for as many of the foods within a category."
>
> (Dietitian 10)
>
> "Pictures . . . . I think pictures are really important . . . particularly with literacy and also non-English speaking patients."
>
> (Dietitian 7)

4. Discussion

Renal dietitians were aware that encouraging a plant-based diet could benefit individuals with CKD and translate to favourable clinical outcomes. Successful implementation of plant-based diets was perceived to require frequent patient contact and ongoing education and support by a dietitian. Common concerns regarding the use of plant-based diets such as high potassium intakes and protein inadequacy could be alleviated with regular dietetic input. Several studies involving people with or without CKD have demonstrated more than adequate levels of total protein consumption in various types of plant-based diets, including those adhering to a vegan dietary pattern [24]. Although, plant proteins may have insufficient levels of one or more essential amino acids. A recent modelling study [25] found that low-protein, plant-based diets did not meet the recommended dietary allowance for all essential amino acids, reinforcing the need for careful planning and dietetic supervision to ensure the adequacy of all nutrients in the diet. However, some barriers to translation into current practice were identified. The lack of staffing, capacity, and time outlined by renal dietitians in this study is consistent with previous research in North America [26–28], Australia [29], and the United Kingdom [30]. New models of dietetic care may be necessary, as it has been demonstrated that patients seen by renal dietitians have fewer hospitalisations and are associated with delays in dialysis commencement [31]. In this study, dietitians highlighted that a coordinated multidisciplinary team approach was essential for implementing plant-based diets into clinical practice and achieving improved patient outcomes [32,33], particularly in conveying safe and consistent dietary messages. A concerted effort was also required to help harmonise differing nutritional philosophies and contradictory nutritional advice. This is especially important given the rapidly changing field, with recent developments suggesting that higher dietary potassium intake is not associated with hyperkalaemia or death in patients treated with haemodialysis [5]. When prescribing plant-based diet advice, additional attention to comorbid conditions is required. Confusion and miscommunication are commonly heightened in the case of comorbid conditions such as diabetes, which are prevalent in this population group [33], and can lead to increased feelings of anxiety amongst patients [33] as well as uncertainty around the value of diet in CKD management.

There is extensive research to support the need for individualised interventions for patients receiving dietetic care [34]. Dietitians in this study described the nuanced approach that may be beneficial when educating people with CKD about a plant-based diet. Approximately 25% of renal patients have limited health literacy [35] and high rates of cognitive impairment, including those in the pre-dialysis stages [36]. Despite this, nutrition resources for CKD management are often not designed to accommodate these deficits [37]. To help accommodate these challenges, dietitians in this study recommended using pictorial resources with limited text to aid comprehension of technical information. As recommended in previous research, strategies encompassing graduated goal-setting and collaborative decision-making [34] were also suggested to support patient adherence and inspire independence [38]. To overcome the sense of frustration described by patients with CKD when receiving didactic nutrition advice [2], the dietitians in the present study suggested an additional explanation of the benefits of plant-based diets, with explicit details on the anticipated benefits for the individual, were needed at the time of education. This may also help alleviate feelings of uncertainty that have been outlined in previous studies when patients are unclear of the reason for making dietary changes [39]. This idea to promote anticipated benefits is similar to findings from a review [40] in the field of diabetes that summarised the literature relating to the barriers and facilitators identified for implementing plant-based diets.

There are several strengths and limitations to this research. The mixed-methods approach incorporates a survey and semi-structured interviews with two exclusive participant groups, enabling a rich collection of data. The use of open-ended and semi-structured questions in both the survey and interview allowed participants to articulate their perceptions to a greater extent and mitigates the risk of researcher bias. The major limitation of this research is the relatively small sample size for the survey. Furthermore, the perspectives gained may not represent all Australian renal dietitians or those working outside Australia. Hence, the findings offer general theoretical concepts that require further research to verify their applicability to other dietitians working in renal care.

5. Relevance to Practice

Interest in plant-based diets as a therapeutic option for CKD continues to grow globally, and recent studies have explored dietitians [41,42] and nephrology professionals' [43] perceptions of plant-based eating. However, to our knowledge, this is the first study to provide explicit information on what might be required in practice to implement plant-based diets and strategies dietitians could use to support their patients with CKD to consume more plant foods.

Dietitians agreed that plant-based diets are beneficial for patients with chronic kidney disease. The successful implementation of plant-based diets was perceived to require extensive contact and education of patients in conjunction with ongoing support from a dietitian. In addition to educating patients, dietitians also need to consider buy-in from the MDT. Identified strengths of the hypothetical dietary prescription that are translatable to practice included shifting the focus of dietary advice to whole foods and overall healthy eating patterns rather than nutrients; positive framing of nutrition messages that encourage the inclusion of healthy foods; and practical supplementary resources such as recipes. Increasing knowledge about the benefits of the plant-based approach and starting with small incremental dietary changes were also recommended to increase patient acceptance.

Supplementary Materials: The following supporting information can be downloaded at: https://www.mdpi.com/article/10.3390/nu14010216/s1; Item S1. Consolidated criteria for reporting qualitative studies (COREQ): a 32-item checklist; Item S2. Online survey questions; Item S3. Interview Guide; Item S4. Example of the seven-day food-target template; Item S5. Example of the foods swap list.

Author Contributions: Conceptualization, J.S. and A.S.-N.; Methodology, J.S., A.S.-N., K.L., and K.C.; Validation, A.S.-N. and K.C.; Formal analysis, J.S., M.Z., and K.L.; Resources, M.Z., J.S., and A.S.-N.; Data curation, J.S., M.Z., and K.L.; Writing—original draft preparation, J.S.; Writing—review and editing, M.Z., A.S.-N., K.L., and K.C.; Visualisation, J.S.; Supervision, A.S.-N., K.L., and K.C.; Project administration, J.S. and A.S.-N.; Funding acquisition, J.S. and K.L. All authors have read and agreed to the published version of the manuscript.

Funding: This work was conducted with financial support from the Australian Government Research Training Program Scholarship, King and Amy O'Malley Trust, and the Faculty of Science, Medicine and Health, University of Wollongong.

Institutional Review Board Statement: The study was conducted in accordance with the Declaration of Helsinki, and approved by the joint University of Wollongong Human Research Ethics Committee (2019/ETH00397 and 2019/STE00428, approved 6 March 2019).

Informed Consent Statement: Informed consent was obtained from all subjects involved in the study. Tacit consent was implied by online survey completion, and written consent forms were obtained from interview participants.

Data Availability Statement: The data presented in this study are not publicly available in accordance with the type of consent obtained about the use of confidential data.

Acknowledgments: The authors wish to thank all participants involved in this study; their valuable contribution to this research is deeply appreciated.

Conflicts of Interest: The authors declare no conflict of interest. The funders had no role in the design of the study; in the collection, analyses, or interpretation of data; in the writing of the manuscript, or in the decision to publish the results.

References

1. Chan, M.; Kelly, J.; Tapsell, L. Dietary Modeling of Foods for Advanced CKD Based on General Healthy Eating Guidelines: What Should Be on the Plate? *Am. J. Kidney Dis.* **2017**, *69*, 436–450. [CrossRef]
2. Lambert, K.; Mansfield, K.; Mullan, J. How do patients and carers make sense of renal dietary advice? A qualitative exploration. *J. Ren. Care* **2018**, *44*, 238–250. [CrossRef]
3. Naismith, D.J.; Braschi, A. An investigation into the bioaccessibility of potassium in unprocessed fruits and vegetables. *Int. J. Food Sci. Nutr.* **2008**, *59*, 438–450. [CrossRef] [PubMed]
4. Picard, K. Potassium Additives and Bioavailability: Are We Missing Something in Hyperkalemia Management? *J. Ren. Nutr.* **2019**, *29*, 350–353. [CrossRef] [PubMed]
5. Bernier-Jean, A.; Wong, G.; Saglimbene, V.; Ruospo, M.; Palmer, S.C.; Natale, P.; Garcia-Larsen, V.; Johnson, D.W.; Tonelli, M.; Hegbrant, J.; et al. Dietary Potassium Intake and All-Cause Mortality in Adults Treated with Hemodialysis. *Clin. J. Am. Soc. Nephrol.* **2021**, *16*, 1851–1861. [CrossRef]
6. Kelly, J.T.; Palmer, S.C.; Wai, S.N.; Ruospo, M.; Carrero, J.J.; Campbell, K.L.; Strippoli, G.F. Healthy Dietary Patterns and Risk of Mortality and ESRD in CKD: A Meta-Analysis of Cohort Studies. *Clin. J. Am. Soc. Nephrol.* **2017**, *12*, 272–279. [CrossRef] [PubMed]
7. Banerjee, T.; Liu, Y.; Crews, D.C. Dietary Patterns and CKD Progression. *Blood Purif* **2016**, *41*, 117–122. [CrossRef]
8. Guldris, S.C.; Parra, E.G.; Amenós, A.C. Gut microbiota in chronic kidney disease. *Nefrología* **2017**, *37*, 9–19. [CrossRef]
9. Koppe, L.; Fouque, D.; Soulage, C.O. The Role of Gut Microbiota and Diet on Uremic Retention Solutes Production in the Context of Chronic Kidney Disease. *Toxins* **2018**, *10*, 155. [CrossRef] [PubMed]
10. Rossi, M.; Johnson, D.W.; Xu, H.; Carrero, J.J.; Pascoe, E.; French, C.; Campbell, K.L. Dietary protein-fiber ratio associates with circulating levels of indoxyl sulfate and p-cresyl sulfate in chronic kidney disease patients. *Nutr. Metab. Cardiovasc. Dis.* **2015**, *25*, 860–865. [CrossRef] [PubMed]
11. Yang, C.-Y.; Tarng, D.-C. Diet, gut microbiome and indoxyl sulphate in chronic kidney disease patients. *Nephrology* **2018**, *23*, 16–20. [CrossRef] [PubMed]
12. D'Alessandro, C.; Piccoli, G.B.; Cupisti, A. The "phosphorus pyramid": A visual tool for dietary phosphate management in dialysis and CKD patients. *BMC Nephrol.* **2015**, *16*, 9. [CrossRef] [PubMed]
13. Eysenbach, G. Improving the Quality of Web Surveys: The Checklist for Reporting Results of Internet E-Surveys (CHERRIES). *J. Med. Internet Res.* **2004**, *6*, e34. [CrossRef]
14. Tong, A.; Sainsbury, P.; Craig, J. Consolidated criteria for reporting qualitative research (COREQ): A 32-item checklist for interviews and focus groups. *Int. J. Qual. Health Care* **2007**, *19*, 349–357. [CrossRef] [PubMed]
15. Jager, J.; Putnick, D.L.; Bornstein, M.H. More than just convenient: The scientific merits of homogeneous convenience samples. *Monogr. Soc. Res. Child. Dev.* **2017**, *82*, 13–30. [CrossRef] [PubMed]
16. McDonald, D.; Hyde, E.; Debelius, J.W.; Morton, J.T.; Gonzalez, A.; Ackermann, G.; Aksenov, A.A.; Behsaz, B.; Brennan, C.; Chen, Y.; et al. American Gut: An Open Platform for Citizen Science Microbiome Research. *mSystems* **2018**, *3*, e00031-18. [CrossRef] [PubMed]
17. National Health and Medical Research Council. Australian Dietary Guidelines. 2013. Available online: https://www.nhmrc.gov.au/_files_nhmrc/file/publications/n55_australian_dietary_guidelines1.pdf (accessed on 4 January 2022).
18. Ash, S.; Campbell, K.; MacLaughlin, H.; McCoy, E.; Chan, M.; Anderson, K.; Corke, K.; Dumont, R.; Lloyd, L.; Meade, A.; et al. Evidence based practice guidelines for the nutritional management of chronic kidney disease. *Nutr. Diet.* **2006**, *63*, S33–S45. [CrossRef]
19. Ash, S.; Campbell, K.L.; Bogard, J.; Millichamp, A. Nutrition prescription to achieve positive outcomes in chronic kidney disease: A systematic review. *Nutrients* **2014**, *6*, 416–451. [CrossRef]

20. R Core Team. *R: A Language and Environment for Statistical Computing*; R Foundation for Statistical Computing: Vienna, Austria, 2018.
21. Elo, S.; Kyngas, H. The qualitative content analysis process. *J. Adv. Nurs.* **2008**, *62*, 107–115. [CrossRef] [PubMed]
22. *Dedoose Version 8.3.17, w.a.f.m., Analyzing, and Presenting Qualitative and Mixed Method Research Data*; SocioCultural Research Consultants, LLC: Los Angeles, CA, USA, 2020.
23. Braun, V.; Clarke, V. Using thematic analysis in psychology. *Qual Res. Psychol.* **2006**, *3*, 77–101. [CrossRef]
24. Joshi, S.; Shah, S.; Kalantar-Zadeh, K. Adequacy of Plant-Based Proteins in Chronic Kidney Disease. *J. Ren. Nutr.* **2019**, *29*, 112–117. [CrossRef] [PubMed]
25. Khor, B.-H.; Tallman, D.A.; Karupaiah, T.; Khosla, P.; Chan, M.; Kopple, J.D. Nutritional Adequacy of Animal-Based and Plant-Based Asian Diets for Chronic Kidney Disease Patients: A Modeling Study. *Nutrients* **2021**, *13*, 3341. [CrossRef] [PubMed]
26. Hand, R.K. Workforce needs and estimated costs/savings for nutrition care in chronic kidney disease–stage 3 through maintenance dialysis. In *Seminars in Dialisis*; Wiley-Blackwell Publishing Ltd: Hoboken, NJ, USA, 2021; pp. 1–9. [CrossRef]
27. Vergili, J.M.; Wolf, R.L. Nutrition practices of renal dietitians in hemodialysis centers throughout the United States: A descriptive study. *J. Ren. Nutr.* **2010**, *20*, e1–e8. [CrossRef]
28. Trudel, T.; McCune, A.; Donahue, K.; Zuberbuhler, L.; Farmer, A.; Mager, D. Variables influencing adoption of practice-based guidelines in Canadian renal dietetic practice. *J. Ren. Nutr.* **2010**, *20*, 235–242. [CrossRef]
29. Hall-McMahon, E.J.; Campbell, K.L. Have renal dietitians successfully implemented evidence-based guidelines into practice? A survey of dietitians across Australia and New Zealand. *J. Ren. Nutr.* **2012**, *22*, 584–591. [CrossRef] [PubMed]
30. British Renal Society. A Multi-Professional Renal Workforce Plan for Adults and Children with Kidney Disease. 2020. Available online: https://britishrenal.org/wp-content/uploads/2020/10/FINAL-WFP-OCT-2020.pdf (accessed on 15 December 2021).
31. Notaras, S.; Galea, L.; Lee, P.; Mak, M.; Lambert, K.; Makris, A. The association between dietetic consultation and time to dialysis for patients attending a pre-dialysis clinic: A retrospective cohort study. *Nephrology* **2020**, *25*, 390–397. [CrossRef]
32. Goldstein, M.; Yassa, T.; Dacouris, N.; McFarlane, P. Multidisciplinary predialysis care and morbidity and mortality of patients on dialysis. *Am. J. Kidney Dis.* **2004**, *44*, 706–714. [CrossRef]
33. Lee, W.-C.; Lee, Y.-T.; Li, L.-C.; Ng, H.-Y.; Kuo, W.-H.; Lin, P.-T.; Liao, Y.-C.; Chiou, T.T.-Y.; Lee, C.-T. The Number of Comorbidities Predicts Renal Outcomes in Patients with Stage 3-5 Chronic Kidney Disease. *J. Clin. Med.* **2018**, *7*, 493. [CrossRef]
34. Sladdin, I.; Chaboyer, W.; Ball, L. Patients' perceptions and experiences of patient-centred care in dietetic consultations. *J. Hum. Nutr. Diet.* **2018**, *31*, 188–196. [CrossRef]
35. Taylor, D.M.; Fraser, S.; Dudley, C.; Oniscu, G.C.; Tomson, C.; Ravanan, R.; Roderick, P.; the ATTOM Investigators. Health literacy and patient outcomes in chronic kidney disease: A systematic review. *Nephrol. Dial. Transplant.* **2017**, *33*, 1545–1558. [CrossRef]
36. Lambert, K.; Mullan, J.; Mansfield, K.; Lonergan, M. Comparison of the extent and pattern of cognitive impairment among predialysis, dialysis and transplant patients: A cross-sectional study from Australia. *Nephrology* **2017**, *22*, 899–906. [CrossRef] [PubMed]
37. Morony, S.; Flynn, M.; McCaffery, K.J.; Jansen, J.; Webster, A.C. Readability of Written Materials for CKD Patients: A Systematic Review. *Am. J. Kidney Dis.* **2015**, *65*, 842–850. [CrossRef] [PubMed]
38. Beto, J.A.; Schury, K.A.; Bansal, V.K. Strategies to promote adherence to nutritional advice in patients with chronic kidney disease: A narrative review and commentary. *Int. J. Nephrol. Renov. Dis.* **2016**, *9*, 21–33. [CrossRef] [PubMed]
39. Walker, R.; James, H.; Burns, A. Adhering to behaviour change in older pre-dialysis populations-what do patients think? A qualitative study. *J. Ren. Care* **2012**, *38*, 34–42. [CrossRef]
40. Rinaldi, S.; Campbell, E.E.; Fournier, J.; O'Connor, C.; Madill, J. A Comprehensive Review of the Literature Supporting Recommendations From the Canadian Diabetes Association for the Use of a Plant-Based Diet for Management of Type 2 Diabetes. *Can. J. Diabetes* **2016**, *40*, 471–477. [CrossRef]
41. Asher, K.E.; Doucet, S.; Luke, A. Registered dietitians' perceptions and use of the plant-based recommendations in the 2019 Canada's Food Guide. *J. Hum. Nutr. Diet.* **2021**, *34*, 715–723. [CrossRef]
42. Janse Van Rensburg, L.M.; Wiles, N.L. The opinion of KwaZulu-Natal dietitians regarding the use of a whole-foods plant-based diet in the management of non-communicable diseases. *S. Afr. J. Clin. Nutr.* **2021**, *34*, 60–64. [CrossRef]
43. Betz, M.V.; Nemec, K.B.; Zisman, A.L. Plant-based Diets in Kidney Disease: Nephrology Professionals' Perspective. *J. Ren. Nutr.* 2021, *in press*. [CrossRef]

nutrients

MDPI

Article

Plant-Based Diets Improve Maternal–Fetal Outcomes in CKD Pregnancies

Rossella Attini [1], Filomena Leone [1], Antoine Chatrenet [2,3], Elisa Longhitano [2,4], Viola Casula [1], Alice Tomasi Cont [1], Gaia Zaccaria [1], Eleonora Dalmasso [1], Ana Maria Manzione [5], Bianca Masturzo [1], Massimo Torreggiani [2], Alberto Revelli [1], Gianfranca Cabiddu [6] and Giorgina Barbara Piccoli [2,*]

[1] Department of Obstetrics and Gynecology SC2U, Città della Salute e della Scienza, Sant'Anna Hospital, 10126 Turin, Italy
[2] Néphrologie et Dialyse, Centre Hospitalier Le Mans, 72037 Le Mans, France
[3] Laboratory "Movement, Interactions, Performance" (EA 4334), Le Mans University, 72000 Le Mans, France
[4] Department of Clinical and Experimental Medicine, Unit of Nephrology and Dialysis, A.O.U. "G. Martino", University of Messina, 98124 Messina, Italy
[5] Nephrology, Dialysis and Renal Transplantation Division, Department of Medical Sciences, "Città della Salute e della Scienza" Hospital, University of Turin, 10100 Turin, Italy
[6] Nephrology, Department of Medical Science and Public Health, San Michele Hospital, G. Brotzu, University of Cagliari, 09047 Cagliari, Italy
* Correspondence: gbpiccoli@yahoo.it

Citation: Attini, R.; Leone, F.; Chatrenet, A.; Longhitano, E.; Casula, V.; Tomasi Cont, A.; Zaccaria, G.; Dalmasso, E.; Manzione, A.M.; Masturzo, B.; et al. Plant-Based Diets Improve Maternal–Fetal Outcomes in CKD Pregnancies. *Nutrients* **2022**, *14*, 4203. https://doi.org/10.3390/nu14194203

Academic Editors: Winston Craig and Ujué Fresán

Received: 27 July 2022
Accepted: 2 October 2022
Published: 9 October 2022

Publisher's Note: MDPI stays neutral with regard to jurisdictional claims in published maps and institutional affiliations.

Abstract: Reducing protein intake in patients with chronic kidney disease (CKD) limits glomerular stress induced by hyperfiltration and can prevent the progression of kidney disease; data in pregnancy are limited. The aim of this study is to analyze the results obtained in CKD patients who followed a plant-based moderately protein-restricted diet during pregnancy in comparison with a propensity-score-matched cohort of CKD pregnancies on unrestricted diets. A total of 52 CKD pregnancies followed up with a protein-restricted plant-based diet (Torino, Italy) were matched with a propensity score based on kidney function and proteinuria with CKD pregnancies with unrestricted protein intake (Cagliari Italy). Outcomes included preterm (<37 weeks) and very preterm (<34 weeks) delivery and giving birth to a small-for-gestational-age baby. The median age in our cohort was 34 years, 63.46% of women were primiparous, and the median body mass index (BMI) was 23.15 kg/m^2 with 13.46% of obese subjects. No statistical differences were found between women on a plant-based diet and women who were not in terms of age, parity, BMI, obesity, CKD stage, timing of referral, or cause of CKD. No differences were found between the two groups regarding the week of delivery. However, the combined negative outcome (birth before 37 completed gestational weeks or birth-weight centile <10) occurred less frequently in women following the diet than in women in the control group (61.54% versus 80.77%; $p = 0.03$). The lower risk was confirmed in a multivariable analysis adjusted for renal function and proteinuria (OR: 0.260 [Q1:0.093–Q3:0.724]; $p = 0.010$), in which the increase in proteinuria from the first to the last check-up before delivery was lower in patients on plant-based diets (median from 0.80 to 1.87 g/24 h; p: ns) than in controls (0.63 to 2.39 g/24 h $p < 0.0001$). Plant-based, moderately protein-restricted diets in pregnancy in patients with CKD are associated with a lower risk of preterm delivery and small-for-gestational-age babies; the effect may be mediated by better stabilization of proteinuria.

Keywords: pregnancy complications; preterm delivery; small for gestational age; preeclampsia; chronic kidney disease; plant-based diets

1. Introduction

Chronic kidney disease (CKD) is defined as a reduction in the kidney function corresponding to an estimated glomerular filtration rate (eGFR) below 60 mL/min or any alteration of renal morphology, including malformations, kidney stones, or kidney scars, or a change in the composition of the urine (such as proteinuria or hematuria) or the blood (such as renal tubular acidosis), independently from kidney function, lasting for at least 3 months [1]. It is estimated that it is present in about 3% of all pregnancies, with about 1:750 occurring in women in an advanced CKD stage (eGFR < 60 mL min, corresponding to CKD stages 3-4-5) [2,3].

Independently from kidney function, the presence of kidney damage, or a reduction in kidney tissue, it is now known to be associated with an increased incidence of adverse pregnancy outcomes, including preterm delivery, preeclampsia, the hypertensive disorders of pregnancy, and giving birth to a small-for-gestational-age (SGA) baby [4–6]. This is also true of those who have a healthy single kidney (as is the case of kidney donors), and is observed, among others, in patients with kidney stones, and in cases with a previous episode of acute kidney injury (AKI), even after complete normalization of renal function [7–10]. The risk of adverse pregnancy outcomes is also higher in patients with kidney damage without hypertension, proteinuria, or loss of kidney function, and are further modulated by these three elements. As expected, risks increase in the presence of pregestational hypertension and proteinuria and in proportion to reduction in kidney function [11–13].

Once a critical reduction in kidney tissue has been reached, a vicious circle is generated, leading to glomerular hypertension and hyperfiltration on the remnant nephrons, causing hypertrophy and ultimately proteinuria and glomerular sclerosis; proteinuria is further associated with glomerular sclerosis in a maladaptive vicious circle [14,15]. On account of this pathogenesis, the main approaches for the preservation of kidney function are aimed at reducing hyperfiltration and intraglomerular pressure and include angiotensin-converting enzyme inhibitors (ACEi) and angiotensin receptor blockers (ARBs), as well as diets with a reduced protein content [16,17].

The definition of low-protein diets has changed over time, in line with the changes in the definition of the recommended daily protein allowances, which are usually taken as a reference of a "normal" protein content in diets. Recommended daily allowances have progressively decreased to the current indication of 0.8 g of proteins per kg of body weight per day, and the current recent guidelines of the Kidney Disease Outcome Quality Initiative (K-DOQI) regarding nutritional management in CKD patients define low-protein diets as those with a protein content of 0.6 to 0.4 g/kg/day, and establish that the energy intake should be 25–35 kcal/kg of ideal body weight per day, according to age and metabolic needs [18].

In recent years there has been growing interest in plant-based diets, given their reduced phosphate content, lower induction of acidosis, and lower probability of triggering hyperfiltration compared to diets with the same amount of animal-derived proteins [19–21].

The recent K-DOQI guidelines on nutritional management in CKD patients underline the interest in plant-based diets, not only citing their potential advantages, but also clearly stating that the "there is insufficient evidence to recommend a particular protein type (plant vs. animal) in terms of the effects on nutritional status, calcium, or phosphorus levels, or the blood lipid profile." This statement (3.2 Statement on Protein Type [18]) acknowledges that well-balanced plant-based diets are no longer considered as dangerous or inferior in CKD patients [18–21].

Plant-based diets are likewise an acknowledged option in pregnancy, provided that they are correctly followed, thus controlling for the complementarity of aminoacids and avoiding nutritional deficits, in particular those of vitamins (B12, vitamin D) and iron [22–26]. Furthermore, some data suggest that plant-based diets in pregnancy may even protect from the development of preeclampsia and hypertensive disorders of pregnancy, possibly through the prevention of excessive weight gain in pregnancy [25,26].

Pregnancy is a well-known cause of physiological renal hyperfiltration; an increase in glomerular filtration is also observed in CKD patients, where it is seen as a sort of "stress test" and is associated with an increase in proteinuria. Given the safety risks posed to the fetus, angiotensin converting enzyme inhibitors (ACEi) and angiotensin receptor blockers (ARBs) are discontinued, which further worsens proteinuria in CKD pregnancies [27,28].

In this context, our team and a few others tried to counterbalance pregnancy-induced hyperfiltration and reduce proteinuria by employing plant-based diets with moderately reduced protein intake during pregnancy [29–33].

The initial results were encouraging both in terms of fetal growth and period of delivery [29,30].

The aim of this study was to compare the results obtained in 52 CKD pregnancies, resulting in singleton deliveries, who followed a plant-based moderately protein-restricted diet during pregnancy, with a propensity-score-matched cohort of CKD pregnancies on unrestricted diets.

2. Materials and Methods

2.1. Settings of the Study

The present study was undertaken in two Italian centers, which began a conjoint study (TOCOS: Torino Cagliari Observational Study) on CKD and pregnancy in 2000: Turin (Piemonte, northern Italy), where the on-diet cases were followed, and Cagliari (the island of Sardinia) where the controls were selected [5].

In Turin, an industrial city with about 800,000 inhabitants, the study was performed at Ospedale Sant'Anna, a tertiary-care hospital that is part of the Città della Salute e della Scienza. With an average of 7000 deliveries per year, Sant'Anna is one of the largest European tertiary-care obstetric facilities. Since 2000, the center has run an outpatient facility for CKD pregnancies and acute kidney problems in pregnancy. The dietary approach with a moderately protein-restricted, plant-based diet has been employed since then [29,30].

The Azienda Ospedaliera Brotzu (AOB), Cagliari, is the largest hospital in Sardinia, an island with about 1.6 million inhabitants. The obstetric ward follows 800–1000 deliveries per year, offering care for high-risk pregnancies (thalassemia, diabetes, and kidney and autoimmune diseases). A nephrology outpatient service specializing in kidney diseases in pregnancy has been operative since 1995.

While both centers follow common protocols of control visits and care [28], the Cagliari center does not employ moderately protein-restricted, plant-based diets during pregnancy. For the sake of this study, the cases followed up in Torino were matched via a propensity score with controls selected in Cagliari, as subsequently specified.

2.2. Definitions Employed

Chronic kidney disease was defined according to the 2002 Kidney Disease Outcomes Quality Initiative (KDOQI) classification and stratification: estimated glomerular filtration rate (eGFR) < 60 mL/min/1.73 m^2 for ≥ 3 months, or kidney damage for ≥ 3 months (Supplementary Table S1). The latter was defined as structural or functional anomalies of the kidneys, with or without decreased GFR, manifested either as pathological abnormalities or markers of kidney damage, including abnormalities in the composition of blood or urine, or abnormalities in imaging tests [1].

Since pre-pregnancy data were available for only a minority of the patients in the study, the definitions of stages were performed using data at referral.

Preeclampsia (PE) and hemolysis, elevated liver enzymes, and low platelets syndrome (HELLP) were defined according to the ACOG guidelines [34]. PE was diagnosed with hypertension (sBP $\geq$ 140 mm Hg and/or dBP$\geq$90 mm Hg, on two occasions at least 4 h apart, with no underlying cause, in a woman with previously normal blood pressure), associated with proteinuria (24 h excretion $\geq$300 mg), diagnosed after 20 weeks of gestation or in the absence of proteinuria, with new onset of any of the following: platelet count <100,000/µL, serum creatinine >1.1 mg/dl, or doubling of its concentration in absence of other renal disease, transaminases to twice normal concentration of liver enzymes, pulmonary edema, cerebral/visual symptoms [34,35].

HELLP syndrome was defined in accordance with the above guidelines (alanine or aspartate transaminase levels $\geq$ twice the upper limit of normal; lactate dehydrogenase $\geq$ 600 U/L; platelet count <100,000/µL) [34]. Superimposed preeclampsia was defined as preeclampsia on already-known treated or untreated pregestational hypertension, or on already-known CKD [35].

Small-for-gestational-age babies were defined in accordance with the two most commonly used cut-points: below the 5th and 10th centiles, following INTERGROWTH standards [36].

Preterm delivery was defined as delivery before 37 completed gestational weeks; early preterm delivery as before 34; and very early preterm delivery as delivery before 28 completed gestational weeks.

Obesity was defined as a pregestational body mass index (BMI) equal to or above 30 kg/m^2; overweight was defined as BMI between 25 and 30 kg/m^2; and underweight was defined as pregestational BMI < 20 kg/m^2.

Gestational hypertension was defined in line with the current guidelines [34–38].

Antihypertensive treatment employed a combination of alphamethyl-dopa and nifedipine, adding doxazosin and small doses of diuretics or clonidine when needed. Treatment was adjusted at every clinical visit with a target of 120–130/60–70 mm Hg [28]. In both settings of care, the frequency of nephrology and obstetric visits, blood and urine tests and biometric and Doppler studies of the uterine and umbilical arteries were tailored to individual needs (visits: 1 week–1 month apart, biometry and fetal Doppler according to fetal growth and the presence of Doppler anomalies), in keeping with the Italian best practices in pregnant CKD patients [28].

Low-dose acetylsalicylate was increasingly prescribed, first to patients with severe proteinuria, and more recently to all women with at-risk pregnancies.

2.3. Indications for the Diet

The main indications for plant-based diets in pregnancy were progressively broadened from subjects with CKD Stages 4–5 and/or nephrotic syndrome to include: patients already on a plant-based diet before pregnancy; CKD Stage 3 to 5 patients not on dialysis; CKD Stage 1 and 2 patients with kidney function impairment during pregnancy; proteinuria above 3 g/day at any time (<30 gestational weeks) of pregnancy, or proteinuria above 1 g/day at referral or in the first trimester; previous nephrotic syndrome, increase in or development of proteinuria in pregnancy; or a combination of any of these elements. The following contraindications were considered: anorexia, hyperemesis gravidarum, language barriers impairing understanding the diet and its aim, psychiatric disorders, low compliance to prescriptions and controls.

2.4. Selection of the Control Group

In line with the criteria used to prescribe a plant-based diet, we selected the controls in the TOCOS database solely for patients being followed up in Cagliari.

The patients of the control group did not undergo a specific nutritional workout in pregnancy, and their protein intake was unrestricted. Patients were referred to the dieticians in the case of excessive weight gain or in the case of hyperemesis. Their follow-up is in keeping with the usual indications of good clinical practice [28,39].

The control patients were chosen using a propensity-match score, based on CKD stage (1, 2, 3, 4, and 5 considered together) and level of proteinuria, dichotomized at 1 g at the first clinical control visit during pregnancy.

2.5. The Plant-Based, Moderately Protein-Restricted Diet

The plant-based diet is based on a simplified schema.

Food is chosen according to a qualitative approach (allowed/forbidden) and the patients are not obliged to regularly weigh food. The dietician controls the quality, quantity, and integration of proteins every four to eight weeks, based upon a 5-day food diary. The use of food diary (5 days instead of the 3 recommended ones, according to the KDOQI guidelines, to improve the quality of the reporting) is chosen, in the absence of other agreed methods, to indirectly evaluate nutrient intake in pregnancy. The frequency of the dietary controls is personalized, according to the clinical status, and the dietitian sees the patients in the occasion of their conjoint nephrology and obstetric control.

When the diet is first prescribed, a short-term consultation (1–2 weeks) is organized, either in person or by phone. Subsequently, the frequency ranges from every 8 weeks in stable patients, with regular weight gain and fetal growth, who follow the diet without declaring difficulties, doubts or problems; to every 2–4 weeks, in particular in cases with either no weight gain, or important weight gain in which—as it is usual in pregnancy—the discrimination between lean weight and fluid overload is not simple, in the absence of overt oedema, and without being able to rely on bioimpedance, whose use is not standardized in pregnancy.

The goal is to control protein intake to 0.8 g/kg/day (real weight) with over 80% of protein coming from plant-derived sources. In normal individuals, 0.8 g/kg/day of proteins is now considered the reference intake (daily allowances); in pregnancy, in patients on plant-based diets, it is advised to increase by 20%, i.e., to 1 g/kg/day [25]. In CKD patients, the recommended intake of 0.6 g/kg/day, taking as reference the recent K-DOQI guidelines, was likewise increased; to be on the safe side, the increase was established at around 30% (rounded at 0.8 g/kg/day) [28–30].

While the overall approach remained the same throughout the period of study, partly on account of changes in definitions of diets with little or no animal-derived proteins in the overall population, we now use the term "plant-based" rather than "vegan-vegetarian" [28–30,40].

The most recent version of the diet is reported in the supplementary material.

The energy intake followed the usual indications in pregnancy. In accordance with the current recommendations, the target was a weight increase ranging between 11.5 kg and 16 kg, adapted for underweight and overweight women [41]. The energy intake followed the usual indications in pregnancy. No variation was advised in plant-based diets in pregnancy [25]. Based on the kidney function, proteinuria levels, and the individual patient's needs and preferences, we allowed 1–3 unrestricted meals per week (without protein restriction but limited in unsaturated fats and short-chain sugars, as indicated in pregnancy). To facilitate compliance with a plant-based diet without the need to eat pulses and cereals at each meal, as it is necessary to avoid deficits of specific aminoacids (in particular lysine), in most cases, supplementation of alpha-keto analogues and aminoacids (Alpha-Kappa or Ketosteril) was added (1 tablet per each 8–10 kg of pregestation body weight) [28–33]. This choice was indeed made to remain "on the safe side" and to avoid the risk of deficits of specific aminoacids in patients who were not used to a plant-based regiment before pregnancy. However, in some patients in which the pill burden was felt to be too high or the supplementation was not tolerated (gastrointestinal discomfort, 1 case) a standard plant-based diet, or a lower pill number, was prescribed, under strict nutritional surveillance [42].

The food distribution and the choice of specific food is highly individualized. For example, for energy intake, a wide choice is discussed, including choosing among the main sources of long-chain carbohydrates; bread and pasta, the bases of the Italian cuisine, are usually widely employed. However, rice, couscous, potatoes, sweet potatoes, and polenta may be either occasional alternatives or the basis of the diet for patients who prefer them.

Olive oil, once more in line with the Mediterranean cuisine and widely available in our country, is preferred; butter and margarine are avoided, but sunflower oil or colza oil are occasionally used by patients who do not like olive oil. Likewise, the choice between the pulses is personalized, and all efforts are made to make the "plant-based" menu as varied as possible. Furthermore, the allowance of unrestricted meals, likewise with personalized frequency, increases the variety, and once again, the choice between meat, fish, poultry, eggs, or cheese, or a combination of animal-derived food, is left free in order to also reduce the psychological constrain of such a demanding diet.

An example of a plant-based diet used in our setting is shown in Supplementary Table S4.

Given the lack of indications on salt restriction in CKD pregnancy, we did not restrict salt intake, but we recommended moderate sodium reduction to patients with severe edema.

Iron status, B12, and 25-OH vitamin D were controlled at baseline and up to monthly tests if needed; vitamins and iron supplements were employed on the basis of blood test results. Erythropoietin was used when needed, with a hemoglobin target of 10 g/dL on account of the physiological hemodilution found in pregnancy.

2.6. Statistical Evaluation

Source of data: in each setting, data on all patients referred with known, newly-diagnosed, or suspected CKD in pregnancy were prospectively recorded in dedicated databases, which were periodically merged, with a final coherence control performed by a trained statistician. Prescription of the moderately protein-restricted plant-based diet is recorded in the database. Data about multiple pregnancies and miscarriages and pregnancy terminations were gathered, but not considered in the statistical analysis. The full list of data gathered is available on request.

The propensity match considered proteinuria (dichotomized at 1 g at baseline) and CKD stage, by means of a greedy 1:1 algorithm. Matching was performed using the "Matchit" R package v 4.2.0 [43].

Continuous series were tested for normality using the Shapiro–Wilk test, and homoscedasticity with Leven's test. According to the conditions of application, for comparing two groups (e.g., seen in nephrology, not seen in nephrology), the independent Student *t*-test and Wilcoxon rank-sum test were used. To compare 3 or more groups, one-way ANOVA and the Kruskal–Wallis test were used.

The comparison of proportions was made with the Chi-squared or the Fisher test, depending on the size of subsample involved. Results are displayed with the median and the interquartile range (IQR, or Q1–Q3 quartiles) or as mean and standard deviation, as appropriate.

The following outcomes were tested by univariable and multivariable methods: birth centile <10 and <5, preterm delivery (all: <37, early: <34 and very early: <28 complete gestational weeks); the outcomes were combined into severe (birth weight<5 centile or delivery <28 weeks) and general (birth weight<10 centile or preterm delivery <37 weeks). The choice of the covariates to include in the multivariate model was based on either statistical significance at the univariate analysis, or well-acknowledged clinical relevance, for instance, diet (plant-based vs. unrestricted); CKD stage (1, 2, 3, 4, and 5 considered together); proteinuria at the first control visit during pregnancy (<1 g vs. $\geq$1 g/24 h), and hypertension (present/absent).

The model employed a backward deletion method and standardized residuals were verified.

Temporal series (e.g., weeks at delivery) were visually analyzed using inverse Kaplan–Meier curves and differences were tested using the log-rank test.

The statistical analysis was performed with JASP version 0.14.1 (Armstrong, The Netherlands, EU) and RStudio version 3.3.0 (Rstudio Project, Boston, MA, USA).

Alpha error was fixed at 5%.

2.7. Ethical Issues

This observational study on current clinical practice was approved by the Ethics Committee of the OIRM Sant'Anna (n° practice 335; n° protocol 11551/c28.2 del 4/3/2011). All patients signed a dedicated informed consent at the first control visit during pregnancy.

Availability of data and materials: TOCOS is a dynamic database updated in real time; the most recent update can be obtained by sending a motivated request to the contact author.

3. Results

3.1. Baseline Data

The 52 pregnancies on the plant-based diet considered for the analysis were selected from 61 pregnancies, after the exclusion of one twin pregnancy, and of 8 cases with miscarriages or pregnancy terminations (the flow chart of the study is reported in Figure 1).

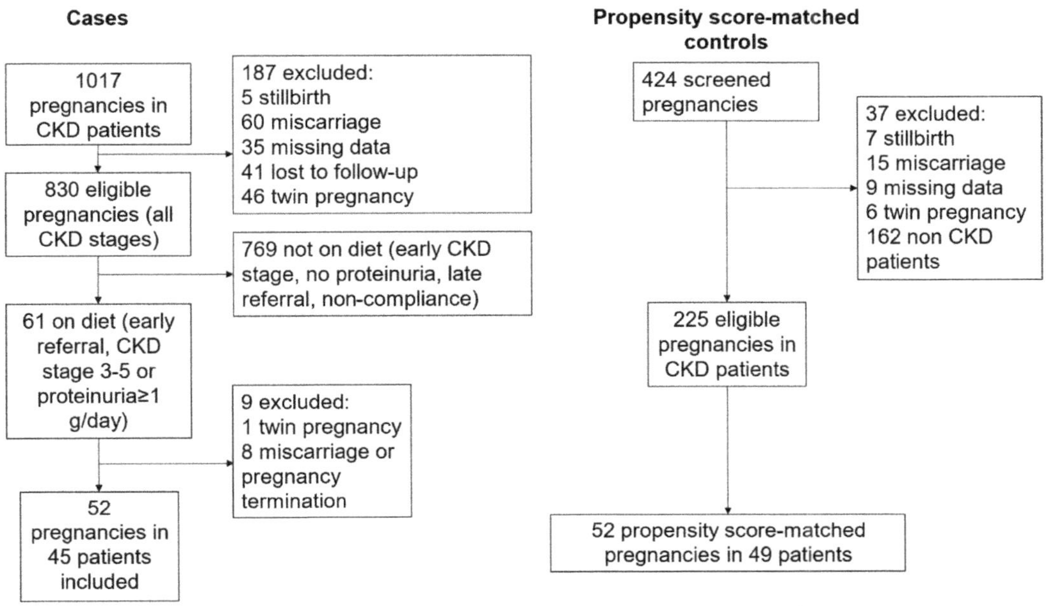

Figure 1. Study flow chart.

Table 1 reports the baseline data in the propensity-matched cohorts selected from the Cagliari database.

Table 1. Baseline data in CKD pregnancies, according to prescription of a plant-based diet.

	All	No Diet	Plant-Based Diet	*p*-Values
Overall data CKD				
N	104	52	52	
Baseline data				
Age (years), median [Q1–Q3]	34 [31.75–38]	34.5 [33–38]	34 [30.75–38]	0.533
Parity (primiparous), *n* (%)	66 (63.46%)	36 (69.23%)	30 (57.69%)	0.222
BMI (kg/m^2), median [Q1–Q3]	23.15 [20.9–26.62]	22.9 [20.19–26.04]	23.63 [21.48–26.62]	0.485
BMI ≥ 30 kg/m^2, *n* (%)	14 (13.46%)	7 (13.46%)	7 (13.46%)	1
Ethnicity (non-Caucasian), *n* (%)	95 (91.35%)	1 (1.92%)	8 (15.39%)	**0.015**
Baseline kidney function data				
Serum creatinine, median [Q1–Q3]	1.02 [0.73–1.39]	0.99 [0.65–1.31]	1.04 [0.79–1.41]	0.301
eGFR CKD-EPI (mL/min), median [Q1–Q3]	71.21 [40.42–106.40]	72.04 [50.75–113.48]	69.72 [48.68–98.37]	0.435
Stage 1, *n* (%)	31 (29.81%)	16 (30.77%)	15 (28.85%)	
Stage 2, *n* (%)	33 (31.73%)	16 (30.77%)	17 (32.69%)	
Stage 3, *n* (%)	19 (18.27%)	11 (21.15%)	8 (15.39%)	0.514
Stage 4, *n* (%)	15 (14.42%)	8 (15.39%)	7 (13.46%)	
Stage 5, *n* (%)	6 (5.775)	1 (1.92%)	5 (9.62%)	
Proteinuria (g/24 h), median [Q1–Q3]	0.705 [0.24–2.06]	0.63 [0.21–1.76]	0.80 [0.29–2.18]	0.196
Proteinuria < 0.5 g/24 h, *n* (%)	38 (36.54%)	20 (38.46%)	18 (34.62%)	
Proteinuria 0.5–1 g/24 h, *n* (%)	24 (23.08%)	12 (23.08%)	12 (23.08%)	0.739
Proteinuria 1–3 g/24 h, *n* (%)	26 (25%)	14 (26.92%)	12 (23.08%)	
Proteinuria ≥ 3 g/24 h, *n* (%)	16 (15.39%)	6 (11.54%)	10 (19.23%)	

Table 1. *Cont.*

	All	No Diet	Plant-Based Diet	p-Values
Timing of referral				
Week at referral, median [Q1–Q3]	8 [6–12]	9 [7–12.25]	7.5 [6–12]	0.201
<12 gestational weeks, n (%)	73 (70.195)	36 (69.23%)	37 (71.15%)	
13–23 gestational weeks, n (%)	8 (7.69%)	5 (9.62%)	3 (5.77%)	0.757
≥24 gestational weeks, n (%)	23 (22.12%)	11 (21.15%)	12 (23.08%)	
Cause of CKD				
Glomerular (primary and secondary GN), n (%)	54 (51.92%)	27 (51.92%)	27 (51.92%)	1
Single kidney, n (%)	3 (2.89%)	2 (3.85%)	1 (1.93%)	0.558
Diabetic nephropathy, n (%)	14 (13.46%)	5 (9.62%)	9 (17.31%)	0.250
ADPKD, n (%)	6 (5.77%)	4 (7.69%)	2 (3.85%)	0.400
Kidney graft, n (%)	9 (8.65%)	5 (9.62%)	4 (7.69%)	0.727
Interstitial (includes interstitial nephropathies, kidney stones, CAKUT and urologic malformations), n (%)	10 (9.62%)	5 (9.62%)	5 (9.62%)	1
Other, n (%)	8 (7.96%)	4 (7.69%)	4 (7.69%)	1

Legend: N, cohort size; BMI, body mass index; eGFR, estimated glomerular filtration rate; GN, glomerulonephritis; ADPKD, autosomal dominant polycystic kidney disease; CAKUT, congenital anomalies of the kidneys and urinary tract; APN, acute pyelonephritis. In bold, significant differences.

Even though the matching considered only CKD stage and proteinuria, the two cohorts were perfectly matched for all the parameters considered, except for ethnicity (non-Caucasian ethnicity was more frequent in Torino, a large industrial city, than in Cagliari, a smaller city in a prevalently agricultural area). In both settings, more than half of the patients were affected by glomerulonephritis, and over one-third were in CKD Stage 3 or higher at their first control visit during pregnancy.

3.2. Pregnancy Outcomes, According to Diet Prescription

Table 2 reports on the main pregnancy outcomes found in the two groups (on diet/not on diet).

While a tendency towards longer pregnancy duration was observed in on-diet patients (preterm delivery: 42.31% on diet vs. 28.85% control group), statistical significance was reached only for delivery <28 weeks (5.77% on-diet patient vs. 11.54% controls, $p = 0.012$).

The combined general outcome (birth before 37 completed gestational weeks and birth centile <10) was observed in 61.54% of the pregnancies in the on-diet group versus 80.77% in the control group ($p = 0.030$).

No on-diet patient and no patient in the control group died, and no mother in either group started dialysis in the first 3 months after delivery. Three neonatal deaths were recorded in the control group, all linked to prematurity, while none were observed in the on-diet group.

None of the surviving children had severe malformations: no malformations were recorded in the control group, while in the on-diet group two newborns presented interatrial septal defects, linked to prematurity, which in both cases spontaneously resolved; and one baby presented ankyloglossia. In addition, one case of complex cardiac anomaly was recorded in a twin pregnancy (excluded from the present analysis) in the on-diet group; the mother was affected by type 1 diabetes, and the child died in the first week of life.

The Kaplan–Meier curve of timing of delivery according to whether or not a diet was prescribed is shown in Figure 2. The curve supports the noninferiority of plant-based diets for the duration of gestation and seems to indicate that they contribute to reducing the incidence of early preterm delivery.

Table 2. Main outcomes according to the diet prescribed.

	All	No Diet	Plant-Based Diet	*p*-Values
Overall data CKD				
N	104	52	52	
Renal data at last control visit				
Serum creatinine, median [Q1–Q3]	1.1 [0.74–1.64]	1 [0.7–1.56]	1.18 [0.81–1.65]	0.194
Proteinuria (g/24 h), median [Q1–Q3]	1.97 [0.58–4.46]	2.39 [0.56–5.6]	1.87 [0.70–3.45]	0.338
eGFR CKD-EPI (mL/min), median [Q1–Q3]	65.62 [41.66–105.01]	72.93 [42.8–113.22]	62.16 [40.65–95.20]	0.302
Stage shift (increase of at least 1 CKD stage), *n* (%)	21 (20.19%)	11 (21.15%)	10 (19.23%)	0.807
Delivery				
Week of delivery, median [Q1–Q3]	36 [33–37]	34.5 [32–37]	36 [33–37]	0.164
Term $\geq$ 37 gw, *n* (%)	37 (35.58%)	15 (28.85%)	22 (42.31%)	0.152
Term < 34 gw, *n* (%)	36 (34.62%)	22 (42.31%)	14 (26.92%)	0.099
Term < 32 gw, *n* (%)	14 (13.46%)	9 (17.31%)	5 (9.62%)	0.250
Term < 28 gw, *n* (%)	6 (5.77%)	6 (11.54%)	0	**0.012**
Offspring data				
Weight at delivery, median [Q1–Q3]	2380 [1797–2820]	2350 [1737.5–2727.5]	2537.5 [1957.5–2872.5]	0.254
Weight < 2500 g, *n* (%)	54 (51.92%)	29 (55.77%)	25 (48.08%)	0.432
Weight < 1500 g, *n* (%)	15 (14.42%)	10 (19.23%)	5 (9.62%)	0.163
Centile, median [Q1–Q3]	36.30 [9.45–59.03]	32.84 [6.29–57.33]	38.81 [14.74–62.08]	0.270
Centile < 10, *n* (%)	28 (26.92%)	18 (34.62%)	10 (19.23)	0.077
Centile < 5, *n* (%)	19 (18.27%)	12 (23.08%)	7 (13.46%)	0.205
Pregnancy-related outcomes				
PE, *n* (%)	3 (2.89%)	3 (5.77)	0	0.079
Combined outcomes				
Term < 37 gw or Centile < 10, *n* (%)	74 (71.15%)	42 (80.77%)	32 (61.54%)	**0.030**
Term < 34 gw or Centile < 10, *n* (%)	50 (48.08%)	29 (55.77%)	21 (40.39%)	0.116
Term < 34 gw or Centile < 5, *n* (%)	43 (41.35%)	24 (46.15%)	19 (36.54%)	0.319
Term < 28 gw or Centile < 5, *n* (%)	20 (19.23%)	13 (25%)	7 (13.46%)	0.135

Legend: N, cohort size; eGFR, estimated glomerular filtration rate; PE, preeclampsia; gw, gestational week. In bold, significant differences.

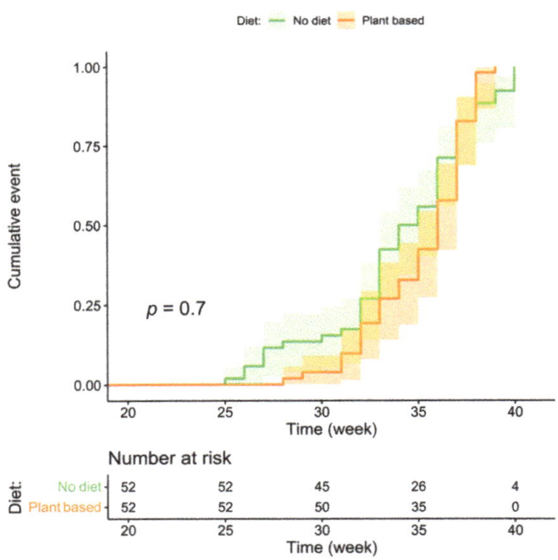

Figure 2. Kaplan–Meier curve for week of delivery according to the patient's diet.

3.3. Variations in Proteinuria and eGFR from the First to Last Control Visit during Pregnancy

Figures 3 and 4 summarize differences in eGFR and proteinuria from referral to delivery, in on-diet cases, and controls. The single-patient graphs are available in Supplementary Figure S1.

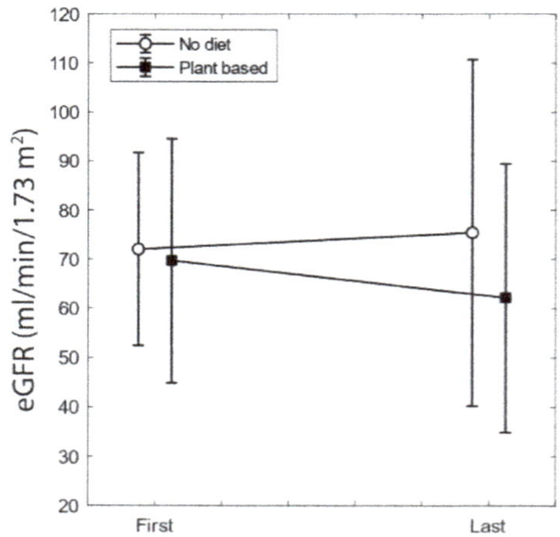

eGFR CKD-EPI (ml/min)	First	Last	*p*-value post-hoc First vs. Last
Controls, median [Q1-Q3]	72.04 [50.75-113.48]	75.45 [42.8-113.22]	0.629
Plant-based diet, median [Q1-Q3]	69.72 [48.68-98.37]	62.16 [40.65-95.20]	0.077

Figure 3. eGFR at the first and last control visit in pregnancy.

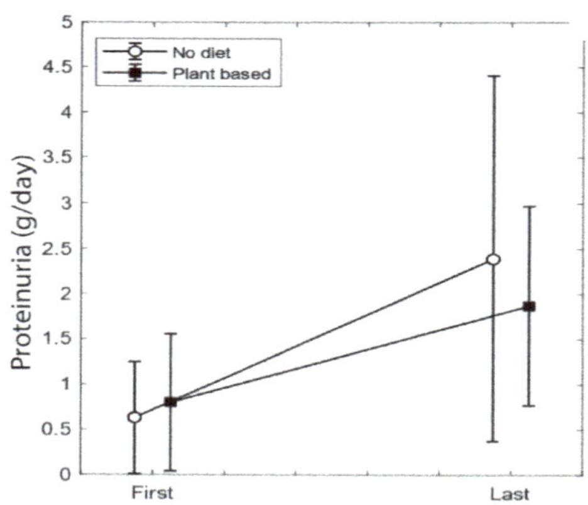

Proteinuria (g/day)	First	Last	*p*-value post-hoc First vs. Last
Controls, median [Q1-Q3]	0.63 [0.21-1.76]	2.39 [0.56-5.60]	<0.001
Plant-based diet, median [Q1-Q3]	0.80 [0.29-2.18]	1.87 [0.70-3.45]	0.115

Figure 4. Proteinuria variation during pregnancy.

While differences from referral to the last control visit during pregnancy were not significant in either group, the increase in proteinuria from the first to the last visit is statistically significant in controls (from 0.63 to 2.39 g/24 h $p < 0.0001$), but nonsignificant in patients on plant-based diets, despite a higher baseline value (median from 0.80 to 1.87 g/24 h).

3.4. Logistic Regression Analysis

The univariable analysis for the different outcomes (preterm delivery and small for gestational age newborns) is reported in the supplementary material (Table S2).

Following the plant-based diet was associated with a lower incidence of all adverse outcomes, which was statistically significant for early preterm delivery (delivery before 34 completed gestational weeks (Table 3); statistical significance was also reached for two general combined outcomes (delivery before 37 gestational weeks or centile <10 and delivery before 34 gestational weeks or centile <10), with a strong protective effect of having been on a plant-based diet (delivery before 37 gestational weeks or centile <10: OR: 0.260 [0.093–0.724], $p = 0.010$) found after adjustment for stage, proteinuria, and hypertension (Table S4). Following a plant-based diet was instead not associated with the single outcome centile <10 or centile <5, delivery before 37 or 28 completed gestational weeks or with the combined outcomes of delivery before 34 gestational weeks or centile <5 and delivery before 28 gestational weeks or centile <5 (Supplementary Table S3).

Table 3. Multivariable logistic regression.

	Preterm Delivery: Gestational Weeks <34		
		OR [CI 95%]	*p* Value
First step	CKD stage	1.495 [0.928–2.406]	0.098
	Plant-based diet	**0.320 [0.122–0.843]**	**0.021**
	Proteinuria > 1 g	1.761 [0.698–4.443]	0.231
	Hypertension	**5.739 [2.180–15.106]**	**<0.001**
Last step	CKD stage	1.455 [0.909–2.328]	0.118
	Plant-based diet	**0.336 [0.129–0.873]**	**0.025**
	Hypertension	**5.697 [2.188–14.837]**	**<0.001**
	Combined Outcome: Preterm Delivery at Week <37 or <10 Centile		
		OR [CI 95%]	*p* value
One step only	**CKD stage**	**2.685 [1.494–4.828]**	**<0.001**
	Plant-based diet	**0.260 [0.093–0.724]**	**0.010**
	Proteinuria > 1 g	2.720 [0.936–7.905]	0.066
	Hypertension	2.294 [0.827–6.367]	0.111
	Combined Outcome: Preterm Delivery at Week <34 or <10 Centile		
One step only	CKD stage	1.460 [0.935–2.280]	0.096
	Plant-based diet	**0.383 [0.158–0.928]**	**0.034**
	Proteinuria > 1 g	2.118 [0.879–5.104]	0.094
	Hypertension	**4.056 [1.679–9.798]**	**0.002**

In bold, significant values.

4. Discussion

The main result of this study is to highlight the potential benefits of plant-based, moderately protein-restricted diets in pregnancy for patients with forms of CKD at high risk of adverse pregnancy outcomes, characterized either by a reduction in kidney function, a high level of proteinuria at baseline, or both.

While previous studies already suggested that such a nutritional approach is safe and is associated with better pregnancy outcomes, the limits of the previous analyses were linked to the small number of cases studied and the lack of a well-matched control group [29–33]. This is the first time that we are reporting the results of a study involving a larger group of on-diet patients, compared with a control group that was propensity-matched for the two main elements considered for diet prescription (CKD stage and proteinuria). It should be noted that the two matched groups of 52 patients each were superimposable for age, BMI, prevalence of primiparity, type of disease, and week of referral. The only difference (higher prevalence of women of non-Caucasian origin in the on-diet group) reflects demographic differences between the two settings (Table 1).

The results suggest that on-diet patients have an advantage for the main outcomes considered (preterm delivery and small-for-gestational-age babies), which reaches statistical significance for the general combined outcome (delivery at <37 completed gestational weeks and birth centile <10), which occurred in 61.54% of on-diet patients and in 80.77% of controls ($p = 0.030$) (Table 2). The multivariable analysis, after adjustment for the main potential confounders (CKD stage, hypertension, and proteinuria), indicates an odds ratio of 0.260 (0.093–0.724) of reaching the combined outcome (weeks < 37 or centile < 10) for on-diet patients (Table 3). The visual analysis of the Kaplan–Meier curve suggests that the advantage occurs early in gestation: no case of delivery <28 gestational age was recorded in the on-diet group, while the prevalence of delivery <34 weeks was 26.92% in the on-diet group and 42.31% in controls (Table 2, Figure 2).

In both cases and controls, probably as a reflection of diligent follow-up targeting normalization of blood pressure and avoiding excessive weight gain, eGFR remained stable from referral to the last check-up before delivery. Instead, proteinuria increased sharply in controls (from 0.63 to 2.39 g/24 h; $p < 0.001$), while the increase was far lower in on-diet patients (0.8 to 1.87 g/24 h; $p = 0.115$) (Figures 2 and 3).

While the small, but potentially relevant, gain in duration of gestation (36 weeks in on-diet patients versus 34.5 in controls) and in birth centile (38.81 in on-diet patients versus 32.84 in controls; Table 2) is not statistically significant, possibly due to the small number of cases, the difference in proteinuria is significant, and suggests less hyperfiltration on the remnant nephrons, which is in turn associated with lower oxidative stress and endothelial damage [14,19,21,44]. Furthermore, plant-based diets are rich in antioxidants and are less diabetogenic in pregnancy [23–25]. All these "endothelial protecting effects" may also play a role in preserving placental function, often impaired in CKD, as has been highlighted in the most recent guidelines on nutrition in CKD [18].

It is difficult to discuss our data in the context of other studies on diets in CKD pregnancies, since, as recently reviewed, there is a cruel lack of such information, specifically in the predialysis phase [45,46]. Plant-based diets are, however, increasingly advised in other situations at risk for adverse pregnancy outcomes that also share a higher risk for endothelial damage, including diabetes, hypertension, and obesity. Of note, all these conditions are risk factors for the development of chronic kidney disease, and may be associated with subclinical renal damage, hence suggesting that common factors are at play in this situations, possibly including—besides endothelial protection—modulation of gut microbiota, and avoidance of excessive weight gain [25,47,48].

More specifically regarding the kidney function, at the same levels of protein intake, plant-based diets are known to induce a lower degree of hyperfiltration compared to mixed protein diets [19–21]. Accordingly, moderate protein restriction, as in our diet plan, in which the diet recommended in pregnancy corresponds to the indications for a healthy, nonpregnant individual, the plant-based regimen is expected to blunt pregnancy-induced renal hyperfiltration.

The potential implications of our results are wide: if the advantage in CKD pregnancies is confirmed on a larger scale, then plant-based diets, with a protein intake adapted to the degree of kidney function and proteinuria, could become the standard indication for women with CKD in pregnancy, at least in severe cases, which is estimated as present in 1:500–1:1000 pregnancies. Furthermore, our experience further underlines the interest of plant-based diets in other high-risk pregnancies, including pre-existing diabetes and hypertension.

This study, which has the advantage of reporting on one treatment proven to be beneficial in CKD pregnancies, and of correlating it for the first time with a lower increase in proteinuria during gestation, has several limitations. First, it involves only two centers. Secondly, as the data were gathered retrospectively, over a long period of time, we were not able to control for variations in care, including pharmacologic treatments, occurring during that period. However, as the Italian best practices in pregnancy in CKD were based on care at these two centers, this is evidence of concordant approaches [28]. Furthermore, although this is the largest CKD cohort followed up with plant-based diets in pregnancy so far published, the number of patients involved is still relatively small, and further stratification, for example, concerning kidney disease, was not possible. The lack of specific nutritional management in CKD pregnancies in the control group did not allow for a detailed comparison of the diet composition to be performed; however, the patients in Cagliari were followed up according to the usual standard of care, in the context of a predominantly Mediterranean nutrition pattern [49,50].

Lastly, and once more due to the only recent introduction in clinical practice of placental biomarkers, the levels of s-flt1 and PlGF were not available, neither for cases nor controls.

Future research is absolutely needed in the field of nutrition in CKD pregnancies. While randomized controlled studies are difficult to propose in the delicate and heterogeneous population with CKD in pregnancy, we need at least large cohort studies comparing different dietary patterns and approaches in these patients. Analysis of placental morphology and correlation with placental biomarkers in CKD patients, as well as in comparison with other diseases sharing similar challenges, such as hypertension, diabetes, and morbid obesity, may allow for a better understanding of the protective effects exerted by plant-based diets, and help us to more finely tune prescriptions.

5. Conclusions

Plant-based, moderately protein-restricted diets are feasible in pregnancy in patients with CKD, and are associated with a lower risk of preterm delivery and of delivery of small-for-gestational-age babies, as compared to mothers on unrestricted omnivorous diets. The favorable effect observed might be mediated by a better stabilization of proteinuria throughout pregnancy, probably resulting from a lower hyperfiltration challenge.

Further prospective studies, systematically employing biomarkers of placental health as well as exploring placental pathology in CKD patients on different diets, are needed if we are to better understand the reasons for these promising results, so that plant-based diets can eventually be proposed also in other high-risk pregnancies.

Supplementary Materials: The following are available online at https://www.mdpi.com/article/10.3390/nu14194203/s1, Table S1: definition and stages of chronic kidney disease according to Kidney Disease Outcome Quality Initiative (KDOQI) guideline 2002; Table S2: univariate logistic regression for different outcomes; Table S3: multivariable logistic regressions for different outcomes; Table S4: The rationale of the plant-based diet the diet and an example for one day of plant-based meals and further advice for pregnant women; Figure S1: eGFR and proteinuria at referral and delivery. First point, referral; second point, delivery.

Author Contributions: Conceptualization, R.A., G.C. and G.B.P.; Data curation, F.L., E.L., V.C., A.T.C., G.Z., A.M.M., B.M. and A.R.; Formal analysis, A.C. and E.D.; Writing—original draft, G.B.P. and M.T. All authors have read and agreed to the published version of the manuscript.

Funding: This research received no external funding.

Institutional Review Board Statement: The study was conducted in accordance with the Declaration of Helsinki. The observational study on current clinical practice was approved by the Ethics Committee of the OIRM Sant'Anna (n° practice 335; n° protocol 11551/c28.2 del 4/3/2011).

Informed Consent Statement: All patients signed a dedicated informed consent at the first control visit during pregnancy.

Data Availability Statement: The data presented in this study are available on request from the corresponding author.

Acknowledgments: We thank Susan Finnel for her careful language editing.

Conflicts of Interest: The authors declare no conflict of interest.

References

1. Kidney Disease: Improving Global Outcomes (KDIGO) CKD Work Group. KDIGO 2012 Clinical Practice Guideline for the Evaluation and Management of Chronic Kidney Disease. *Kidney Inter. Suppl.* **2013**, *3*, 1–150.
2. Williams, D.; Davison, J. Chronic kidney disease in pregnancy. *BMJ* **2008**, *336*, 211–215. [CrossRef] [PubMed]
3. Webster, P.; Lightstone, L.; McKay, D.B.; Josephson, M.A. Pregnancy in chronic kidney disease and kidney transplantation. *Kidney Int.* **2017**, *91*, 1047–1056. [CrossRef] [PubMed]
4. Nevis, I.F.; Reitsma, A.; Dominic, A.; McDonald, S.; Thabane, L.; Akl, E.A.; Hladunewich, M.; Akbari, A.; Joseph, G.; Sia, W.; et al. Pregnancy outcomes in women with chronic kidney disease: A systematic review. *Clin. J. Am. Soc. Nephrol.* **2011**, *6*, 2587–2598. [CrossRef]
5. Piccoli, G.B.; Cabiddu, G.; Attini, R.; Vigotti, F.N.; Maxia, S.; Lepori, N.; Tuveri, M.; Massidda, M.; Marchi, C.; Mura, S.; et al. Risk of Adverse Pregnancy Outcomes in Women with CKD. *J. Am. Soc. Nephrol.* **2015**, *26*, 2011–2022. [CrossRef]
6. Zhang, J.J.; Ma, X.X.; Hao, L.; Liu, L.J.; Lv, J.C.; Zhang, H. A Systematic Review and Meta-Analysis of Outcomes of Pregnancy in CKD and CKD Outcomes in Pregnancy. *Clin. J. Am. Soc. Nephrol.* **2015**, *10*, 1964–1978. [CrossRef]
7. Tangren, J.S.; Wan Md Adnan, W.A.H.; Powe, C.E.; Ecker, J.; Bramham, K.; Hladunewich, M.A.; Ankers, E.; Karumanchi, S.A.; Thadhani, R. Risk of Preeclampsia and Pregnancy Complications in Women With a History of Acute Kidney Injury. *Hypertension* **2018**, *72*, 451–459. [CrossRef]
8. Sebastian, N.; Czuzoj-Shulman, N.; Spence, A.R.; Abenhaim, H.A. Maternal and fetal outcomes of urolithiasis: A retrospective cohort study. *J. Gynecol. Obstet. Hum. Reprod.* **2021**, *50*, 102161. [CrossRef]
9. Garg, A.X.; Nevis, I.F.; McArthur, E.; Sontrop, J.M.; Koval, J.J.; Lam, N.N.; Hildebrand, A.M.; Reese, P.P.; Storsley, L.; Gill, J.S.; et al. Gestational hypertension and preeclampsia in living kidney donors. *N. Engl. J. Med.* **2015**, *372*, 124–133. [CrossRef]
10. Davis, S.; Dylewski, J.; Shah, P.B.; Holmen, J.; You, Z.; Chonchol, M.; Kendrick, J. Risk of adverse maternal and fetal outcomes during pregnancy in living kidney donors: A matched cohort study. *Clin. Transplant.* **2019**, *33*, e13453. [CrossRef]
11. Piccoli, G.B.; Attini, R.; Vasario, E.; Conijn, A.; Biolcati, M.; D'Amico, F.; Consiglio, V.; Bontempo, S.; Todros, T. Pregnancy and chronic kidney disease: A challenge in all CKD stages. *Clin. J. Am. Soc. Nephrol.* **2010**, *5*, 844–855. [CrossRef] [PubMed]
12. Alsuwaida, A.; Mousa, D.; Al-Harbi, A.; Alghonaim, M.; Ghareeb, S.; Alrukhaimi, M.N. Impact of early chronic kidney disease on maternal and fetal outcomes of pregnancy. *J. Matern. Fetal. Neonatal Med.* **2011**, *24*, 1432–1436. [CrossRef] [PubMed]
13. Imbasciati, E.; Gregorini, G.; Cabiddu, G.; Gammaro, L.; Ambroso, G.; Del Giudice, A.; Ravani, P. Pregnancy in CKD stages 3 to 5: Fetal and maternal outcomes. *Am. J. Kidney Dis.* **2007**, *49*, 753–762. [CrossRef]
14. Hoy, W.E.; Hughson, M.D.; Bertram, J.F.; Douglas-Denton, R.; Amann, K. Nephron number, hypertension, renal disease, and renal failure. *J. Am. Soc. Nephrol.* **2005**, *16*, 2557–2564. [CrossRef] [PubMed]
15. Luyckx, V.A.; Brenner, B.M. The clinical importance of nephron mass. *J. Am. Soc. Nephrol.* **2010**, *21*, 898–910. [CrossRef]
16. Kalantar-Zadeh, K.; Fouque, D. Nutritional Management of Chronic Kidney Disease. *N. Engl. J. Med.* **2017**, *377*, 1765–1776. [CrossRef] [PubMed]
17. Naber, T.; Purohit, S. Chronic Kidney Disease: Role of Diet for a Reduction in the Severity of the Disease. *Nutrients* **2021**, *13*, 3277. [CrossRef]
18. Ikizler, T.A.; Burrowes, J.D.; Byham-Gray, L.D.; Campbell, K.L.; Carrero, J.J.; Chan, W.; Fouque, D.; Friedman, A.N.; Ghaddar, S.; Goldstein-Fuchs, D.J.; et al. KDOQI Clinical Practice Guideline for Nutrition in CKD: 2020 Update. *Am. J. Kidney Dis.* **2020**, *76*, S1–S107. [CrossRef]
19. Carrero, J.J.; Gonzalez-Ortiz, A.; Avesani, C.M.; Bakker, S.J.L.; Bellizzi, V.; Chauveau, P.; Clase, C.M.; Cupisti, A.; Espinosa-Cuevas, A.; Molina, P.; et al. Plant-based diets to manage the risks and complications of chronic kidney disease. *Nat. Rev. Nephrol.* **2020**, *16*, 525–542. [CrossRef]
20. Chauveau, P.; Koppe, L.; Combe, C.; Lasseur, C.; Trolonge, S.; Aparicio, M. Vegetarian diets and chronic kidney disease. *Nephrol. Dial. Transplant.* **2019**, *34*, 199–207. [CrossRef]
21. Joshi, S.; McMacken, M.; Kalantar-Zadeh, K. Plant-Based Diets for Kidney Disease: A Guide for Clinicians. *Am J Kidney Dis* **2021**, *77*, 287–296. [CrossRef] [PubMed]
22. Craig, W.J.; Mangels, A.R.; American Dietetic, A. Position of the American Dietetic Association: Vegetarian diets. *J. Am. Diet. Assoc.* **2009**, *109*, 1266–1282. [CrossRef] [PubMed]
23. Procter, S.B.; Campbell, C.G. Position of the Academy of Nutrition and Dietetics: Nutrition and lifestyle for a healthy pregnancy outcome. *J. Acad. Nutr. Diet.* **2014**, *114*, 1099–1103. [CrossRef] [PubMed]
24. Kaiser, L.L.; Campbell, C.G.; Academy Positions Committee, W. Practice paper of the Academy of Nutrition and Dietetics abstract: Nutrition and lifestyle for a healthy pregnancy outcome. *J. Acad. Nutr. Diet.* **2014**, *114*, 1447. [CrossRef] [PubMed]
25. Sebastiani, G.; Herranz Barbero, A.; Borras-Novell, C.; Alsina Casanova, M.; Aldecoa-Bilbao, V.; Andreu-Fernandez, V.; Pascual Tutusaus, M.; Ferrero Martinez, S.; Gomez Roig, M.D.; Garcia-Algar, O. The Effects of Vegetarian and Vegan Diet during Pregnancy on the Health of Mothers and Offspring. *Nutrients* **2019**, *11*, 557. [CrossRef]
26. Perry, A.; Stephanou, A.; Rayman, M.P. Dietary factors that affect the risk of pre-eclampsia. *BMJ Nutr. Prev. Health* **2022**, *5*, 118–133. [CrossRef]
27. Helal, I.; Fick-Brosnahan, G.M.; Reed-Gitomer, B.; Schrier, R.W. Glomerular hyperfiltration: Definitions, mechanisms and clinical implications. *Nat. Rev. Nephrol.* **2012**, *8*, 293–300. [CrossRef]

28. Cabiddu, G.; Castellino, S.; Gernone, G.; Santoro, D.; Moroni, G.; Giannattasio, M.; Gregorini, G.; Giacchino, F.; Attini, R.; Loi, V.; et al. A best practice position statement on pregnancy in chronic kidney disease: The Italian Study Group on Kidney and Pregnancy. *J. Nephrol.* **2016**, *29*, 277–303. [CrossRef]
29. Piccoli, G.B.; Leone, F.; Attini, R.; Parisi, S.; Fassio, F.; Deagostini, M.C.; Ferraresi, M.; Clari, R.; Ghiotto, S.; Biolcati, M.; et al. Association of low-protein supplemented diets with fetal growth in pregnant women with CKD. *Clin. J. Am. Soc. Nephrol.* **2014**, *9*, 864–873. [CrossRef]
30. Attini, R.; Leone, F.; Parisi, S.; Fassio, F.; Capizzi, I.; Loi, V.; Colla, L.; Rossetti, M.; Gerbino, M.; Maxia, S.; et al. Vegan-vegetarian low-protein supplemented diets in pregnant CKD patients: Fifteen years of experience. *BMC Nephrol.* **2016**, *17*, 132. [CrossRef]
31. Attini, R.; Leone, F.; Montersino, B.; Fassio, F.; Minelli, F.; Colla, L.; Rossetti, M.; Rollino, C.; Alemanno, M.G.; Barreca, A.; et al. Pregnancy, Proteinuria, Plant-Based Supplemented Diets and Focal Segmental Glomerulosclerosis: A Report on Three Cases and Critical Appraisal of the Literature. *Nutrients* **2017**, *9*, 770. [CrossRef] [PubMed]
32. Orozco-Guillien, A.O.; Munoz-Manrique, C.; Reyes-Lopez, M.A.; Perichat-Perera, O.; Miranda-Araujo, O.; D'Alessandro, C.; Piccoli, G.B. Quality or Quantity of Proteins in the Diet for CKD Patients: Does "Junk Food" Make a Difference? Lessons from a High-Risk Pregnancy. *Kidney Blood Press. Res.* **2021**, *46*, 1–10. [CrossRef] [PubMed]
33. Nava, J.; Moran, S.; Figueroa, V.; Salinas, A.; Lopez, M.; Urbina, R.; Gutierrez, A.; Lujan, J.L.; Orozco, A.; Montufar, R.; et al. Successful pregnancy in a CKD patient on a low-protein, supplemented diet: An opportunity to reflect on CKD and pregnancy in Mexico, an emerging country. *J. Nephrol.* **2017**, *30*, 877–882. [CrossRef] [PubMed]
34. Anonymous. Gestational Hypertension and Preeclampsia: ACOG Practice Bulletin, Number 222. *Obstet. Gynecol.* **2020**, *135*, e237–e260. [CrossRef] [PubMed]
35. Masuyama, H.; Nobumoto, E.; Okimoto, N.; Inoue, S.; Segawa, T.; Hiramatsu, Y. Superimposed preeclampsia in women with chronic kidney disease. *Gynecol. Obstet. Investig.* **2012**, *74*, 274–281. [CrossRef] [PubMed]
36. Villar, J.; Cheikh Ismail, L.; Victora, C.G.; Ohuma, E.O.; Bertino, E.; Altman, D.G.; Lambert, A.; Papageorghiou, A.T.; Carvalho, M.; Jaffer, Y.A.; et al. International standards for newborn weight, length, and head circumference by gestational age and sex: The Newborn Cross-Sectional Study of the INTERGROWTH-21st Project. *Lancet* **2014**, *384*, 857–868. [CrossRef]
37. National Institute for Health and Care Excellence. Hypertension in pregnancy: Diagnosis and management. NICE Guideline [NG133]. Available online: https://www.nice.org.uk/guidance/ng133 (accessed on 22 April 2022).
38. Brown, M.A.; Magee, L.A.; Kenny, L.C.; Karumanchi, S.A.; McCarthy, F.P.; Saito, S.; Hall, D.R.; Warren, C.E.; Adoyi, G.; Ishaku, S.; et al. The hypertensive disorders of pregnancy: ISSHP classification, diagnosis & management recommendations for international practice. *Pregnancy Hypertens.* **2018**, *13*, 291–310. [CrossRef]
39. La gravidanza Fisiologica, Aggiornamento 2011. Available online: https://www.salute.gov.it/imgs/C_17_pubblicazioni_1436_allegato.pdf (accessed on 10 September 2022).
40. Piccoli, G.B.; Attini, R.; Vasario, E.; Gaglioti, P.; Piccoli, E.; Consiglio, V.; Deagostini, C.; Oberto, M.; Todros, T. Vegetarian supplemented low-protein diets. A safe option for pregnant CKD patients: Report of 12 pregnancies in 11 patients. *Nephrol. Dial. Transplant.* **2011**, *26*, 196–205. [CrossRef]
41. Most, J.; Dervis, S.; Haman, F.; Adamo, K.B.; Redman, L.M. Energy Intake Requirements in Pregnancy. *Nutrients* **2019**, *11*, 1812. [CrossRef]
42. Attini, R.; Montersino, B.; Leone, F.; Minelli, F.; Fassio, F.; Rossetti, M.M.; Colla, L.; Masturzo, B.; Barreca, A.; Menato, G.; et al. Dialysis or a Plant-Based Diet in Advanced CKD in Pregnancy? A Case Report and Critical Appraisal of the Literature. *J. Clin. Med.* **2019**, *8*, 123. [CrossRef]
43. Ho, D.E.; Imai, K.; King, G.; Stuart, E.A. Matching as Nonparametric Preprocessing for Reducing Model Dependence in Parametric Causal Inference. *Political Anal.* **2007**, *15*, 199–236. [CrossRef]
44. Brenner, B.M.; Lawler, E.V.; Mackenzie, H.S. The hyperfiltration theory: A paradigm shift in nephrology. *Kidney Int.* **1996**, *49*, 1774–1777. [CrossRef] [PubMed]
45. Esposito, P.; Garibotto, G.; Picciotto, D.; Costigliolo, F.; Viazzi, F.; Conti, N.E. Nutritional Challenges in Pregnant Women with Renal Diseases: Relevance to Fetal Outcomes. *Nutrients* **2020**, *12*, 873. [CrossRef] [PubMed]
46. Reyes-Lopez, M.A.; Piccoli, G.B.; Leone, F.; Orozco-Guillen, A.; Perichart-Perera, O. Nutrition care for chronic kidney disease during pregnancy: An updated review. *Eur. J. Clin. Nutr.* **2020**, *74*, 983–990. [CrossRef]
47. Chen, Z.; Qian, F.; Liu, G.; Li, M.; Voortman, T.; Tobias, D.K.; Ley, S.H.; Bhupathiraju, S.N.; Li, L.J.; Chavarro, J.E.; et al. Prepregnancy plant-based diets and the risk of gestational diabetes mellitus: A prospective cohort study of 14,926 women. *Am. J. Clin. Nutr.* **2021**, *114*, 1997–2005. [CrossRef]
48. Craig, W.J.; Mangels, A.R.; Fresan, U.; Marsh, K.; Miles, F.L.; Saunders, A.V.; Haddad, E.H.; Heskey, C.E.; Johnston, P.; Larson-Meyer, E.; et al. The Safe and Effective Use of Plant-Based Diets with Guidelines for Health Professionals. *Nutrients* **2021**, *13*, 4144. [CrossRef]
49. Nieddu, A.; Vindas, L.; Errigo, A.; Vindas, J.; Pes, G.M.; Dore, M.P. Dietary Habits, Anthropometric Features and Daily Performance in Two Independent Long-Lived Populations from Nicoya peninsula (Costa Rica) and Ogliastra (Sardinia). *Nutrients* **2020**, *12*, 1621. [CrossRef]
50. Tessier, S.; Gerber, M. Comparison between Sardinia and Malta: The Mediterranean diet revisited. *Appetite* **2005**, *45*, 121–126. [CrossRef]

nutrients

Article

Modeling the Effect of Environmentally Sustainable Food Swaps on Nutrient Intake in Pregnant Women

Tian Wang [1,2], Allison Grech [2], Hasthi U. Dissanayake [2], Sinead Boylan [2,3] and Michael R. Skilton [1,2,4,*]

[1] Boden Collaboration for Obesity, Nutrition, Exercise and Eating Disorders, The University of Sydney, Sydney, NSW 2006, Australia; tian.wang@sydney.edu.au
[2] Sydney Medical School, The University of Sydney, Sydney, NSW 2006, Australia; agre2034@uni.sydney.edu.au (A.G.); hasthi.dissanayake@sydney.edu.au (H.U.D.); sinead.boylan@sydney.edu.au (S.B.)
[3] School of Life and Environmental Sciences, The University of Sydney, Sydney, NSW 2006, Australia
[4] Sydney Institute for Women, Children and Their Families, Sydney Local Health District, Camperdown, NSW 2050, Australia
* Correspondence: michael.skilton@sydney.edu.au; Tel.: +61-2-8627-1916

Abstract: Food production greatly contributes to greenhouse gas emissions (GHG), but there remain concerns that consuming environmentally sustainable foods can increase the likelihood of nutritional deficiencies during pregnancy. We identified commonly consumed foods of pregnant women and determined the effect of their replacement with environmentally sustainable alternatives on nutrient intake and measures of environmental sustainability. Dietary intake data from 171 pregnant women was assessed and foods that contributed the most to energy and protein intake were identified. Of these, foods producing the highest GHG emissions were matched with proposed environmentally sustainable alternatives, and their impact on nutrient provision determined. Meats, grains, and dairy products were identified as important sources of energy and protein. With the highest GHG emissions, beef was selected as the reference food. Proposed alternatives included chicken, eggs, fish, tofu, legumes, and nuts. The most pronounced reductions in CO_2 emissions were from replacing beef with tofu, legumes, and nuts. Replacing one serve per week of beef with an isocaloric serve of firm tofu during pregnancy could reduce GHG emissions by 372 kg CO_2 eq and increase folate (+28.1 µg/serve) and fiber (+3.3 g/serve) intake without compromising iron (+1.1 mg/serve) intake. Small dietary substitutions with environmentally sustainable alternatives can substantially reduce environmental impact without compromising nutrient adequacy.

Keywords: nutrition; sustainability; pregnancy; nutritional requirements; food production system; environment; diet

Citation: Wang, T.; Grech, A.; Dissanayake, H.U.; Boylan, S.; Skilton, M.R. Modeling the Effect of Environmentally Sustainable Food Swaps on Nutrient Intake in Pregnant Women. *Nutrients* **2021**, *13*, 3355. https://doi.org/10.3390/nu13103355

Academic Editor: Ujué Fresán

Received: 13 July 2021
Accepted: 22 September 2021
Published: 24 September 2021

1. Introduction

The food production system is a major contributor to global warming and environmental change, through greenhouse gas (GHG) emissions, freshwater use, land use, acidification, and eutrophication [1]. According to the EAT-Lancet Commission and the Food and Agricultural Organization, there is an urgent need to shift to environmentally sustainable diets on a global scale, by moving towards greater consumption of plant-based foods and reducing the production of less environmentally sustainable animal-derived products [1,2].

Requirements for certain nutrients are elevated during pregnancy to support maternal needs and optimize fetal growth and development [1,3]. Animal-derived foods are an important source of some of these nutrients, such as iron and zinc, and replacement with more environmentally sustainable plant-based alternatives may negatively impact the intake of these nutrients [4]. Health messaging aimed at pregnant women emphasizes the importance of animal-derived foods, particularly red meat, to meet the nutritional requirements of pregnancy [5]. While women consuming a vegetarian or vegan diet are considered to be at higher risk of iron deficiency during pregnancy [6,7], appropriately planned vegetarian diets are nutritionally adequate and suitable for pregnant women to consume [8]. However, it is not clear whether replacement of a small portion of animal-derived foods (e.g., one serve/week) with plant-based alternatives without other dietary modifications will meaningfully affect nutrient intakes among pregnant women consuming a mixed diet, who make up the majority of pregnant women in Western countries [9,10]. Accordingly, we sought to model the net nutritional and environmental effects of partial replacement of commonly consumed animal-derived foods with more environmentally sustainable alternatives within the context of a mixed diet during pregnancy.

2. Materials and Methods

2.1. Maternal Demographics and Identification of Commonly Consumed Foods

A total of 224 mothers and their babies were recruited between April 2015 and September 2016 at the Royal Prince Alfred Hospital in Sydney, Australia, from the Newborn Body Fatness study [11]. Eligibility criteria included: gestational age $\geq$ 34 weeks; singleton born at the Royal Prince Alfred Hospital; and completed assessment of infants' body fatness within 24 h of birth using air-displacement plethysmography. Infants who required respiratory support or had major congenital abnormalities were excluded. Pregnant women with diabetes and preeclampsia were excluded from this dietary analysis as they would have received medical nutritional advice to manage these conditions, leaving 171 women for inclusion in this analysis (Figure S1).

Dietary intake data of pregnant women were collected using a validated food frequency questionnaire (FFQ; Dietary Questionnaire for Epidemiological Studies, version 2) [12]. The median intake per day of 96 food items was recorded and categorized according to the same grouping as Poore et al. [13]. For example, full-cream milk, reduced-fat milk, skim-milk, and flavored-milk drinks were combined into the "milk" food group. For each food group, the energy and protein content of individual food items were analyzed using the AUSNUT 2011-13 AHS Food Nutrient Database [14] (Table S1). Food groups with more than one food item were calculated based on the median intake of each food items. Food items were then ranked by energy and protein intake.

Ethics approval was granted by the Sydney Local Health District (HREC/14/RPAH/478), with written informed consent provided by the participating mothers.

2.2. Environmentally Sustainable Food Alternatives

The commonly consumed foods items, those with a relatively high mean GHG emissions per kilogram (kg) retail weight, based on data from Poore et al. [13], were selected as reference foods. These include beef, chicken, white fish, and milk. Foods with lower mean GHG emissions per kg retail weight were proposed as more environmentally sustainable alternatives. These include beans and legumes, tofu, and mixed nuts. Using the Australian Guide to Healthy Eating [15] as a basis, the standard serve size of reference foods (cooked-size) was determined, and the energy and protein content were calculated using the AUSNUT 2011-13 AHS Food Nutrient Database [14]. The portion size of the proposed food substitutes was determined by matching the energy and protein content of reference foods.

Nutrients included in the analysis were protein, calcium, iron, zinc, iodine, folic acid, saturated fat, and dietary fiber on the basis of these being essential nutrients during pregnancy according to the National Health and Medical Research Council [16] and the American College of Obstetricians and Gynecologists [17], and/or associated with non-communicable disease risk. Estimated absorbed iron was calculated based on the following assumptions: 100% of iron in plant-based foods was non-heme iron; 60% of iron in animal-derived foods was non-heme and 40% was heme iron [18]; 16.8% of non-heme iron and 25% of heme iron are absorbed within the context of a mixed diet [19].

Mean global measures of environmental sustainability, specifically GHG emissions (kg CO_2 eq), land use (m^2 by years occupied), acidifying emissions (g SO_2 eq), eutrophying emissions (g PO_4^{3-} eq), and stress-weighted water use (L) of the food production, were derived from Poore and Nemecek [13]. The total GHG emissions (kg CO_2 eq), land use (m^2), acidifying emissions (g SO_2 eq), eutrophying emissions (g PO_4^{3-} eq), and stress-weighted water use (L) of consuming one serve of an individual food item per week for the entire duration of pregnancy (270 days) were calculated. GHG emissions of an average passenger vehicle were derived from the United States Environmental Protection Agency GHG equivalencies calculator [20].

Furthermore, we modeled the effect on nutrient reference values (NRVs) of replacing one serve of beef with an isoenergetic serve of firm tofu in this population. The NRVs of important nutrients (protein, dietary fiber, iron, zinc, calcium, and folate) during pregnancy were acquired from the National Health and Medical Research Council [16] and the number (%) of people meeting NRVs before and after this replacement were calculated.

2.3. Statistical Analyses

Data are presented as mean for continuous variables and proportions for categorical variables, unless otherwise noted. Histograms were generated for visual assessment of normality. Statistical analyses were performed with IBM® SPSS Statistics 26 Software (IBM, Armonk, NY, USA).

3. Results

3.1. Participant Dietary Characteristics: Nutrition and Commonly Consumed Foods

A total of 171 pregnant women without diabetes and/or preeclampsia were included in the dietary intake analysis. Their mean age was 33.5 $\pm$ 4.5 years. In total, 22% of them had pre-pregnancy BMI $\geq$ 25 kg/m^2, classified as having overweight or obesity.

Table 1 lists the food items that were commonly consumed and contributed the most to pregnant women's energy and protein intake. Rice, pasta, full-cream milk, yogurt, chicken, beef, and mixed dishes with cereal as the major ingredient were important sources of both energy and protein. The dietary intake analysis can be found in Table S1.

Table 1. Commonly consumed food items that contributed most to the energy and protein intake of pregnant women.

Food Items	Energy (kJ/Day)	Food Items	Protein (g/Day)
Rice	539	Beef	6.4
Pasta	518	Chicken	4.6
Full-cream milk	281	Pasta	4.5
Chocolate	271	Yogurt	3.9
Yogurt	236	Full-cream milk	3.4
Chicken	199	Rice	2.3
Cakes	148	Fish	2.2
Beef	142	Mixed dishes *	1.9
Mixed dishes *	131	Lamb	1.9
Tropical fruits	123	Eggs	1.6

* Mixed dishes with cereal as the major ingredient.

3.2. Environmentally Sustainable Food Alternatives

From the identified commonly consumed foods, four food items with relatively high GHG emissions were selected as reference foods: beef (99.5 kg CO_2 eq per 1 kg), chicken (9.9 kg CO_2 eq per 1kg), white fish (13.6 kg CO_2 eq per 1 kg), and milk (3.2 kg CO_2 eq per 1 L) [12]. Food items with lower GHG emissions than each reference food, including the other reference foods, were proposed as more environmentally sustainable alternatives. Beef was selected as the primary reference food given its contributions to average energy intake, protein intake, and GHG emissions, and its proposed more environmentally sustainable alternatives include other red meats (pork and lamb), chicken, egg, fish, beans and legumes, tofu, and nuts. The proposed more environmentally sustainable alternatives for other reference foods could be found in the Supplementary Material (Tables S2–S10).

3.2.1. Replacement of Less Environmentally Sustainable Foods with Energy-Matched Serves of More Environmentally Sustainable Alternatives

The nutrient analysis and measures of environmental sustainability for the beef and isoenergetic serves of more environmentally sustainable alternatives are shown in Tables 2 and 3. The net difference in nutrients and measures of environmental sustainability of replacing one serve of a reference food for an isoenergetic serve of a more environmentally sustainable alternative are shown in Table S2 (beef), Table S3 (chicken), Table S4 (white fish), and Table S5 (milk). In general, animal-derived alternatives have a similar weight (mean cooked weight: 69 g) to beef (65 g cooked weight) whereas an isoenergetic serve of a plant-based alternative (except nuts) weighs approximately 1.5 to 2 times as much. Protein content was generally higher in animal-derived options than isoenergetic serves of plant-based alternatives, with the exception of nuts. Animal-derived foods were important sources of iodine, particularly eggs and white fish. Plant-based foods contain fiber and were generally rich sources of calcium and folate compared to animal-derived foods.

Table 2. Nutrient analysis of beef and isoenergetic serves of more environmentally sustainable options #.

Foods	Serve Size, † g	Protein, g	SFA, g	Dietary Fiber, g	Fe, mg	Estimated Fe Absorption, mg	Zn, mg	Ca, mg	Iodine, µg	Folate, * µg
Beef	65	20.2	1.2	0.0	1.7	0.33	5.1	5.2	1.1	0.0
Pork	83	23.7	0.6	0.0	0.8	0.16	2.0	3.3	0.7	36.6
Lamb	49	14.2	2.5	0.0	1.1	0.22	2.1	4.5	0.2	7.9
Chicken	74	21.4	0.9	0.0	0.4	0.07	0.6	6.7	0.4	2.2
Egg	81	10.0	2.1	0.0	1.3	0.26	1.0	31.5	38.2	67.0
Salmon	39	11.4	1.6	0.0	0.6	0.11	0.2	3.9	3.8	0.0
White fish	89	23.4	0.5	0.0	0.5	0.10	0.6	38.3	39.6	1.8
Beans, mixed	110	7.1	0.1	6.8	2.2	0.38	0.9	47.5	0.6	74.0
Chickpeas	101	6.4	0.2	4.7	1.8	0.31	1.0	45.4	0.5	63.6
Lentils	133	9.0	0.1	4.9	2.7	0.45	1.2	22.6	0.7	26.6
Baked beans	133	6.5	0.1	6.9	1.3	0.23	0.7	51.7	2.0	66.3
Tofu, firm	94	11.3	0.9	3.3	2.7	0.46	1.6	300.0	2.7	28.1
Tofu, silken	210	11.3	0.7	4.8	3.8	0.64	1.1	50.4	2.5	27.3
Mixed nuts	18	3.9	1.4	1.1	0.5	0.09	0.7	16.2	0.1	10.2

Energy content of foods was matched to one serve of beef (471 kJ). † Cooked weight. * Dietary folate equivalents. Ca, calcium; Fe, iron; SFA, saturated fatty acid; Zn, zinc.

Table 3. Environmental sustainability of beef and isoenergetic serves of more environmentally sustainable options #.

Foods	Estimated Effect of Consuming One Serve					Overall Estimated Effect of Consuming One Serve per Week for Duration of Pregnancy					
	GHG Emissions, kg CO_2 eq	Land Use, m^2	Acid., g SO_2 eq	Eutroph., g PO_4^{3-} eq	Stress-Weighted Water Use, L	kg CO_2 eq	Equiv km ^	Land Use, m^2	Acid., g SO_2 eq	Eutroph., g PO_4^{3-} eq	Stress-Weighted Water Use, L
Beef	10.0	32.6	31.9	30.1	3473	383.8	1545	1258.2	1229.7	1162.5	133,968.2
Pork	1.4	2.0	16.5	8.8	7722	54.8	221	77.5	635.6	340.3	297,842.7
Lamb	2.7	25.4	9.6	6.7	9753	105.2	424	980.2	368.4	257.4	376,202.6
Chicken	0.9	1.1	9.5	4.5	1310	35.3	142	43.5	364.9	173.5	50,519.1
Egg	0.4	0.5	4.3	1.8	1452	14.6	59	19.6	167.2	67.9	55,988.9
Salmon	0.6	0.4	3.0	10.6	1871	23.6	95	14.6	114.4	408.1	72,157.5
White fish	1.4	0.9	6.8	24.1	4259	53.7	216	33.2	260.4	929.0	164,278.5
Beans, mixed	0.2	1.7	2.4	1.9	2483	7.7	31	66.5	94.2	72.9	95,775.9
Chickpeas	0.2	1.6	2.2	1.7	2270	7.0	28	60.8	86.1	66.6	87,556.9
Lentils	0.2	2.1	2.9	2.3	2988	9.2	37	80.0	113.3	87.7	115,257.0
Baked beans	0.2	2.1	2.9	2.3	2980	9.2	37	79.8	113.0	87.4	114,927.6
Tofu, firm	0.3	0.3	0.6	0.6	479	11.6	47	12.7	24.2	22.4	18,489.7
Tofu, silken	0.7	0.7	1.4	1.3	1074	25.9	104	28.4	54.3	50.2	41,434.5
Mixed nuts	0.0	0.2	0.6	0.3	2598	1.2	5	7.6	23.3	11.4	100,188.6

Energy content of foods were matched to one serve of beef (471 kJ). ^ Greenhouse gas emissions equivalent to the driving distance by an average passenger vehicle with average fuel economy. Acid., acidifying emissions; Equiv, equivalent; Eutroph., eutrophying emissions; GHG, greenhouse gas. All measures of environmental sustainability were calculated from Poore and Nemecek [13].

Not accounting for bioavailability, the iron content provided by one serve of beef is 1.7 mg (Table 2) and, of the food items analyzed, was the richest animal-derived iron source. The overall iron content of plant-based alternatives was high, with mixed beans, lentils, and tofu (firm and silken) being the richest sources. The estimated absorbed iron from isoenergetic serves of mixed beans, lentils, and tofu (firm and silken) was slightly higher than from beef.

Using data from Poore et al. [13], the production of beef produced far greater GHG emissions (10.0 kg CO_2 eq per serve) than other animal-derived alternatives and plant-based alternatives (Table 3). Extrapolating these results, if pregnant women replace one serve of beef with one isoenergetic serve of firm tofu each week throughout their pregnancy, GHG emissions would be reduced by 372.2 kg CO_2 eq, equivalent to the emissions produced by a typical passenger vehicle driven for 1498 km (Table S2). For other measures of environmental sustainability, all food alternatives have land use less than 2.1 m^2, except for lamb (25.4 m^2), per isoenergetic serve (Table 3). Most food alternatives have acidifying emissions of lower than 10 g SO_2 eq per isoenergetic serve. White fish produced the highest eutrophying emissions (24.1 g PO_4^{3-} eq). Lastly, pork, lamb, and white fish have higher stress-weighted water use (L) than beef (3473.3 L) per serve. Modeling of this replacement of one serve of beef with one isoenergetic serve of firm tofu indicates a small positive impact or no impact on nutrient intakes (Table 4). For example, it would not meaningfully impact iron intake (+1.1 mg/serve), whilst folate (+28.1 μg/serve) and dietary fiber (+3.3 g/serve) would both increase. The exception is zinc, which decreases by 0.5 mg per serve. The net results of this substitution would be that among women who consume a mixed diet, the proportion who meet NRVs for zinc fall by 11%, and the proportion of pregnant women meeting NRVs for calcium and fiber increase by 5% and 2%, respectively.

Table 4. Modeling nutrient intake of pregnant women replacing one serve of beef per week with an isoenergetic serve of firm tofu #.

		Original Intake						Modeled Intake					
Nutrients	**NRVs**	**Mean**	**SD**	**Median**	**IQR**	**Met NRVs (n)**	**Met NRVs (%)**	**Mean**	**SD**	**Median**	**IQR**	**Met NRVs (n)**	**Met NRVs (%)**
Protein, g	EAR 49	97.8	45.3	87.9	72.2–109.4	112	98	96.5	45.3	86.6	70.9–108.1	112	98
Dietary fiber, g	AI 28	23.3	8.8	22.2	17.6–27.9	27	24	23.7	8.8	22.7	18.1–28.3	29	26
Iron, mg	EAR 22	14.2	6.9	12.7	10.0–16.0	5	5	14.3	6.9	12.9	10.1–16.1	6	5
Zinc, mg	EAR 9.0	12.6	5.0	11.5	9.4–14.2	93	83	12.1	5.0	11.0	8.9–13.7	81	72
Calcium, mg	EAR 840	984.1	332.9	930.8	759.5–1114.7	71	63	1026.2	332.9	972.9	801.6–1156.8	77	69
Dietary Folate equivalents, μg	EAR 520	280.2	118.6	254.6	204.4–323.7	4	4	284.2	118.6	258.7	208.4–327.8	4	4

Modeling undertaken in 112 pregnant women (65.5%) consuming >1 serve of beef per week. AI, Adequate Intake; EAR, Estimated Average Requirement; NRV, Nutrient Reference Value.

Further analyses of replacing reference foods with environmentally sustainable alternatives can be found in the Supplementary Material. For example, replacing one serve of beef with one isoenergetic serve of mixed beans reduces protein, saturated fat, and zinc content as expected, while increasing dietary fiber, iron, calcium, and folate content, as well as lowering all measures of environmental sustainability (Table S2). In addition, replacing one serve of milk with an isoenergetic serve of soy milk does not negatively impact on calcium content (+68.4 mg/serve) but reduces the impact on all measures of environmental sustainability (Table S5). The net effect of this replacement per day over the course of an entire pregnancy would reduce GHG emissions by 138.9 kg CO_2 eq, equivalent to the emissions produced by a typical passenger vehicle driven for 559 km (Table S5).

3.2.2. Protein-Matching Environmentally Sustainable Alternatives

The nutrient analysis and measures of environmental sustainability for the reference foods and protein-matched serves of more environmentally sustainable alternatives are shown in Table S6. The net differences in nutrients and measures of environmental sustainability of replacing one serve of a reference food for a protein-matched serve of a more environmentally sustainable alternative are shown in Table S7 (beef), Table S8 (chicken), Table S9 (white fish), and Table S10 (milk).

The weight of protein-matched portion sizes for all of the more environmentally sustainable alternatives was markedly greater than that of beef, with some plant-based alternatives weighing more than four times as much as one serve of beef (Table S4). Similar to the isoenergetic serves, animal-derived foods were rich in iodine, whilst plant-based alternatives were rich sources of calcium, folate, and dietary fiber.

In general, plant-based alternatives were high in iron when matching protein content, with the richest sources from silken tofu, mixed beans, and lentils (Table S4). The estimated iron absorption of these three foods was three times as much as beef in one protein-matched serve. The average zinc content provided by plant-based alternatives was lower than that of beef but higher than other animal-derived alternatives.

Using protein-matched serves did not markedly alter the results regarding the net benefits to environmental sustainability, particularly GHG emissions. For example, if pregnant women substitute one serve of beef with a protein-matched serve of firm tofu each week throughout the course of pregnancy, GHG emissions would be reduced by 363 kg CO_2 eq, equivalent to 1461 km of typical driving distance by an average passenger vehicle. The net results of this replacement would decrease the proportion of women who meet NRVs for zinc by 5% and increase the proportion of pregnant women meeting NRVs for calcium and fiber by 10% and 2%, respectively (Table 5). For other measures of environmental sustainability, all food alternatives have land use less than 6.4 m^2, except for lamb (36.3 m^2), per protein-matched serve (Table S6). Most food alternatives have acidifying emissions of lower than 15 g SO_2 eq per protein-matched serve. White fish produced the highest eutrophying emissions (20.8 g PO_4^{3-} eq). Lastly, protein-matched serves of more environmentally sustainable alternatives have relatively high stress-weighted water use on average, although this varied greatly, ranging from 861L per serve for firm tofu through to over 13,000 L per serve for mixed nuts and lamb.

Table 5. Modeling the nutrient intake of pregnant women replacing one serve of beef per week with a protein-matched serve of firm tofu #.

Nutrients	NRVs	Original Intake						Modeled Intake					
		Mean	SD	Median	IQR	Met NRVs (n)	Met NRVs (%)	Mean	SD	Median	IQR	Met NRVs (n)	Met NRVs (%)
Dietary fiber, g	AI 28	23.3	8.8	22.2	17.6–27.9	27	24	24.1	8.8	23.0	18.5–28.7	29	26
Iron, mg	EAR 22	14.2	6.9	12.7	10.0–16.0	5	5	14.6	6.9	13.2	10.4–16.4	6	5
Zinc, mg	EAR 9.0	12.6	5.0	11.5	9.4–14.2	93	83	12.3	5.0	11.2	9.1–13.9	87	78
Calcium, mg	EAR 840	984.1	332.9	930.8	759.5–1114.7	71	63	1060.3	332.9	1007.1	835.7–1190.9	82	73
Dietary Folate equivalents, µg	EAR 520	280.2	118.6	254.6	204.4–323.7	4	4	287.4	118.6	261.9	211.6–331.0	4	4

Modeling undertaken in 112 pregnant women (65.5%) consuming >1 serve of beef per week. AI, Adequate Intake; EAR, Estimated Average Requirement; NRV, Nutrient Reference Value.

4. Discussion

The Food and Agriculture Organization of the United Nations and others has identified an urgent need to shift to environmentally sustainable diets on a global scale [1,2]. Concerns over nutrient adequacy in populations with high nutrient demands, such as pregnant women, are potential challenges to the broad implementation of environmentally sustainable diets. Our findings indicate these concerns are likely misplaced within the context of a mixed diet. Modeled replacement of animal-derived food with more environmentally sustainable plant-based alternatives has only a small effect on overall nutrient intake but a considerable positive effect on environmental sustainability.

There remain concerns among many practitioners and community members regarding the potential risk of nutrition inadequacy of pregnant women consuming plant-based diets [21,22]. Our focus was not plant-based diets but rather environmentally sustainable foods within the context of mixed diets. Focusing on mixed diets enables our findings to be relevant to a large proportion of the population for whom consuming a purely vegetarian or plant-based diet is neither practicable nor desirable [9,10].

In general, our results support animal-derived foods as a rich source of zinc, and plant-based foods as being rich in calcium, folate, and dietary fiber. A specific swap replacing one serve per week of beef with firm tofu reduces zinc and protein levels, while calcium, folate, and dietary fiber increases. The absolute differences in the nutrient intake of this swap were small. In this modeling, the largest differences were for calcium (raised by about 13% of a standard deviation), zinc (reduced by about 10% of a standard deviation), and fiber (raised by about 5% of a standard deviation), resulting in an increase in the proportion of pregnant women who meet NRVs of calcium, dietary fiber, and iron, but a decrease for zinc. Maternal zinc deficiency during pregnancy may increase the risk of low birth weight and small for gestational age infants [23], although severe zinc deficiency is rare. Indeed, in our population, the majority of women met the NRVs for zinc, based on both actual and modeled intakes. Furthermore, zinc is a common ingredient in pregnancy multivitamins, which are used by approximately 70–80% of women in the USA, Europe, and Australia [7,24–26]. The amount of zinc in such multivitamins (typically 11 mg per day) [27] exceeds the NRV for zinc.

Alongside folate, the public is perhaps most aware of concerns regarding sufficient iron intake during pregnancy [3,7]. Plant-based foods are a good source of overall dietary iron, but this does not account for differences in the bioavailability of heme and non-heme iron. Heme iron is only found in animal-derived meat products. Heme iron constitutes approximately 40% of total iron from animal-derived meat products and it is more readily absorbed by humans than non-heme iron [3]. To account for this, we estimated the amount of absorbed iron. This estimation did not account for the increased absorption of non-heme iron during pregnancy [28], and as such is a conservative estimate of absorbed iron from plant-based foods. Nonetheless, both total dietary iron intake and estimated absorbed iron were slightly higher after replacing a serve of beef with an isoenergetic serve of firm tofu (Table S2), although the magnitude did not appear to be clinically meaningful on an individual basis. It is notable that the proportion of women meeting NRVs for iron intake by diet alone was low in our population (about 5%). This is consistent with other studies of pregnant women in Australia [7], and with dietary modeling undertaken as part of the development of the Australian Guide to Healthy Eating, in which no dietary models could provide sufficient iron to meet the needs of pregnant women [29].

The low prevalence of participants meeting the NRVs for folate, iron, and fiber, for both actual and modeled intakes, highlights the necessity of appropriately planned diets, by health professionals, such as dietitians or individuals with nutrition training, to fulfill the nutritional needs of women during pregnancy. The use of dietary supplements during pregnancy may at least partially alleviate these deficiencies, irrespective of the background diet. There is currently limited publicly available information concerning the environmental sustainability of pregnancy supplements.

Iron and zinc absorption can be affected by other factors. Within the context of a mixed diet, non-heme iron and zinc absorption can be enhanced by other components of the diet, including meat, poultry, fish, and other seafood [30,31], alongside vitamin C-rich foods, e.g., citrus fruits and green leafy vegetables [32]. Therefore, one way to implement a one-serve per week replacement within the context of a mixed diet whilst maintaining the effect of iron and zinc absorption enhancers (e.g., meat products, green leafy vegetables) would be to replace half the portion of a less environmentally sustainable animal-derived meat product with a more sustainable plant-based food twice per week.

The most environmentally sustainable alternatives produced approximately 98% less GHG emissions than one serve of beef when matched for energy or protein content. To facilitate a broader understanding of the impact of incorporating more environmentally sustainable foods, we compared GHG emissions generated in food production to those produced by typical passenger vehicle usage. Using the example above, replacing one serve of beef with an isoenergetic serve of firm tofu per week during pregnancy could reduce GHG emissions by the equivalent to those produced by a typical passenger vehicle driven for 1498 km. Similarly, most of the proposed environmentally sustainable alternatives have a lesser environmental impact when assessed by other measures of environmental sustainability (including land use, acidifying emissions, and eutrophying emissions), although water use in the production of some plant-based alternatives (e.g., legumes and beans) appears to be similar to that of some animal-derived foods.

When matching the energy and protein content to reference foods, the portion size of plant-based alternatives is two to four times heavier than animal-derived foods, consistent with the density of energy and protein being notably higher in animal-derived foods. Plant-based alternatives are rich in dietary fiber with lower energy density, which increase satiety, helping to maintain a healthy weight by limiting calorie intake [33,34], and optimize weight gain during pregnancy [35,36]. Excessive protein intake from animal sources, primarily meat products, may also increase the risk of overweight and obesity in offspring [37].

Our study has a number of limitations. We used an FFQ administered in the immediate postpartum period. Women were asked to recall their habitual diet during pregnancy, which we validated against dietary biomarkers in a subgroup [38]. FFQs are well described as tools for assessing habitual diet over 6–12-month periods, although we cannot rule out that there may have been a greater emphasis on third trimester intake due to recency bias. Furthermore, detailed information of dietary intake (e.g., ingredients of mixed dishes) is difficult to assess but will include meat or other food items as an ingredient. As such, the values for meat and other food items will have been underestimated. Nonetheless, previous research has shown dietary patterns during pregnancy remain relatively stable when compared to pre-pregnancy intake [39–41] and are not significantly different to those of non-pregnant women of reproductive age [42]. Our study population from which we identified the commonly consumed food items during pregnancy has a relatively low prevalence of overweight and obesity (22%). This is less than in the general population in Western countries, where up to 50% of women have overweight or obesity before pregnancy [43,44], and is likely due at least in part to the exclusion of women with diabetes and preeclampsia from our analyses. Future research may seek to determine the environmental sustainability of foods consumed by representative samples of pregnant women. We mainly focused on GHG emissions given their contribution to global warming and did not describe the impact of environmentally sustainable foods on the economy and society. A range of indicators of economic and societal aspects [45] (e.g., affordability, employment, and food insecurity) can be used to assess the effects of improving environmental sustainability and should be the topic of future research. We acquired measures of environmental sustainability from a global dataset by Poore and Nemecek [13], consisting of data derived from 570 studies in 119 countries to ensure that our findings can be broadly generalizable. Future studies could employ country-specific measures of environmental sustainability to enable a more geographically accurate indication of the environmental impact of these food swaps. The role of food–food interactions that influence absorption was beyond the scope of our current study; however, future research should look to model these interactions within the context of dietary changes to promote environmental sustainability. Finally, to translate our findings into practice, the acceptability and popularity of proposed environmentally sustainable options need to be taken into consideration. Future research should identify whether there are unique challenges or opportunities for promoting environmentally sustainable foods during pregnancy. Nutrition communicators, dietitians, and practitioners may need to focus on the promotion of health benefits of environmentally sustainable plant-based foods, and provide practical advice (e.g., design recipes) in incorporating these replacements into their individual diets.

5. Conclusions

Our research highlights simple dietary substitutions that can substantially reduce environmental impact without compromising essential nutrient intake during pregnancy. Moving forward, environmentally sustainable food replacements should be the focus of applied clinical research and inform nutrition practice and policy development.

Supplementary Materials: The following are available online at https://www.mdpi.com/article/10.3390/nu13103355/s1, Figure S1: Flow diagram of participant selection; Table S1: Dietary intake analysis of pregnant women in the Newborn Body Fatness study; Table S2: Food swaps—beef. Differences in nutrients and measures of environmental sustainability between beef and more sustainable isoenergetic options; Table S3: Food swaps—chicken. Differences in nutrients and measures of environmental sustainability between chicken and more environmentally sustainable isoenergetic options; Table S4: Food swaps—white fish. Differences in nutrients and measures of environmental sustainability between white fish and more environmentally sustainable isoenergetic options; Table S5: Food swaps—milk. Differences in nutrients and measures of environmental sustainability between milk and more environmentally sustainable isoenergetic options; Table S6: Nutrient analysis and measures of environmental sustainability of the reference food (beef) and more environmentally sustainable protein-matched options; Table S7: Food swaps—beef. Differences in nutrients and measures of environmental sustainability between beef and more environmentally sustainable protein-matched options; Table S8: Food swaps—chicken. Differences in nutrients and measures of environmental sustainability between chicken and more environmentally sustainable protein-matched options; Table S9: Food swaps—white fish. Differences in nutrients and measures of environmental sustainability between white fish and more environmentally sustainable protein-matched options; Table S10: Food swaps—milk. Differences in nutrients and measures of environmental sustainability between milk and more environmentally sustainable protein-matched options.

Author Contributions: Conceptualization, M.R.S., T.W., S.B. and A.G.; Data curation, T.W.; Formal analysis, T.W.; Investigation, T.W. and H.U.D.; Methodology, M.R.S., T.W., S.B. and A.G.; Project administration, M.R.S. and T.W.; Resources, H.U.D. and M.R.S.; Software, T.W.; Supervision, M.R.S.; Validation, T.W., M.R.S. and A.G.; Visualization, T.W.; Writing—original draft preparation, T.W.; Writing—review and editing, T.W., M.R.S., A.G., S.B. and H.U.D. All authors have read and agreed to the published version of the manuscript.

Funding: This research received no external funding.

Institutional Review Board Statement: The study was approved by the Ethics Committee of the Sydney Local Health District (HREC/14/RPAH/478, date of approval: 20 February 2015), with written informed consent provided by the participating mothers.

Informed Consent Statement: Informed consent was obtained from all subjects involved in the study.

Data Availability Statement: The data presented in this study are available on request from the corresponding author. The data are not publicly available for ethical reasons.

Acknowledgments: The authors thank Graham Giles, Cancer Epidemiology Centre, Cancer Council Victoria for permission to use the Dietary Questionnaire for Epidemiological Studies (Version 2).

Conflicts of Interest: Skilton is employed by The University of Sydney as the Maurice Blackmore Principal Research Fellow in Integrative Medicine. This position was established through a gift from the Blackmores Institute. Skilton receives no research funding or in kind support from Blackmores Limited. The authors have no other conflicts of interest to declare.

References

1. Willett, W.; Rockström, J.; Loken, B.; Springmann, M.; Lang, T.; Vermeulen, S.; Garnett, T.; Tilman, D.; DeClerck, F.; Wood, A.; et al. Food in the Anthropocene: The EAT–Lancet Commission on healthy diets from sustainable food systems. *Lancet* **2019**, *393*, 447–492. [CrossRef]
2. Fischer, C.G.; Garnett, T. *Plates, Pyramids, and Planets: Developments in National Healthy and Sustainable Dietary Guidelines: A State of Play Assessment*; Food and Agriculture Organization of the United Nations: Rome, Italy, 2016.
3. Sebastiani, G.; Herranz Barbero, A.; Borrás-Novell, C.; Alsina Casanova, M.; Aldecoa-Bilbao, V.; Andreu-Fernández, V.; Pascual Tutusaus, M.; Ferrero Martínez, S.; Gómez Roig, M.D.; García-Algar, O. The Effects of Vegetarian and Vegan Diet during Pregnancy on the Health of Mothers and Offspring. *Nutrients* **2019**, *11*, 557. [CrossRef] [PubMed]
4. Queensland Government. Iron for Pregnant Women. Queensland 2020. Available online: https://www.health.qld.gov.au/__data/assets/pdf_file/0023/150089/antenatal-iron.pdf (accessed on 7 October 2020).
5. Piccoli, G.B.; Clari, R.; Vigotti, F.N.; Leone, F.; Attini, R.; Cabiddu, G.; Mauro, G.; Castelluccia, N.; Colombi, N.; Capizzi, I.; et al. Vegan-vegetarian diets in pregnancy: Danger or panacea? A systematic narrative review. *BJOG Int. J. Obstet. Gynaecol.* **2015**, *122*, 623–633. [CrossRef] [PubMed]
6. Pistollato, F.; Cano, S.S.; Elio, I.; Vergara, M.M.; Giampieri, F.; Battino, M.; Sumalla Cano, S.; Masias Vergara, M. Plant-Based and Plant-Rich Diet Patterns during Gestation: Beneficial Effects and Possible Shortcomings. *Adv. Nutr. (Bethesda)* **2015**, *6*, 581–591. [CrossRef] [PubMed]
7. Slater, K.; Rollo, M.E.; Szewczyk, Z.; Ashton, L.; Schumacher, T.; Collins, C. Do the Dietary Intakes of Pregnant Women Attending Public Hospital Antenatal Clinics Align with Australian Guide to Healthy Eating Recommendations? *Nutrients* **2020**, *12*, 2438. [CrossRef]
8. Melina, V.; Craig, W.; Levin, S. Position of the Academy of Nutrition and Dietetics: Vegetarian Diets. *J. Acad. Nutr. Diet* **2016**, *116*, 1970–1980. [CrossRef] [PubMed]
9. Mendez, M.A.; Kogevinas, M. A comparative analysis of dietary intakes during pregnancy in Europe: A planned pooled analysis of birth cohort studies. *Am. J. Clin. Nutr.* **2011**, *94* (Suppl. 6), 1993S–1999S. [CrossRef]
10. Forbes, L.E.; Graham, J.E.; Berglund, C.; Bell, R.C. Dietary Change during Pregnancy and Women's Reasons for Change. *Nutrients* **2018**, *10*, 1032. [CrossRef]

11. McKenzie, K.M.; Dissanayake, H.U.; McMullan, R.; Caterson, I.D.; Celermajer, D.S.; Gordon, A.; Hyett, J.; Meroni, A.; Phang, M.; Raynes-Greenow, C.; et al. Quantity and Quality of Carbohydrate Intake during Pregnancy, Newborn Body Fatness and Cardiac Autonomic Control: Conferred Cardiovascular Risk? *Nutrients* **2017**, *9*, 1375. [CrossRef]
12. Dissanayake, H.U.; McMullan, R.L.; Kong, Y.; Caterson, I.D.; Celermajer, D.S.; Phang, M.; Raynes-Greenow, C.; Polson, J.W.; Gordon, A.; Skilton, M.R. Body Fatness and Cardiovascular Health in Newborn Infants. *J. Clin. Med.* **2018**, *7*, 270. [CrossRef]
13. Poore, J.; Nemecek, T. Reducing food's environmental impacts through producers and consumers. *Science* **2018**, *360*, 987–992. [CrossRef]
14. Food Standards Australia New Zealand. AUSNUT 2011–2013 Food Nutrient Database. Available online: https://www.foodstandards.gov.au/science/monitoringnutrients/ausnut/foodnutrient/Pages/default.aspx (accessed on 7 October 2020).
15. National Health and Medical Research Council. *Serve Sizes*; Australian Government: Canberra, Australia, 2015.
16. National Health and Medical Research Council. *Nutrient Reference Values for Australia and New Zealand*; Department of Health and Ageing, National Health and Medical Research Council: Canberra, Australia, 2017.
17. The American College of Obstetricians and Gynecologists. Nutrition during Pregnancy 2020 [Updated June 2020; Cited 2020]. Available online: https://www.acog.org/womens-health/faqs/nutrition-during-pregnancy (accessed on 7 October 2020).
18. Beard, J.L.; Dawson, H.; Piñero, D.J. Iron metabolism: A comprehensive review. *Nutr. Rev.* **1996**, *54*, 295–317. [CrossRef]
19. Institute of Medicine Panel on Micronutrients. *Dietary Reference Intakes for Vitamin A, Vitamin K, Arsenic, Boron, Chromium, Copper, Iodine, Iron, Manganese, Molybdenum, Nickel, Silicon, Vanadium, and Zinc*; National Academies Press (US): Washington, DC, USA, 2001; Copyright 2001 by the National Academy of Sciences. All rights reserved.
20. United States Environmental Protection Agency. Greenhouse Gas Equivalencies Calculator 2020 [cited 7 October 2020]. Available online: https://www.epa.gov/energy/greenhouse-gas-equivalencies-calculator (accessed on 7 October 2020).
21. Victoria State Government. Pregnancy and Diet. 2019. Available online: https://www.betterhealth.vic.gov.au/health/healthyliving/pregnancy-and-diet (accessed on 7 October 2020).
22. Queensland Government. Healthy Eating for Vegetarian or Vegan Pregnant and Breastfeeding Mothers. 2018. Available online: https://www.health.qld.gov.au/__data/assets/pdf_file/0024/726063/antenatal-veganveget.pdf (accessed on 7 October 2020).
23. Wang, H.; Hu, Y.-F.; Hao, J.-H.; Chen, Y.-H.; Su, P.-Y.; Wang, Y.; Yu, Z.; Fu, L.; Xu, Y.-Y.; Zhang, C.; et al. Maternal zinc deficiency during pregnancy elevates the risks of fetal growth restriction: A population-based birth cohort study. *Sci. Rep.* **2015**, *5*, 11262. [CrossRef]
24. Bailey, R.L.; Pac, S.G.; Fulgoni, V.L., III; Reidy, K.C.; Catalano, P.M. Estimation of Total Usual Dietary Intakes of Pregnant Women in the United States. *JAMA Netw. Open* **2019**, *2*, e195967. [CrossRef] [PubMed]
25. Oliver, E.M.; Grimshaw, K.E.C.; Schoemaker, A.A.; Keil, T.; McBride, D.; Sprikkelman, A.B.; Ragnarsdottir, H.S.; Trendelenburg, V.; Emmanouil, E.; Reche, M.; et al. Dietary Habits and Supplement Use in Relation to National Pregnancy Recommendations: Data from the EuroPrevall Birth Cohort. *Matern. Child Health J.* **2014**, *18*, 2408–2425. [CrossRef] [PubMed]
26. Shand, A.W.; Walls, M.; Chatterjee, R.; Nassar, N.; Khambalia, A.Z. Dietary vitamin, mineral and herbal supplement use: A cross-sectional survey of before and during pregnancy use in Sydney, Australia. *ANZJOG* **2016**, *56*, 154–161. [CrossRef] [PubMed]
27. BLACKMORES Pregnancy & Breast-Feeding Gold 2021. Available online: https://www.blackmores.com.au/products/pregnancy-breast-feeding-gold (accessed on 15 March 2021).
28. Barrett, J.F.R.; Whittaker, P.G.; Williams, J.G.; Lind, T. Absorption of non-haem iron from food during normal pregnancy. *BMJ* **1994**, *309*, 79–82. [CrossRef]
29. National Health and Medical Research Council. *A Modelling System to Inform the Revision of the Australian Guide to Healthy Eating*; Australia Co.: Canberra, Australia, 2011.
30. Hallberg, L.; Hulthén, L. Prediction of dietary iron absorption: An algorithm for calculating absorption and bioavailability of dietary iron. *Am. J. Clin. Nutr.* **2000**, *71*, 1147–1160. [CrossRef]
31. Lonnerdal, B. Dietary Factors Influencing Zinc Absorption. *J. Nutr.* **2000**, *130*, 1378S–1383S. [CrossRef]
32. Hunt, J.R. Bioavailability of iron, zinc, and other trace minerals from vegetarian diets. *Am. J. Clin. Nutr.* **2003**, *78* (Suppl. 3), 633s–639s. [CrossRef]
33. Ledikwe, J.H.; Blanck, H.M.; Kettel Khan, L.; Serdula, M.K.; Seymour, J.D.; Tohill, B.C.; Rolls, B.J. Dietary energy density is associated with energy intake and weight status in US adults. *Am. J. Clin. Nutr.* **2006**, *83*, 1362–1368. [CrossRef]
34. De Castro, J.M. Dietary Energy Density Is Associated with Increased Intake in Free-Living Humans. *J. Nutr.* **2004**, *134*, 335–341. [CrossRef] [PubMed]
35. Deierlein, A.L.; Siega-Riz, A.M.; Herring, A. Dietary energy density but not glycemic load is associated with gestational weight gain. *Am. J. Clin. Nutr.* **2008**, *88*, 693–699. [CrossRef] [PubMed]
36. Hirko, K.A.; Comstock, S.S.; Strakovsky, R.S.; Kerver, J.M. Diet during Pregnancy and Gestational Weight Gain in a Michigan Pregnancy Cohort. *Curr. Dev. Nutr.* **2020**, *4*, nzaa121. [CrossRef]
37. Maslova, E.; Rytter, D.; Bech, B.H.; Henriksen, T.B.; Rasmussen, M.A.; Olsen, S.F.; Halldorsson, T.I. Maternal protein intake during pregnancy and offspring overweight 20 y later. *Am. J. Clin. Nutr.* **2014**, *100*, 1139–1148. [CrossRef] [PubMed]
38. Phang, M.; Dissanayake, H.U.; McMullan, R.L.; Hyett, J.; Gordon, A.; Garg, M.L.; Skilton, M.R. Increased α-Linolenic Acid Intake during Pregnancy is Associated with Higher Offspring Birth Weight. *Curr. Dev. Nutr.* **2019**, *3*, nzy081. [CrossRef]
39. Blumfield, M.L.; Hure, A.J.; MacDonald-Wicks, L.K.; Smith, R.; Simpson, S.J.; Giles, W.B.; Raubenheimer, D.; Collins, C.E. Dietary balance during pregnancy is associated with fetal adiposity and fat distribution. *Am. J. Clin. Nutr.* **2012**, *96*, 1032–1041. [CrossRef]

40. Lee, Y.Q.; Collins, C.E.; Schumacher, T.L.; Weatherall, L.J.; Keogh, L.; Sutherland, K.; Gordon, A.; Rae, K.M.; Pringle, K.G. Disparities exist between the dietary intake of Indigenous Australian women during pregnancy and the Australian dietary guidelines: The Gomeroi gaaynggal study. *J. Hum. Nutr. Diet* **2018**, *31*, 473–485. [CrossRef]
41. Crozier, S.R.; Robinson, S.M.; Godfrey, K.M.; Cooper, C.; Inskip, H.M. Women's dietary patterns change little from before to during pregnancy. *J. Nutr.* **2009**, *139*, 1956–1963. [CrossRef]
42. Hure, A.; Young, A.; Smith, R.; Collins, C. Diet and pregnancy status in Australian women. *Public Health Nutr.* **2009**, *12*, 853–861. [CrossRef]
43. Australian Institute of Health and Welfare. *Australia's Mothers and Babies 2018: In Brief*; Australian Institute of Health and Welfare: Canberra, Australia, 2020.
44. Driscoll, A.K.; Gregory, E.C.W. *Increases in Prepregnancy Obesity: United States, 2016–2019*; NCHS Data Brief, no 392; Hyattsville, M.D., Ed.; National Center for Health Statistics: Hyattsville, MD, USA, 2020.
45. Boylan, S.M.; Thow, A.-M.; Tyedmers, E.K.; Malik, A.; Salem, J.; Alders, R.; Raubenheimer, D.; Lenzen, M. Using Input-Output Analysis to Measure Healthy, Sustainable Food Systems. *Front. Sustain. Food Syst.* **2020**, *4*, 93.

nutrients

MDPI

Article

Comparison of Plate Waste between Vegetarian and Meat-Containing Meals in a Hospital Setting: Environmental and Nutritional Considerations

Andrew Berardy [1,*], Brianna Egan [1], Natasha Birchfield [2], Joan Sabaté [1] and Heidi Lynch [3]

[1] Center for Nutrition, Lifestyle and Disease Prevention, School of Public Health, Loma Linda University, Loma Linda, CA 92350, USA; began@students.llu.edu (B.E.); jsabate@llu.edu (J.S.)
[2] Roche Diagnostic Solutions, Clinical Development Core, Clinical Development Medical Affairs, Tucson, AZ 85755, USA; birchfin@roche.com
[3] Department of Kinesiology, College of Health Sciences, Point Loma Nazarene University, San Diego, CA 92106, USA; hlynch@pointloma.edu
* Correspondence: andrewberardy@llu.edu; Tel.: +1-909-558-1000 (ext. 15312)

Abstract: Vegetarian diets can satisfy nutritional requirements and have lower environmental impacts than those containing meat. However, fruits and vegetables are wasted at higher rates than meat. Reducing both food waste (FW) and the environmental impacts associated with food production is an important sustainability goal. Therefore, the aim of this study was to examine potential tradeoffs between vegetarian meals' lower impacts but potentially higher FW compared to meat-containing meals. To examine this, seven consecutive days of plate FW data from Loma Linda University Medical Center (LLUMC) patients were collected and recorded from 471 meals. Mean total FW and associated greenhouse gas emissions (GHGE) were higher among meat-containing meals (293 g/plate, 604 g CO_2-eq/plate) than vegetarian meals (259 g/plate, 357 g CO_2-eq/plate) by 34 g ($p = 0.05$) and 240 g CO_2-eq ($p < 0.001$), respectively. Statistically significant differences were observed in both FW and associated GHGE across major food categories, except fruit, when comparing vegetarian and meat-containing meals. Overall, vegetarian meals were preferable to meat-containing meals served at LLUMC both in terms of minimizing FW and lowering environmental impacts. Other institutions serving vegetarian meal options could expect similar advantages, especially in reduced GHGE due to the high CO_2 embodied in meat.

Keywords: food waste; global warming; vegetarian meals; hospital setting; plant based; sustainability; public health

Citation: Berardy, A.; Egan, B.; Birchfield, N.; Sabaté, J.; Lynch, H. Comparison of Plate Waste between Vegetarian and Meat-Containing Meals in a Hospital Setting: Environmental and Nutritional Considerations. *Nutrients* **2022**, *14*, 1174. https://doi.org/10.3390/nu14061174

Academic Editor: Lindsay Brown

Received: 5 January 2022
Accepted: 7 March 2022
Published: 11 March 2022

1. Introduction

Human activities cause global environmental changes that threaten to disrupt the stability of the Earth's systems, leading to potentially disastrous consequences [1]. This recognition has prompted a widespread call for emergency action to limit global temperature increases, restore biodiversity, and protect health [2]. Food systems are responsible for between 19 and 37% of global anthropogenic greenhouse gas emissions (GHGE), depending on what is included in the estimate [3,4]. A recent estimate attributed 34% of global GHGE to food systems, with 71% coming from agriculture and land use, and the rest from downstream supply chain activities [5].

Yet, current practices of food production and distribution are insufficient, as there are 815 million people globally, or one in nine, who are undernourished [6]. In order to end hunger, different scenarios predict that between 3 and 20% more food production will be necessary, depending upon the approach, increasing the associated environmental impacts [7]. This challenge will only become more difficult as the global population continues to expand to approximately 9 billion people [8].

Reducing the consumption of animal-based foods is a possible measure to reduce environmental impacts while improving health outcomes, with the potential to reduce diet-related GHGE by between 33 and 51% in the United States [9,10]. A systematic review found that vegan diets could reduce GHGE by up to 70%, land use by up to 86%, and water use by up to 70% [11]. Another review found that along with improved health, shifting from current omnivorous dietary patterns to vegetarian or vegan diets increases environmental sustainability while also improving health [12].

The Academy of Nutrition and Dietetics considers appropriately planned vegetarian diets to be healthful and nutritionally adequate for all stages of the life cycle [13]. Consuming vegetarian or vegan diets has been shown to lower risk for developing obesity [14], cardiovascular diseases [15], hypertension [16], type 2 diabetes [17], and metabolic syndrome [18]. These health-protective effects may be due to the higher nutrient quality typical of plant-based diets [19]. Notably, vegetarian and vegan diets tend to be lower in total fat, saturated fat, monounsaturated fat, dietary cholesterol, protein, alcohol, and sodium, and higher in polyunsaturated fat, fiber, and iron [19]. This is likely because plant-based diets tend to be higher in fruits, greens, and pulses; subcategories of vegetables [19]. In addition, plant-based diets have been found to sufficiently support athletic performance while also contributing to better overall health and reducing environmental impacts [20–22].

Nonetheless, certain nutrients are less bioavailable or less frequently consumed on a vegetarian or vegan diet. For example, as non-heme iron (found in plants) is less bioavailable compared to heme iron (from animals), the Recommended Dietary Allowance for iron for vegetarians and vegans is 1.8-fold greater than that for omnivores [23]. Additionally, vegans (who exclude all animal products) must be mindful to consume foods fortified with vitamin B12 or take a vitamin B12 supplement as this vitamin is not present in plant foods [13]. Lacto-ovo vegetarians typically consume at least the recommended intake for calcium, while vegans may risk insufficiency. Furthermore, vitamin D is not abundant in food and is a nutrient for which the use of supplements is frequently advised, regardless of dietary pattern [24].

Higher diet quality, as measured by the Healthy Eating Index, is associated with higher food waste (FW), primarily in the form of fruits and vegetables [25]. FW is a significant challenge, as 32% of all food produced in the world by weight or 24% by kilocalories (kcal) is wasted [26]. If global FW were treated as its own country, it would be the third largest emitter of GHGE, behind China and the United States, occupy 30% of the world's agricultural land area, and use the equivalent water of the annual discharge of the Volga river in Russia (i.e., 250 km^3) [27]. These FW statistics represent wasted resources and wasted opportunities to eat health-promoting foods, which comprise a large portion of total waste. In fact, the average global FW per capita per year could fulfill a person's dietary recommended intake (DRI) of 25 nutrients for 18 days [28]. Based on the types of food wasted, that amount of FW contains between 25 and 50% of the DRI for vitamin C, K, zinc, copper, manganese, and selenium for a person [28].

It is important to understand possible tradeoffs when promoting a solution to one problem to ensure it does not exacerbate another. For example, given the relatively high proportion of fruit and vegetable waste compared to meat waste, and the small proportion of vegetarians in the general population, could there be higher FW as a result of reducing meat-containing meals? Moreover, would the environmental impacts associated with that FW be substantial enough to negate the benefits of serving vegetarian meals as the default in large institutional settings? Although there are publications that assess hospital FW, its environmental impacts, and techniques for FW reduction, no literature has previously examined these questions [29–32].

Loma Linda University Medical Center (LLUMC) provides a unique setting to examine the potential tradeoffs associated with serving lower environmental impact foods with potentially higher FW compared to higher environmental impact foods with lower FW. Unlike many hospitals, LLUMC serves lacto-ovo vegetarian meals to patients by default for the first 24 h upon admission. However, patients have the option to reject the default and can choose their preferred meal items from standard menus, which include meat, after 24 h. As such, the aim of this case study was to examine the differences in FW and GHGE between vegetarian meals and meat-containing meals served in a hospital setting.

2. Materials and Methods

A plate waste audit was performed by Loma Linda University dietetics graduate students across seven consecutive days, from September 6 to 12, 2020 at LLUMC. Plates audited included those served at breakfast, lunch, and dinner and were provided by meal services on three hospital floors, which housed patients with the fewest special dietary orders (e.g., liquid diets or "nil per os" (NPO, nothing by mouth)). At least 20 plates were audited, upon tray return and prior to disposal, after each meal service. Each tray was assigned a de-identifiable number and meal type (i.e., "meat-containing" or "vegetarian") based on the food items listed on the tray ticket, which reflected the patient's menu order. Trays returned without tray tickets and no remaining meat items were categorized as unknown meal type. Floor number and diet order (regular or therapeutic) were also noted for each tray. Institutional review board approval was not needed since no patient-identifying information was collected. For each tray, all remaining individual food items were removed and individually weighed in grams before being discarded. Liquid diet trays were excluded from measurement due to the high proportion of total weight from liquid.

Data processing included removing container weight values from FW measured in containers. LLUMC also provided recipes for composite foods such as cooked entrees and soups, which were used to determine the proportional weights for individual ingredients (e.g., spinach, cheese, and egg white for spinach quiche). In addition to FW weight, GHGE were estimated using a combination of SimaPro life cycle assessment software and published literature used to fill any gaps where SimaPro did not have appropriate data [33–37]. The life cycle assessment (LCA) studies used for GHGE estimates all had cradle to farm or manufacturer gate system boundaries and reported results using a weight-based functional unit. The parameters of LCA studies included were system boundaries from cradle to farm or manufacturer gate or distributor, excluding retail, consumption, and disposal, and used a weight-based functional unit and attributional assumptions.

Descriptive and statistical analyses were conducted using IBM SPSS Statistics for Windows, version 28 (IBM Corp., Armonk, NY, USA). Tests for assumptions of normality and homogeneity of variance were performed using Kolmogorov–Smirnov and Levene's test, respectively. To examine between-group differences in total FW and total GHGE, independent *t*-tests were performed. Values that were ± 2.5 standard deviations from the mean were considered outliers. Visual assessment using boxplots indicated that there were no outliers, defined as values that were ± 2.5 standard deviations from the mean. Post hoc exploratory analyses were also conducted using independent t-tests to compare between-group differences for FW and GHGE by primary food categories. The exploratory analyses were considered secondary analyses, which were not driven by hypothesis testing; therefore, the significance level was not adjusted for multiple comparisons. In addition, effect size was calculated using Hedges' g for all primary and secondary outcomes. Data are reported as the mean $\pm$ standard deviation and the level of statistical significance was set at $p = 0.05$.

3. Results

Plate data were analyzed for 447 patient trays of the 471 that were collected. Twenty-four patient trays were excluded from analysis due to unknown meal type and absence of identifying characteristics (e.g., leftover meat or a tray ticket). Key findings of this study were that mean total plate waste was higher among meat-containing meals, and that the associated GHGE was lower among vegetarian meals.

3.1. Food Waste

3.1.1. Descriptive Statistics

The corresponding means and standard error of mean are presented graphically in Figure 1. The data for total FW were not normally distributed for either group ($p < 0.05$). Skewness of variables prevented transformation to a normal distribution. Non-parametric tests, such as the Mann–Whitney U test, resulted in unacceptable values (U > 10,000) and are most appropriate for analyzing ordinal data. Thus, non-parametric testing was excluded from the analytical approach. However, based on the central limit theorem, with adequate sample sizes ($n \geq 30$), violation of the normality assumption is unlikely to affect statistical findings. Therefore, parametric tests were acceptable due to the large sample size. Homogeneity of variances was observed ($p = 0.64$). Descriptive statistics for FW (g) by each food category and meal type are provided in Table A1 in Appendix A.

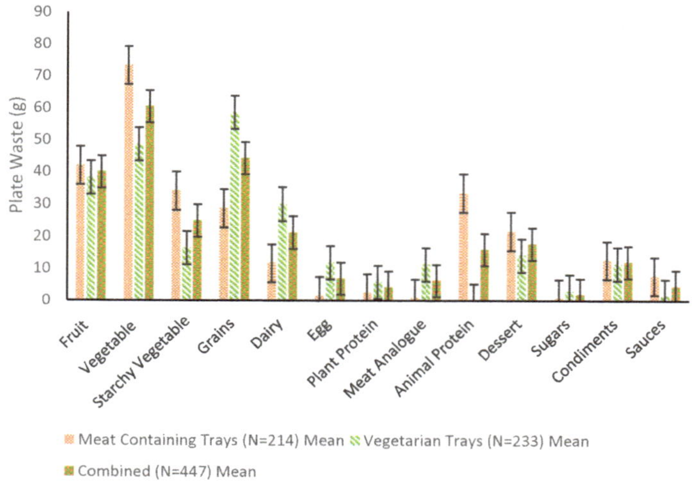

Figure 1. Categories of foods and their respective amounts of waste differentiated by meal type. Error bars represent the standard error of mean.

Total mean FW was greater among meat-containing meals (292.51 ± 180.77 g/plate) compared to vegetarian meals (258.46 ± 186.09 g/plate), with a mean difference of 34.05 g/plate, $t(445) = 1.96$, $p = 0.05$, $g = 0.19$ (Figure 2). The largest FW source for meat-containing meals was vegetables and fruit, while vegetarian meals had the most FW from grains and vegetables.

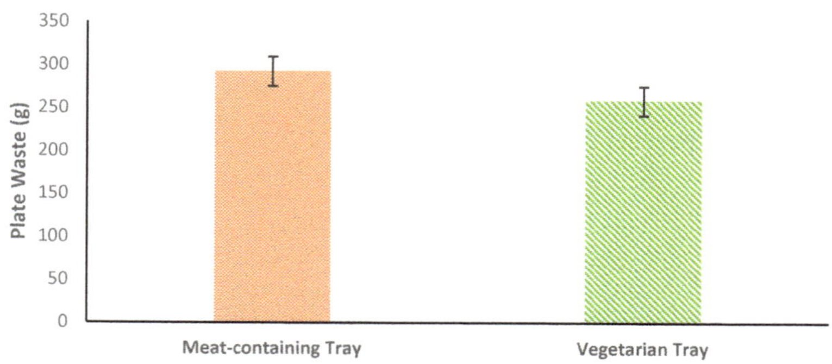

Figure 2. Plate waste from meat-containing and vegetarian meals. Vegetarian meals had less FW than meat-containing meals. Error bars represent the standard error of mean.

3.1.2. Exploratory Analyses

Exploratory analyses revealed significant differences in FW and GHGE between groups for analyzed food categories except fruit (Table 1). Vegetable and dessert waste were significantly greater among the meat-containing meals, while grains, dairy, egg, and plant protein waste were significantly greater among the vegetarian meals.

Table 1. Exploratory comparison analyses for food waste (g/plate) and GHGE (g CO_2 eq/plate) between meat-containing and vegetarian meal types by food category.

Food Waste (g/plate)		Total (N = 447)					
	M	±	SD	SE	*t*-Statistic	*p*-Value	Hedges' g
Fruit	40.03	±	50.19	2.37	0.80	0.43	0.08
Vegetable [1]	140.06	±	65.51	3.10	4.60	<0.001	0.44
Grains	44.27	±	54.73	2.59	6.10	<0.001	0.57
Dairy	21.26	±	45.69	2.16	4.40	<0.001	0.41
Egg	6.90	±	17.79	0.84	6.62	<0.001	0.61
Plant Protein [2]	4.20	±	17.96	0.85	2.03	0.049	0.19
Dessert	17.74	±	35.54	1.68	2.22	0.03	0.21
GHGE (g CO_2 eq/plate)							
Fruit	19.56	±	29.32	1.39	0.90	0.38	0.09
Vegetable [1]	30.27	±	33.16	1.57	4.17	<0.001	0.39
Grains	41.61	±	54.44	2.57	4.38	<0.001	0.41
Dairy	77.84	±	161.04	7.62	3.82	<0.001	0.36
Egg	23.40	±	60.18	2.85	6.14	<0.001	0.61
Plant Protein [2]	5.32	±	19.93	0.94	2.42	0.008	0.23
Dessert	108.08	±	323.83	15.32	2.67	0.004	0.25

[1] Includes vegetables and starchy vegetables; [2] Plant protein items consist of peanut butter, tofu, black beans, brown lentils, and hummus.

There were statistically significant differences between meat-containing meals and vegetarian meals for every major food category shared by both meal types except fruit.

3.2. Global Warming Potential

Descriptive statistics for GHGE by food category and meal type are provided in Table A2. The difference in total GHGE was also compared between meal types. The data were not normally distributed ($p < 0.001$) and homogeneity of variance was not observed ($p < 0.001$). The ratio of the meat-containing meals to the vegetarian meals is 1.1; thus, this violation is unlikely to affect statistical findings. Total GHGE was significantly greater for meat-containing meals (604.20 ± 643.45 g CO_2 eq) compared to vegetarian meals (356.66 ± 376.98 g CO_2 eq), t(445) = 4.995, $p < 0.001$, g = 0.47 (Figure 3).

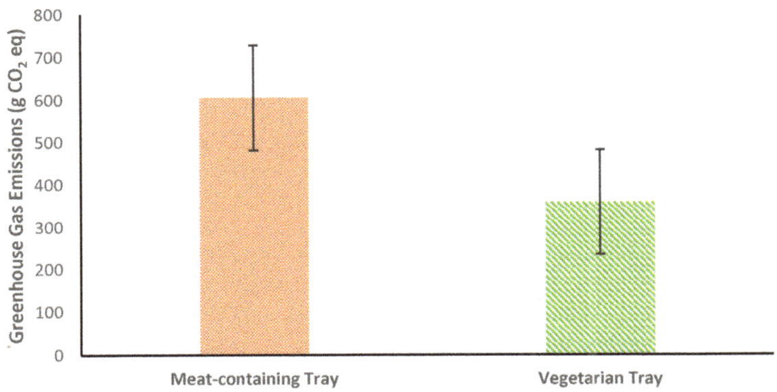

Figure 3. Mean total GHGE (g CO_2 eq) by meal type. Error bars represent the standard error of mean.

Total GHGE were significantly higher for FW from meat-containing meals than for vegetarian. The highest contributor to GHGE was animal protein, followed by dessert. The highest contributor to FW from vegetarian meals was dairy, followed by dessert.

GHGE from both meat-containing and vegetarian meals' waste had a high standard error of means. GHGE from vegetarian meals' waste was much lower than that from meat-containing meals' waste.

GHGE was significantly greater among meat-containing meals for the vegetable and dessert food categories compared to vegetarian meals. GHGE was significantly greater among vegetarian meals for grains, dairy, egg, and plant protein.

GHGE associated with plate waste showed statistically significant differences across all food categories except fruit when comparing plate waste from meat-containing meals to plate waste from vegetarian meals.

4. Discussion

The objective of this study was to examine the differences in FW and GHGE between vegetarian meals and meat-containing meals to determine if greater FW among vegetarian meals offset the associated environmental benefits when compared to meat-containing meals. Analysis of plate FW failed to demonstrate evidence that vegetarian meals are associated with more FW or corresponding GHGE. Therefore, there does not appear to be a tradeoff or downside to providing vegetarian meals to patients by default for the first 24 h following their admission to a hospital setting from this perspective.

Previous work has not investigated the possibility that extra FW would be generated by providing vegetarian meals by default, which could potentially negate the environmental benefit of doing so, when compared to serving meat-containing meals by default. Only a couple of studies have reported actual FW in hospitals at the item level [29,32]. Change in meal service style from traditional foodservice to room service can reduce FW by approximately one-third [30]. GHGE from meals in a hospital setting were estimated to be approximately 5 kg CO_2-eq per day for a 2000 kcal diet, with a range between approximately 0.5 and 8 kg CO_2-eq for liquid diets and high protein diets, respectively [31]. GHGE from plate waste itself amounted to an average of approximately 1 kg waste per patient per day, which was associated with approximately 1.8 kg CO_2-eq [32]. Plate waste refers to food that was served to a patient but not consumed, as opposed to tray waste, which includes other non-food waste, such as packaging [38]. Numerous studies indicate that the GHGE from animal-based foods are higher than those from plant-based foods [9,10,33,35].

However, FW from vegetarian meals in this study was approximately 11% lower than that from meat-containing meals, which represents a difference that is approximately half the reduction in FW observed in another study that examined FW reduction from a transition to room service rather than traditional foodservice [30]. Additionally, the average GHGE from daily plate waste per patient reported here for meat-containing and vegetarian meals was approximately 36% and 21%, respectively, of the average GHGE per day for a 2000 kcal diet in a hospital setting reported in another study [31]. In addition, the GHGE per patient per day in this study of approximately 1.8 kg CO_2 eq for meat-containing meals matches the value reported in another study of hospital FW and emissions of 1.8 kg CO_2 eq per patient per day [32].

Meal provision is considered an "environmental hot spot" in hospitals [39]. To address this, it has been proposed to list vegetarian meal choices first on menus and to offer more vegetarian meal options in hospitals [39]. The European Society for Clinical Nutrition and Metabolism (ESPEN) affirms the importance of providing vegetarian meals and other specialized dietary patterns to be respective of religious and dietary preferences to patients as well, noting the increased demand for vegetarian meals by patients [40]. Providing vegetarian meals in hospital settings may have synergistic benefits beyond reducing FW and environmental impacts by also promoting health.

California licensed health care facilities and state prisons are required by law to make available "wholesome, plant-based meal options" to meet patient needs and follow physicians' diet orders according to CA Senate Bill No. 1138 [41]. Additional California law (Senate Bill No. 1383) sets targets for statewide organics recycling to reduce short-lived climate pollutants, such as methane from food waste sent to landfill [42]. The American Medical Association passed a resolution in 2017 (H-150.949) calling on US hospitals to "improve the health of patients, staff, and visitors by providing a variety of healthy food, including plant-based meals" [43]. As hospitals work to comply with such laws and resolutions, this study demonstrates that serving plant-based or vegetarian meals may provide overall reductions in FW and GHGE generated from meal service.

United States federal regulations require that hospitals provide "a nourishing, palatable, well-balanced diet that meets the daily nutritional and special dietary needs" of patients (42 Code of Federal Regulations 483.35), informed by the recommendations of a qualified registered dietitian, and that menus meet nutritional needs as recommended by the Food and Nutrition Board of the National Research Council, National Academy of Sciences [44]. While maintaining compliance with such regulations, as well as specific state regulations, there may be particular advantages conveyed by providing vegetarian meals. For example, there is a clear connection between proper nutrition and a healthy immune system to protect against infections [45]. Of particular relevance currently, healthy diets as measured by the Plant-Based Diet Score are associated with lower risk and severity of COVID-19 [46]. Health care workers (who often eat meals provided by the hospital cafeteria) who reported following plant-based diets and low-meat diets also had lower odds of moderate to severe COVID-19 [47].

There were some limitations to this study. Some food categories were excluded from exploratory statistical analysis due to inherent differences between meal types (e.g., vegetarian meals contained no animal protein). Some additional food categories were excluded due to having near negligible mean values. The food categories excluded were meat analogues, animal protein, sugars, condiments, and sauces. The larger amount of plant protein waste from vegetarian meals was expected, as these trays were more likely to contain higher amounts of plant proteins including peanut butter, tofu, black beans, brown lentils, and hummus.

Future research should include measurements of initial food weights to understand the proportion of each meal wasted and facilitate comparison across meals with different starting weights. It may also be useful to explore differences when correcting for kcal content of meals. Additional research could also examine correlations between meal type (e.g., liquid, dysphagia, cardiac, and low sodium), patient ward (e.g., surgery and intensive care), and outcomes (e.g., length of stay), as well as explore differences based on demographic factors such as sex or age.

Generalizability of the findings from this research is likely most applicable to other hospitals and similar settings where food is provided, but from fairly limited options and with few if any alternatives. In a hospital setting, there are often limited choices and the consumer may be feeling unwell, both of which increase the likelihood of them wasting food. In contrast, consumers are normally able to choose from a wide array of foods in a variety of settings, reducing the likelihood that they will waste the food they choose to consume. Therefore, it is unlikely that similar levels of food waste would be observed outside a hospital setting. It is unclear whether or not a proportional difference in food waste between vegetarian and meat-containing meals would be maintained outside a hospital setting. However, it is well known that the environmental impacts associated with meat are greater than those associated with most vegetarian foods, so it is reasonable to expect that food waste from meat-containing meals would still have higher GHGE for a similar amount of food wasted.

5. Conclusions

It is important both to reduce the GHGE associated with food provision and reduce the proportion of food that goes to waste as part of efforts to limit the negative environmental consequences of food systems. Fortunately, the case study examined here provides an example where one choice—serving vegetarian meals to patients by default for their first 24 h in a hospital setting—improves both outcomes. Food waste from vegetarian meals was lower in both total weight and associated GHGE than food waste from meat-containing meals.

Author Contributions: Conceptualization, A.B., J.S. and H.L.; methodology A.B., B.E. and H.L.; validation, A.B.; formal analysis, N.B.; investigation, B.E.; data curation, B.E. and N.B.; writing—original draft preparation, A.B., B.E. and N.B.; writing—review and editing, A.B., B.E., J.S. and H.L.; visualization, A.B. and N.B.; supervision, H.L. and J.S.; project administration, H.L.; funding acquisition, H.L. All authors have read and agreed to the published version of the manuscript.

Funding: This research was funded by Point Loma Nazarene University's Wesleyan Center (internal) grant.

Institutional Review Board Statement: Not applicable.

Informed Consent Statement: Not applicable.

Data Availability Statement: Data supporting results are available in Appendix A.

Acknowledgments: The authors thank Jean Sellars and Kalvin Lazcano for their cooperation and allowing the collection of food waste at LLUMC, as well as Anna Salisbury for her role in data collection and analysis.

Conflicts of Interest: The authors declare no conflict of interest. The funders had no role in the design of the study; in the collection, analyses, or interpretation of data; in the writing of the manuscript, or in the decision to publish the results.

Appendix A

Table A1. Descriptive statistics for plate waste (g/plate) by meal type and food category, presented as M ± SD.

Food Category	Meat-Containing Meals (N = 214)			Vegetarian Meals (N = 233)			Total (N = 447)		
Fruit	42.00	±	50.78	38.21	±	49.69	40.03	±	50.19
Vegetable	73.33	±	77.09	48.64	±	80.20	60.46	±	79.60
Starchy Vegetable	34.12	±	46.64	16.42	±	37.06	24.90	±	42.80
Grains	28.74	±	36.74	58.54	±	63.97	44.27	±	54.73
Dairy	11.72	±	32.52	30.02	±	53.68	21.26	±	45.69
Egg	1.51	±	8.23	11.86	±	22.24	6.90	±	17.79
Plant Protein	2.45	±	11.75	5.80	±	22.09	4.20	±	17.96
Meat Analogue	0.79	±	9.46	11.32	±	27.24	6.28	±	21.36
Animal Protein	33.39	±	36.41	0.00	±	0.00	15.98	±	30.20
Dessert	21.64	±	38.00	14.16	±	32.79	17.74	±	35.54
Sugars	0.72	±	4.59	2.91	±	9.04	1.86	±	7.33
Condiments	12.67	±	15.22	11.50	±	10.27	12.06	±	12.88
Sauces	7.72	±	14.77	1.50	±	4.53	4.48	±	11.16
Total Plate Waste	292.51	±	180.77	258.46	±	186.09	274.76	±	181.15

Table A2. Descriptive statistics for GHGE (g CO_2 eq/plate) by meal type and food category, presented as M ± SD.

GHGE by Food Category	Meat-Containing Meals (N = 214)			Vegetarian Meals (N = 233)			Total (N = 447)		
Fruit	20.87	±	32.00	18.36	±	49.69	19.56	±	29.32
Vegetable	26.95	±	32.05	18.05	±	28.79	22.31	±	30.74
Starchy Vegetable	9.47	±	13.40	4.66	±	10.94	6.96	±	12.43
Grains	30.04	±	45.03	52.24	±	59.81	41.61	±	54.44
Dairy	47.87	±	122.28	105.37	±	185.29	77.84	±	161.04
Egg	5.12	±	27.79	40.20	±	75.08	23.40	±	60.18
Plant Protein	2.95	±	12.95	7.50	±	24.42	5.32	±	19.93

Table A2. *Cont.*

GHGE by Food Category	Meat-Containing Meals (N = 214)			Vegetarian Meals (N = 233)			Total (N = 447)		
Meat Analogue	1.315	±	15.84	20.92	±	51.04	11.54	±	39.71
Animal Protein	285.01	±	398.24	0.00	±	0.00	136.45	±	310.83
Dessert	150.53	±	380.48	69.09	±	254.16	108.08	±	323.83
Sugars	0.55	±	3.47	2.22	±	7.10	1.42	±	5.72
Condiments	22.21	±	40.88	16.68	±	31.06	19.33	±	1.35
Sauces	1.31	±	2.45	1.50	±	4.53	1.37	±	4.21
Total	604.21	±	643.45	356.66	±	374.98	475.17	±	536.09

References

1. Rockström, J.; Steffen, W.; Noone, K. A safe operating space for humanity. *Nature* **2009**, *461*, 472–475. [CrossRef] [PubMed]
2. Atwoli, L.; Baqui, A.H.; Benfield, T.; Bosurgi, R.; Godlee, F.; Hancocks, S.; Horton, R.; Laybourn-Langton, L.; Monteiro, C.A.; Norman, I.; et al. Call for emergency action to limit global temperature increases, restore biodiversity and protect health. *Allergy* **2022**, *77*, 730–733. [CrossRef] [PubMed]
3. Vermeulen, S.J.; Campbell, B.M.; Ingram, J.S.I. Climate Change and Food Systems. *Annu. Rev. Environ. Resour.* **2012**, *37*, 195–222. [CrossRef]
4. Rosenzweig, C.; Mbow, C.; Barioni, L.G.; Benton, T.G.; Herrero, M.; Krishnapillai, M.; Liwenga, E.T.; Pradhan, P.; Rivera-Ferre, M.G.; Sapkota, T.; et al. Climate change responses benefit from a global food system approach. *Nat. Food* **2020**, *1*, 94–97. [CrossRef]
5. Crippa, M.; Solazzo, E.; Guizzardi, D.; Monforti-Ferrario, F.; Tubiello, F.N.; Leip, A. Food systems are responsible for a third of global anthropogenic GHG emissions. *Nat. Food* **2021**, *2*, 198–209. [CrossRef]
6. FAO; IFAD; UNICEF; WFP; WHO. *The State of Food Security and Nutrition in the World 2017: Building Resilience for Peace and Food Security*; FAO: Rome, Italy, 2017.
7. Hasegawa, T.; Havlík, P.; Frank, S.; Palazzo, A.; Valin, H. Tackling food consumption inequality to fight hunger without pressuring the environment. *Nat. Sustain.* **2019**, *2*, 826–833. [CrossRef]
8. Godfray, H.C.J.; Beddington, J.R.; Crute, I.R.; Haddad, L.; Lawrence, D.; Muir, J.F.; Pretty, J.; Robinson, S.; Thomas, S.M.; Toulmin, C. Food security: The challenge of feeding 9 billion people. *Science* **2010**, *327*, 812–818. [CrossRef] [PubMed]
9. Heller, M.; Keoleian, G.; Rose, D. Implications of Future Us Diet Scenarios on Greenhouse Gas Emissions. 2020, p. 34. Available online: https://habitsofwaste.org/wp-content/uploads/2020/11/2020-CSS20-01-v3.pdf (accessed on 22 February 2022).
10. Stylianou, K.S.; Fulgoni, V.L.; Jolliet, O. Small targeted dietary changes can yield substantial gains for human and environmental health. *Nat. Food* **2021**, *2*, 616–627. [CrossRef]
11. Kustar, A.; Patino-echeverri, D. A Review of Environmental Life Cycle Assessments of Diets: Plant-Based Solutions Are Truly Sustainable, even in the Form of Fast Foods. *Sustainability* **2021**, *13*, 9926. [CrossRef]
12. Fresán, U.; Sabaté, J. Vegetarian Diets: Planetary Health and Its Alignment with Human Health. *Adv. Nutr.* **2019**, *10*, S380–S388. [CrossRef] [PubMed]
13. Melina, V.; Craig, W.; Levin, S. Position of the Academy of Nutrition and Dietetics: Vegetarian Diets. *J. Acad. Nutr. Diet.* **2016**, *116*, 1970–1980. [CrossRef] [PubMed]
14. Barnard, N.D.; Kahleova, H.; Levin, S.M. The Use of Plant-Based Diets for Obesity Treatment. *Int. J. Dis. Reversal Prev.* **2019**, *1*, 12. [CrossRef]
15. Sofi, F.; Dinu, M.; Pagliai, G.; Cesari, F.; Gori, A.M.; Sereni, A.; Becatti, M.; Fiorillo, C.; Marcucci, R.; Casini, A. Low-calorie vegetarian versus mediterranean diets for reducing body weight and improving cardiovascular risk profile. *Circulation* **2018**, *137*, 1103–1113. [CrossRef] [PubMed]
16. Lee, K.W.; Loh, H.C.; Ching, S.M.; Devaraj, N.K.; Hoo, F.K. Effects of vegetarian diets on blood pressure lowering: A systematic review with meta-analysis and trial sequential analysis. *Nutrients* **2020**, *12*, 1604. [CrossRef] [PubMed]
17. Lee, Y.; Park, K. Adherence to a vegetarian diet and diabetes risk: A systematic review and meta-analysis of observational studies. *Nutrients* **2017**, *9*, 603. [CrossRef] [PubMed]
18. Rizzo, N.S.; Sabaté, J.; Jaceldo-Siegl, K.; Fraser, G.E. Vegetarian dietary patterns are associated with a lower risk of metabolic syndrome: The Adventist Health Study 2. *Diabetes Care* **2011**, *34*, 1225–1227. [CrossRef] [PubMed]
19. Clarys, P.; Deliens, T.; Huybrechts, I.; Deriemaeker, P.; Vanaelst, B.; De Keyzer, W.; Hebbelinck, M.; Mullie, P. Comparison of nutritional quality of the vegan, vegetarian, semi-vegetarian, pesco-vegetarian and omnivorous diet. *Nutrients* **2014**, *6*, 1318–1332. [CrossRef] [PubMed]
20. Lynch, H.; Johnston, C.; Wharton, C. Plant-based diets: Considerations for environmental impact, protein quality, and exercise performance. *Nutrients* **2018**, *10*, 1841. [CrossRef] [PubMed]
21. Leischik, R.; Spelsberg, N. Vegan Triple-Ironman (Raw Vegetables/Fruits). *Case Reports Cardiol.* **2014**, *2014*, 317246. [CrossRef]
22. Wirnitzer, K.C.; Kornexl, E. Energy and Macronutrient Intake of a Female Vegan Cyclist During an 8-Day Mountain Bike Stage Race. *Baylor Univ. Med. Cent. Proc.* **2014**, *27*, 42–45. [CrossRef]

23. Trumbo, P.; Yates, A.; Schlicker, S.; Poos, M. Dietary Reference Intakes: Vitamin A, Vitamin K, Arsenic, Boron, Chromium, Copper, Iodine, Iron, Manganese, Molybdenum, Nickel, Silicon, Vanadium, and Zinc. *J. Am. Diet. Assoc.* **2001**, *101*, 294–301. [CrossRef]
24. Holick, M. Vitamin D Deficiency. *N. Engl. J. Med.* **2007**, *357*, 266–281. [CrossRef] [PubMed]
25. Conrad, Z.; Niles, M.T.; Neher, D.A.; Roy, E.D.; Tichenor, N.E.; Jahns, L. Relationship between food waste, diet quality, and environmental sustainability. *PLoS ONE* **2018**, *13*, e0195405. [CrossRef] [PubMed]
26. Lipinski, B.; Hanson, C.; Lomax, J.; Kitinoja, L.; Waite, R.; Searchinger, T. *Toward a Sustainable Food System Reducing Food Loss and Waste*; World Resources Institute: Washington, DC, USA, 2016.
27. Food and Agriculture Organization of the United Nations. *Food Wastage Footprint. Impacts on Natural Resources. Summary Report*; FAO: Rome, Italy, 2013; ISBN 9789251077528.
28. Chen, C.; Chaudhary, A.; Mathys, A. Nutritional and environmental losses embedded in global food waste. *Resour. Conserv. Recycl.* **2020**, *160*, 104912. [CrossRef]
29. Williams, P.; Walton, K. Plate waste in hospitals and strategies for change. *e-SPEN* **2011**, *6*, e235–e241. [CrossRef]
30. McCray, S.; Maunder, K.; Krikowa, R.; MacKenzie-Shalders, K. Room Service Improves Nutritional Intake and Increases Patient Satisfaction While Decreasing Food Waste and Cost. *J. Acad. Nutr. Diet.* **2018**, *118*, 284–293. [CrossRef] [PubMed]
31. Vidal, R.; Moliner, E.; Pikula, A.; Mena-Nieto, A.; Ortega, A. Comparison of the carbon footprint of different patient diets in a Spanish hospital. *J. Health Serv. Res. Policy* **2015**, *20*, 39–44. [CrossRef] [PubMed]
32. Dias-Ferreira, C.; Santos, T.; Oliveira, V. Hospital food waste and environmental and economic indicators-A Portuguese case study. *Waste Manag.* **2015**, *46*, 146–154. [CrossRef] [PubMed]
33. Berardy, A.; Fresán, U.; Matos, R.A.; Clarke, A.; Mejia, A.; Jaceldo-Siegl, K.; Sabaté, J. Environmental impacts of foods in the adventist health study-2 dietary questionnaire. *Sustainability* **2020**, *12*, 267. [CrossRef]
34. Mejia, A.; Harwatt, H.; Jaceldo-Siegl, K.; Sranacharoenpong, K.; Soret, S.; Sabaté, J. Greenhouse Gas Emissions Generated by Tofu Production: A Case Study. *J. Hunger Environ. Nutr.* **2018**, *13*, 131–142. [CrossRef]
35. Clune, S.; Crossin, E.; Verghese, K. Systematic review of greenhouse gas emissions for different fresh food categories. *J. Clean. Prod.* **2017**, *140*, 766–783. [CrossRef]
36. Carlsson-Kanyama, A.; González, A.D. Potential contributions of food consumption patterns to climate change. *Am. J. Clin. Nutr.* **2009**, *89*, 1704S–1709S. [CrossRef] [PubMed]
37. González, A.D.; Frostell, B.; Carlsson-Kanyama, A. Protein efficiency per unit energy and per unit greenhouse gas emissions: Potential contribution of diet choices to climate change mitigation. *Food Policy* **2011**, *36*, 562–570. [CrossRef]
38. Alharbi, N.S.; Qattan, M.Y.; Alhaji, J.H. Towards sustainable food services in hospitals: Expanding the concept of "plate waste" to "tray waste". *Sustainability* **2020**, *12*, 6872. [CrossRef]
39. Keller, R.L.; Muir, K.; Roth, F.; Jattke, M.; Stucki, M. From bandages to buildings: Identifying the environmental hotspots of hospitals. *J. Clean. Prod.* **2021**, *319*, 128479. [CrossRef]
40. Thibault, R.; Abbasoglu, O.; Ioannou, E.; Meija, L.; Ottens-Oussoren, K.; Pichard, C.; Rothenberg, E.; Rubin, D.; Siljamäki-Ojansuu, U.; Vaillant, M.F.; et al. ESPEN guideline on hospital nutrition. *Clin. Nutr.* **2021**, *40*, 5684–5709. [CrossRef] [PubMed]
41. California Senate. Senate Bill No. 1138. 2018. Available online: https://leginfo.legislature.ca.gov/faces/billTextClient.xhtml?bill_id=201720180SB1138 (accessed on 22 February 2022).
42. California Senate. Senate Bill No. 1383. 2016. Available online: https://leginfo.legislature.ca.gov/faces/billNavClient.xhtml?bill_id=201520160SB1383 (accessed on 22 February 2022).
43. American Medical Association Healthy Food Options in Hospitals H-150.949. Available online: https://policysearch.ama-assn.org/policyfinder/detail/HealthyFoodOptionsinHospitalsH-150.949?uri=%2FAMADoc%2FHOD.xml-0-627.xml (accessed on 16 February 2022).
44. *Code of Federal Regulations. Title 42: Public Health*; National Archives and Records Administration: Washington, DC, USA, 2011; pp. 1–871.
45. Calder, P.C.; Carr, A.C.; Gombart, A.F.; Eggersdorfer, M. Optimal nutritional status for a well-functioning immune system is an important factor to protect against viral infections. *Nutrients* **2020**, *12*, 2326. [CrossRef] [PubMed]
46. Merino, J.; Joshi, A.D.; Nguyen, L.H.; Leeming, E.R.; Mazidi, M.; Drew, D.A.; Gibson, R.; Graham, M.S.; Lo, C.-H.; Capdevila, J.; et al. Diet quality and risk and severity of COVID-19: A prospective cohort study. *Gut* **2021**, *70*, 2096–2104. [CrossRef] [PubMed]
47. Kim, H.; Rebholz, C.M.; Hegde, S.; Lafiura, C.; Raghavan, M.; Lloyd, J.F.; Cheng, S.; Seidelmann, S.B. Plant-based diets, pescatarian diets and COVID-19 severity: A population-based case-control study in six countries. *BMJ Nutr. Prev. Health* **2021**, *4*, 257–266. [CrossRef]

nutrients

MDPI

Article

Pea Proteins Have Anabolic Effects Comparable to Milk Proteins on Whole Body Protein Retention and Muscle Protein Metabolism in Old Rats

Jérôme Salles [1], Christelle Guillet [1], Olivier Le Bacquer [1], Carmen Malnero-Fernandez [2], Christophe Giraudet [1], Véronique Patrac [1], Alexandre Berry [1], Philippe Denis [1], Corinne Pouyet [1], Marine Gueugneau [1], Yves Boirie [1,3], Heidi Jacobs [2] and Stéphane Walrand [1,3,*]

[1] Unité de Nutrition Humaine (UNH), Université Clermont Auvergne, INRAE, CRNH Auvergne, 63000 Clermont-Ferrand, France; jerome.salles@inrae.fr (J.S.); christelle.guillet@uca.fr (C.G.); olivier.lebacquer@inrae.fr (O.L.B.); christophe.giraudet@inrae.fr (C.G.); veronique.patrac@inrae.fr (V.P.); alexandre.berry@inrae.fr (A.B.); philippe.denis.2@inrae.fr (P.D.); corinne.pouyet@inrae.fr (C.P.); marine.gueugneau@inrae.fr (M.G.); yves.boirie@inrae.fr (Y.B.)

[2] Cosucra-Groupe S.A., 7740 Warcoing, Belgium; cmalnero-fernandez@cosucra.com (C.M.-F.); hjacobs@cosucra.com (H.J.)

[3] Service de Nutrition Clinique, CHU Clermont-Ferrand, 63000 Clermont-Ferrand, France

* Correspondence: stephane.walrand@inrae.fr; Tel.: +33-4-73-17-82-81; Fax: +33-4-73-17-82-22

Abstract: Plant proteins are attracting rising interest due to their pro-health benefits and environmental sustainability. However, little is known about the nutritional value of pea proteins when consumed by older people. Herein, we evaluated the digestibility and nutritional efficiency of pea proteins compared to casein and whey proteins in old rats. Thirty 20-month-old male Wistar rats were assigned to an isoproteic and isocaloric diet containing either casein (CAS), soluble milk protein (WHEY) or Pisane™ pea protein isolate for 16 weeks. The three proteins had a similar effect on nitrogen balance, true digestibility and net protein utilization in old rats, which means that different protein sources did not alter body composition, tissue weight, skeletal muscle protein synthesis or degradation. Muscle mitochondrial activity, inflammation status and insulin resistance were similar between the three groups. In conclusion, old rats used pea protein with the same efficiency as casein or whey proteins, due to its high digestibility and amino acid composition. Using these plant-based proteins could help older people diversify their protein sources and more easily achieve nutritional intake recommendations.

Keywords: pea proteins; plant proteins; sarcopenia; skeletal muscle; protein digestibility; muscle protein metabolism

Citation: Salles, J.; Guillet, C.; Le Bacquer, O.; Malnero-Fernandez, C.; Giraudet, C.; Patrac, V.; Berry, A.; Denis, P.; Pouyet, C.; Gueugneau, M.; et al. Pea Proteins Have Anabolic Effects Comparable to Milk Proteins on Whole Body Protein Retention and Muscle Protein Metabolism in Old Rats. *Nutrients* **2021**, *13*, 4234. https://doi.org/10.3390/nu13124234

Academic Editors: Winston Craig and Ujué Fresán

Received: 13 October 2021
Accepted: 22 November 2021
Published: 25 November 2021

1. Introduction

Alongside animal proteins, plant proteins are a critical part of the equation to help meet future protein demand and achieve worldwide food security. In the US, demand for plant proteins grew by 20% in both 2018 and 2019 [1]. This growing interest in plant proteins is driven by multiple factors, such as food safety concerns, rising food intolerances, increased accessibility of vegetarian and vegan foods, environmental concerns, sustainability imperatives, and consumer adoption of proactive approaches to health and wellbeing. The nutritional benefits of these new protein sources are still under investigation, with studies looking into their health benefits while also exploring their limits, such as allergenicity or anti-nutritional substance content [2]. Consumer acceptability needs to be carefully defined, as it remains the final bottleneck for developing new protein sources.

Grain legumes are a valuable source of plant food proteins, and so rising protein demand is expected to increase the dietary importance of grain legumes. Pulses generally have a higher nutritional value than other crops, especially since the onset of domestication and genetic selection processes operated by humans. Pea proteins have enough essential amino acid (EAA) content (30%) to meet WHO/FAO/UNU-recommended requirements [3]. Note that EAA requirement is based on a recommended adult protein intake of 0.8 g/kg body weight/day. Note also that peas provide well above the recommended leucine requirements [4]. In addition to providing proteins with suitable EAA profiles, legumes contain digestible carbohydrates, and some of them also contain fat.

There is considerable interest in the potential of using plant-based proteins to support muscle mass maintenance and/or growth, as demonstrated by the number of recent papers studying the impact of intakes of plant-based protein, e.g., pea proteins, on skeletal muscle anabolic response in athletes [2,5]., Dairy whey protein is a shared choice for protein supplementation in athletes because of its leucine content, its digestibility, and its ability to activate muscle protein synthesis. Most extant research on plant proteins in athletes has set out to compare and evaluate the effects of dietary supplementations with whey and pea proteins in conjunction with resistance training on muscle anabolism and strength. Taken together, the data revealed that whey and pea protein treatments led to similar responses to resistance exercise. Whey and pea proteins promote comparable muscle strength, physical performance, and body composition following resistance training [6], especially in beginners or people returning to weight training [7].

These same plant proteins could be equally valuable in other populations, such as older people, to help maintain muscle mass and slow down the aging-related process of sarcopenia. However, despite their reported efficacy in athletes, the effects of pea and other plant proteins in older people suffering from sarcopenia have not yet been disclosed. The fact that pea protein provides well above the recommended leucine requirements points to it playing a potentially valuable role in combating the loss of skeletal muscle mass and function in older subjects. Leucine is an anabolic amino acid with proven effectiveness for the maintenance of muscle mass during aging [8]. Meeting the body's quantitative daily demand for EAA is vitally important; the quality of protein consumed by older people is an equally important factor, and is generally determined by its digestibility and utilizability by the body. Among milk proteins, whey protein digests quickly, while casein digests slowly as it clots at acidic pH in the stomach. Numerous experiments have set out to determine whether fast or slow digestion was better for muscle protein synthesis and muscle building. The bottom line is that rapid digestion is best for stimulating muscle protein synthesis and increasing muscle mass, even in older people [9]. Interestingly, a previous study has shown that pea protein transiently aggregates in the stomach and has an intermediately-fast intestinal bioavailability midway between those of whey and casein [10].

When new sources of dietary proteins are tested for nutritional quality, the first studies are carried out using animal models, as advised by FAO. The second step in such studies is often to evaluate the interest of the protein in some pathophysiological situations characterized by a reduced capacity to assimilate and metabolize proteins, as is the case in older subjects. These animal studies make it possible to precisely assess protein metabolism in certain key tissues such as skeletal muscle. Such a study is difficult to perform in humans. For pea proteins, although its digestibility is high in young rats, there is little data on the nutritional value of pea proteins in old rats as compared to dairy proteins, and particularly in terms of protein digestibility and metabolism. To address this gap, this study used old rats to evaluate the efficiency of pea proteins as compared to dairy proteins, i.e., casein and whey proteins, in terms of protein digestibility, body protein retention, muscle protein synthesis and degradation and muscle protein accretion.

2. Materials and Methods

2.1. Animal Experiment

All animal procedures were approved by the local institutional animal care and use committee (Comité d'Ethique en Matière d'Expérimentation Animale Auvergne: C2EA-02) and conducted in accordance with the European guidelines for the care and use of laboratory animals (2010-63UE) (Authorization number: APAFIS#5329-2016051115541284 v2). Animals were housed in the INRAE's Human Nutrition Research animal facility (Agreement No. D6334515).

A total of thirty 20-month-old male Wistar rats were obtained from Janvier Labs (Le Genest-St-Isle, France). All animals came from the same batch and were bred under the same conditions throughout their lives. The rats were housed in individual cages under controlled environment conditions (12-h light/12-h dark cycle, temperature 22 °C) with free access to water. All of the rats were fed a maintenance diet (A04, Safe, Augy, France) ad libitum for a 2-week acclimatization period. Rats were then randomized into three groups according to body weight, fat mass and lean mass. Animals were assigned (n = 10 per group) to a diet containing either 14% casein (Armorprotéines, Saint-Brice-en-Cogles, France) (CAS rats), 14% soluble milk protein, i.e., Protarmor™ 80, a Whey protein concentrate (Armorprotéines, Saint-Brice-en-Cogles, France) (WHEY rats) or 14% pea proteins, i.e., Pisane™ (Cosucra, Warcoing, Belgium) (PEA rats) for 16 weeks. The three experimental diets were isoproteic and isocaloric (Tables 1 and 2). Different protein to nitrogen conversion factors were used depending on the protein source used. Specifically, the conversion factors used were: 6.15 for casein, 6.08 for whey and 5.36 for pea protein. Dietary AA levels were analyzed by the ABioC laboratory (Arzacq, France) according to EN ISO 13903:2005 standard method (Table 1). Body weight and food intake were measured weekly. At the end of the experiment and after an overnight fast, the remaining CAS (n = 6), WHEY (n = 6) and PEA (n = 8) rats were anesthetized. Blood samples were collected from the abdominal aorta and drawn into precooled ethylenediaminetetraacetic acid (EDTA) tubes. After centrifugation, plasma was removed and frozen at −80 °C until analysis. Liver, heart, adipose tissues and hindlimb skeletal muscles were weighed, snap-frozen in liquid nitrogen, and stored at −80 °C for later analysis.

Table 1. Experimental diet: composition and amino acid content.

	CAS	WHEY	PEA
Diet composition (g/100 g)			
Protein			
Casein	14		
Soluble milk protein		14	
Pea protein			14
Fat (soybean oil)	6	6	6
Carbohydrates	68	68	68
Cellulose	7.5	7.5	7.5
Vitamin and mineral mix	4.5	4.5	4.5
Calculated energy (kcal/100 g)	**412**	**412**	**412**
Amino acid content (g/100 g protein)			
Tryptophan	1.17	2.09	0.87
Threonine	4.18	5.09	3.79
Aspartic acid	6.86	11.47	12.26
Serine	5.57	4.69	5.37
Lysine	7.55	9.84	7.45
Valine	6.16	5.23	5.25
Proline	10.84	4.77	4.30
Alanine	2.87	4.93	4.34
Phenylalanine	4.69	3.62	5.56
Isoleucine	4.79	5.26	4.67
Glycine	1.75	1.83	4.02
Tyrosine	4.19	2.76	3.28
Arginine	3.11	2.54	8.12
Leucine	8.99	12.15	8.51
Histidine	2.69	2.11	2.41
Glutamic acid	21.47	16.86	17.70
Methionine	2.65	2.05	1.03
Cysteine	0.49	2.70	1.07

Table 2. Composition of the protein sources.

	CASEIN Protein	WHEY Protein	PEA Protein
Protein (%)	90.2	80.9	83.6
Fat (%)	<1	4.7	<1
Carbohydrates (%)	<1	4.3	5.6
Moisture (%)	9.0	5.6	4.4
Ash (%)	<2	4.5	5.8

Compositions were obtained from technical data sheets provided by suppliers.

2.2. Whole Body Composition

At the beginning, middle (after 8 weeks) and end (after 16 weeks) of the experiment, fat and lean body mass (g) were measured in non-anesthetized living animals placed in an EchoMRI-100 body composition analyzer (Echo Medical Systems LLC, Houston, TX, USA).

2.3. Protein Quality Evaluation

To collect total urine and feces, rats were placed in metabolic cages (Tecniplast France, Decines-Charpieu, France) for 4 days in the last week of the experimental protocol. Total excreted nitrogen was then determined by the Dumas method at Institut UniLaSalle (Beauvais, France) [11]. Dietary protein quality was evaluated by calculating nitrogen balance (NB), apparent protein digestibility (AD), true protein digestibility (TD), net protein utilization (NPU) and biological value (BV) using the following equations [12]:

$$\mathrm{NB(g)} = \mathrm{NI} - (\mathrm{FN} + \mathrm{UN})$$

$$\mathrm{AD}\ (\%) = \frac{\mathrm{NI} - \mathrm{FN}}{\mathrm{NI}} \times 100$$

$$\mathrm{TD}\ (\%) = \frac{\mathrm{NI} - (\mathrm{FN} - \mathrm{EFN})}{\mathrm{NI}} \times 100$$

$$\mathrm{NPU}\ (\%) = \frac{\mathrm{NI} - (\mathrm{FN} + \mathrm{UN} - \mathrm{EFN} - \mathrm{EUN})}{\mathrm{NI}} \times 100$$

$$\mathrm{BV}\ (\%) = \frac{\mathrm{NPU}}{\mathrm{TD}} \times 100$$

where NI is nitrogen intake, FN is fecal nitrogen, UN is urinary nitrogen, EFN is endogenous fecal nitrogen, and EUN is endogenous urinary nitrogen. A group of old rats that received a nitrogen-free diet during the metabolic cage period was used to deduce fecal and urinary endogenous nitrogen excretions.

2.4. Plasma Analyses

Plasma levels of fasting glucose, triglycerides, and total cholesterol were determined using a Konelab 20 analyzer (Thermo-Electron Corporation, Waltham, MA, USA). ELISA kits were used to determine insulin (Alpco Diagnostics, Salem, NH, USA), leptin, (Biovendor, Bmo, Czech Republic), adiponectin (AssayPro, St Charles, MO, USA), TNFα (Millipore, Molsheim, France) and IL-10 (Diaclone, Besançon, France). Homeostatic model assessment of insulin resistance (HOMA-IR) was calculated to assess insulin sensitivity in old rats, using the formula:

$$\mathrm{HOMA} - \mathrm{IR} = \frac{(\text{fasting glucose} \times \text{fasting insulin})}{22.5}$$

with fasting glucose level expressed as mmol/L and fasting insulin level expressed as mIU/L.

2.5. Protein Synthesis Measurement

To study muscle protein synthesis, we measured rate of incorporation of a stable isotope, i.e., an AA L-[$^{13}C_6$]-labeled phenylalanine (Eurisotop Saint-Aubin, France), into muscle proteins using the flooding dose method. Fasting rats were injected subcutaneously with a large dose of L-[$^{13}C_6$] phenylalanine (50% mol excess, 150 µmol/100 g) to flood the precursor pool of protein synthesis. Incorporation time of labeled phenylalanine was 50 min. A 50-mg piece of plantaris muscle was used to isolate and hydrolyze total mixed proteins as previously described [13]. After derivatization, L-[$^{13}C_6$] phenylalanine enrichments in hydrolyzed proteins and in tissue fluid were assessed using gas chromatography–mass spectrometry (Hewlett-Packard 5971A; Hewlett-Packard Co., Palo Alto, CA, USA). Fractional synthesis rates (FSR) of proteins were calculated using the equation:

$$\mathrm{FSR} = \frac{\mathrm{Ei}}{\mathrm{Ep}\ \times \mathrm{t}} \times 100 \qquad (1)$$

where Ei is enrichment as atom percent excess of L-[$^{13}C_6$] phenylalanine derived from phenylalanine from proteins at time t (minus basal enrichment), Ep is mean enrichment in the precursor pool (tissue fluid L-[$^{13}C_6$] phenylalanine), and t is incorporation time in hours.

2.6. Western-Blot Analysis

Homogenates of frozen plantaris muscles were prepared as previously described [14]. Denatured proteins were separated on a polyacrylamide gel and electrotransferred to a polyvinylidene difluoride membrane (Millipore, Molsheim, France). After blocking with 5% skimmed dry milk in Tris-buffered saline (TBS) + 0.1% Tween-20, membranes were incubated with primary antibodies: p70 S6 kinase (Thr389) and anti-total p70 S6 kinase (Cell Signaling Technology, Ozyme distributor, Saint-Quentin-en-Yvelines, France). After washing with TBS + 0.1% Tween-20, immunoblots were exposed to swine anti-rabbit immunoglobulins conjugated with horseradish peroxidase (HRP) (DAKO, Trappes, France). The antigen/primary antibody/secondary antibody/HRP complexes were visualized by luminescence using ECL Western Blotting Substrate (Pierce, Thermo Fisher Scientific, Courtaboeuf, France) and a Fusion Fx imaging system (Vilber Lourmat, Collegien, France). Quantification of band density was done using MultiGauge 3.2 software (Fujifilm Corporation, Tokyo, Japan). The values represented the ratio of the phosphorylated protein levels to total protein levels, and were expressed in arbitrary units.

2.7. mRNA Analysis

The protocol for total RNA extraction and mRNA analysis has been previously described [14]. Briefly, a piece of plantaris muscle was homogenized in Tri-Reagent (Euromedex, Mundolsheim, France) and total RNA was isolated according to manufacturer's instructions. RNA amount was measured by spectrophotometry at 260 nm. Total RNA was reverse-transcribed using SuperScript III reverse transcriptase and a random hexamer and oligo dT primer combination (Invitrogen, Life Technologies, Saint-Aubin, France). PCR amplification was performed using a Rotor-Gene Q system and 2 × Rotor-Gene SYBR Green PCR master mix (Qiagen, Courtaboeuf, France). Relative concentrations of mRNA corresponding to genes of interest were quantified using Rotor-Gene software and the standard curve method. The primers used for real-time PCR analysis were listed in Table 3. Hypoxanthine-guanine phosphoribosyltransferase (HPRT) was used as housekeeping gene. Data were expressed in arbitrary units.

Table 3. Primer sequences used for quantitative analysis of gene expression.

Gene Name	Forward and Reverse Primers
MAFbx (Muscle atrophy F-box)	*For* 5′-AGTGAAGACCGGCTACTGTGGAA-3′ *Rev* 5′-TTGCAAAGCTGCAGGGTGAC-3′
MuRF1 (Muscle RING finger-1)	*For* 5′-GTGAAGTTGCCCCCTTACAA-3′ *Rev* 5′-TGGAGATGCAATTGCTCAGT-3′
HPRT (Hypoxanthine-guanine phosphoribosyltransferase)	*For* 5′-AGTTGAGAGATCATCTCCAC-3′ *Rev* 5′-TTGCTGACCTGCTGGATTAC-3′

2.8. Mitochondrial Enzymatic Assays

First, 50 mg of frozen rat plantaris muscle was homogenized in homogenization buffer (225 mM mannitol, 75 mM sucrose, 10 mM Tris-HCl, 10 mM EDTA, pH 7.2) and then centrifugated at 650× *g* for 20 min at 4 °C. The supernatant was kept and the pellet was suspended in homogenization buffer and resubmitted to the same procedure. Both supernatants were pooled and used for activity measurements [14–16]. Complex I and 3-hydroxyacyl-CoA dehydrogenase (HAD) activities were spectrophotometrically assayed in the supernatant fraction by following the oxidation of nicotinamide adenine dinucleotide, reduced (NADH). Citrate synthase (CS) activity was measured by following the reduction of 5,5-dithiobis (2-nitrobenzoic acid) (DTNB) [14–17]. Activities were expressed in nmol/min/mg of proteins.

2.9. Statistics

To calculate the sample size, we used published and unpublished data of net protein utilization (NPU) [18]. A difference of 20–25% and a mean variance of 10% were expected for this parameter between CAS group and WHEY group. Based on these data, the setting of type I error (α) at 5% and a power of 90%, a total of 6 rats per group was required. To anticipate potential rat death for the 16-week experimental period, 10 rats were assigned to each diet. All results were presented as means ± SEM. Animals that died or developed tumors during the experiment were excluded from the analysis. In detail, while we had 10 rats per group at baseline, the number of rats remaining at the end of the experiment was 6 CAS rats, 6 WHEY rats, and 8 PEA rats. The data were analyzed for homogeneity of variance and normality. Homogeneous data were analyzed by a one-way analysis of variance (ANOVA) followed by a Tukey-Kramer test to evaluate the significance of inter-group differences. Heterogeneous data were analyzed using Kruskal-Wallis test and the significance of inter-group differences was assessed using a Steel–Dwass test. Differences were considered significant at $p < 0.05$. Statistical analysis was performed using NCSS 2020 software (NCSS LLC., Kaysville, UT, USA).

3. Results

3.1. Caloric Intake, Body Composition Evolution, and Final Tissue Weights

No significant difference in calculated daily caloric intake was observed between experimental groups throughout the study period (86.0 ± 4.6 kcal/day, 92.0 ± 3.2 kcal/day and 94.8 ± 5.9 kcal/day for CAS, WHEY and PEA rats, respectively). Rat groups were purpose-defined at the beginning of the experiment to ensure no significant between-group differences in body weight, fat mass and lean mass. Thereafter, body weight, fat mass and lean mass remained not significantly different between CAS, WHEY and PEA rats at each timepoint (i.e., the middle (week 8) and the end (week 16) of the experiment) (Table 4). In accordance with the body composition measurements, the weights of several lean tissues, (i.e., skeletal muscle, liver and heart) and two different fat tissues (i.e., perirenal adipose tissue and subcutaneous adipose tissue) presented no significant between-group differences at the end of the experiment (Table 5).

Table 4. Body weight, fat mass and lean mass variations over the course of the experimental study.

	CAS	WHEY	PEA
Body weight (g)			
Week 0	582 ± 23	577 ± 14	595 ± 28
Week 8	585 ± 22	607 ± 16	612 ± 34
Week 16	583 ± 20	605 ± 20	590 ± 39
Fat mass (g)			
Week 0	91 ± 8	99 ± 8	108 ± 13
Week 8	104 ± 7	129 ± 18	136 ± 21
Week 16	99 ± 18	127 ± 15	131 ± 26
Lean Mass (g)			
Week 0	442 ± 18	427 ± 15	434 ± 15
Week 8	431 ± 19	424 ± 14	420 ± 15
Week 16	430 ± 17	421 ± 16	403 ± 14

Week 0, week 8 and week 16 mark the beginning, the middle and the end of the experiment, respectively. Data are expressed as means ± SEM.

Table 5. Tissue weights in CAS, WHEY and PEA old rats after 16 weeks of different diets.

	CAS	WHEY	PEA
Plantaris (mg)	309 ± 34	300 ± 23	263 ± 0.17
Soleus (mg)	175 ± 20	173 ± 24	165 ± 13
Gastrocnemius (g)	1.52 ± 0.29	1.19 ± 0.13	1.13 ± 0.06
Quadriceps (g)	1.94 ± 0.26	1.78 ± 0.27	1.72 ± 0.24
Hindlimb muscle mass (g)	8.82 ± 0.51	8.01 ± 0.88	7.36 ± 0.61
Perirenal adipose tissue (g)	11.7 ± 2.6	15.3 ± 1.4	19.7 ± 4.6
Subcutaneous adipose tissue (g)	11.9 ± 2.3	13.3 ± 2.0	12.2 ± 2.4
Liver (g)	13.7 ± 0.9	14.3 ± 0.9	13.2 ± 1.7
Heart (g)	1.91 ± 0.05	1.88 ± 0.08	1.96 ± 0.08

Results are given as means ± SEM. Hindlimb muscle mass is the sum of plantaris, soleus, gastrocnemius, quadriceps and tibialis muscle weights.

3.2. *Protein Quality Evaluation*

Nitrogen intake and fecal and urinary nitrogen contents were evaluated during the metabolic cage period (Table 6). None of these parameters were significantly different between rat groups. Nitrogen balance, which is the difference between nitrogen intake and nitrogen loss by both fecal and urinary routes, was similar between CAS, WHEY and PEA rats (Table 6). There were no significant between-group differences in apparent digestibility, which considers all of the digestive processes involving protein digestion, including endogenous nitrogen losses, or in true digestibility, which considers the specific digestion of dietary protein by subtracting endogenous nitrogen losses. Finally, net protein utilization, which is the ratio of retained nitrogen to ingested nitrogen, and biological value, which is the ratio of retained nitrogen to absorbed nitrogen, were similar between CAS, WHEY and PEA rats (Table 6).

Table 6. Evaluation of the protein quality of the different experimental diets during the 4-day period in metabolic cages.

	CAS	WHEY	PEA
Nitrogen intake (g)	1.47 ± 0.10	1.61 ± 0.12	1.57 ± 0.11
Fecal nitrogen (g)	0.12 ± 0.01	0.13 ± 0.01	0.14 ± 0.02
Urinary nitrogen (g)	0.86 ± 0.07	0.88 ± 0.11	0.91 ± 0.08
Nitrogen balance (g)	0.49 ± 0.08	0.60 ± 0.20	0.61 ± 0.08
Apparent digestibility (%)	91.6 ± 0.7	92.1 ± 0.7	91.8 ± 0.8
True digestibility (%)	99.9 ± 0.5	101.2 ± 0.6	100.5 ± 0.7
Net protein utilization (%)	66.3 ± 6.7	74.7 ± 6.1	81.3 ± 6.8
Biological value (%)	66.4 ± 6.9	73.8 ± 6.0	80.8 ± 6.6

Results are given as means ± SEM.

3.3. Plasma Metabolic Parameters and Cytokines

Fasting levels of lipid metabolic markers, i.e., triglycerides and total cholesterol, were not significantly different between CAS, WHEY and PEA rats (Table 7). There were no significant dietary source-protein effects on parameters related to insulin sensitivity, i.e., fasting glucose and insulin concentrations and calculated HOMA-IR. Circulating leptin concentrations were similar between experimental groups, while adiponectin levels tended to be higher in PEA rats compared to CAS rats and WHEY rats ($p = 0.07$). After 16 weeks of feeding with dietary treatment, rats showed similar plasma concentrations of pro-inflammatory cytokines such as IL-1β and TNFα, and the anti-inflammatory cytokine IL-10 (Table 7). To evaluate inflammatory status, we calculated the ratios of the inflammatory markers TNF-α and IL-1β to the anti-inflammatory marker IL-10. TNFα/IL-10 and IL-1β/IL-10 ratios did not differ between groups (Table 7).

Table 7. Fasting metabolic parameters in plasma of old rats after the 16 weeks of different diets.

	CAS	WHEY	PEA
Insulin sensitivity			
Glucose (g/L)	0.955 ± 0.106	1.010 ± 0.075	0.970 ± 0.108
Insulin (ng/mM)	1.285 ± 0.585	0.678 ± 0.213	0.553 ± 0.102
HOMA-IR	6.055 ± 1.884	4.232 ± 1.392	3.202 ± 0.761
Lipids			
Triglycerides (g/L)	0.789 ± 0.088	0.994 ± 0.364	0.604 ± 0.176
Total cholesterol (g/L)	0.843 ± 0.073	0.878 ± 0.067	0.833 ± 0.167
Adipokines			
Adiponectin (µg/mL)	5.145 ± 1.240	6.355 ± 0.764	8.751 ± 1.109
Leptin (ng/mL)	4.547 ± 0.416	5.172 ± 1.170	6.730 ± 2.935
Cytokines			
TNFα (pg/mL)	11.93 ± 5.90	6.28 ± 2.79	11.29 ± 2.86
IL-1β (pg/mL)	155.4 ± 67.9	158.1 ± 63.9	133.5 ± 29.6
IL-10 (pg/mL)	58.56 ± 22.95	57.26 ± 22.74	54.10 ± 10.93
TNFα / IL-10 ratio	0.264 ± 0.099	0.185 ± 0.033	0.254 ± 0.071
IL-1β / IL-10 ratio	2.540 ± 0.090	2.580 ± 0.111	2.429 ± 0.049

Results are given as means ± SEM.

3.4. Markers of Muscle Protein Anabolism and Catabolism

Fractional synthesis rates (FSR) were measured in plantaris muscles of old rats (Figure 1A). According to skeletal muscle mass measurements, muscle FSR was similar between CAS, WHEY and PEA rats. Associated with these data, protein quality did not affect the phosphorylation rates of p70 S6 kinase (an intermediate of the translation initiation step) in plantaris muscles of old rats (Figure 1B). The involvement of the ubiquitin-proteasome pathway in the regulation of skeletal muscle mass in the three experimental groups was assessed by measuring mRNA expressions of MuRF1 and MAFbx. Gene expressions of both E3 ubiquitin ligases were also unchanged by experimental diets in rat skeletal muscles (Figure 1C,D).

3.5. Muscle Mitochondrial Activity

To explore the effect of protein quality on muscle mitochondrial function in old rats, we measured the maximal activity of citrate synthase, which is a mitochondrial matrix enzyme often used as a marker of mitochondrial density. CAS, WHET and PEA rats showed similar citrate synthase activities in plantaris muscles (Figure 2A). Likewise, the activities of muscle complex 1 and 3-hydroxyacyl-CoA dehydrogenase (HAD), i.e., one of the electron transport chain complexes and a key enzyme of the mitochondrial β-oxidation cycle, respectively, were not affected by the different experimental diets (Figure 2B,C).

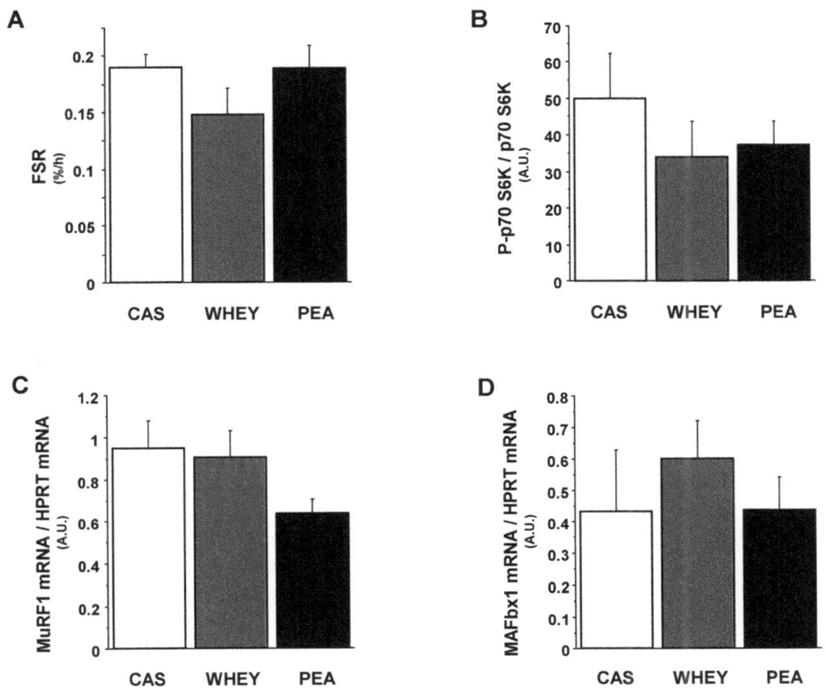

Figure 1. Effects of different experimental diets on protein synthesis and expression of ubiquitin-proteasome pathway markers in plantaris muscles of old rats. Fractional synthesis rate (**A**) was measured by tracer enrichment in plantaris muscles after a 50-min incubation with L-[$^{13}C_6$] phenylalanine. In the same muscles, the phosphorylation states of p70 S6 kinase (**B**) were determined by Western-blotting, and the gene expressions of the two ubiquitin E3 ligases MuRF1 (**C**) and MAFbx (**D**) were analyzed by quantitative RT-PCR analysis. Statistical significance was assessed by ANOVA, followed by a Tukey-Kramer test or a Kruskal-Wallis test followed by a Steel–Dwass test depending on homogeneity of variance and normality. Data are expressed as means ± SEM. A.U.: Arbitrary units.

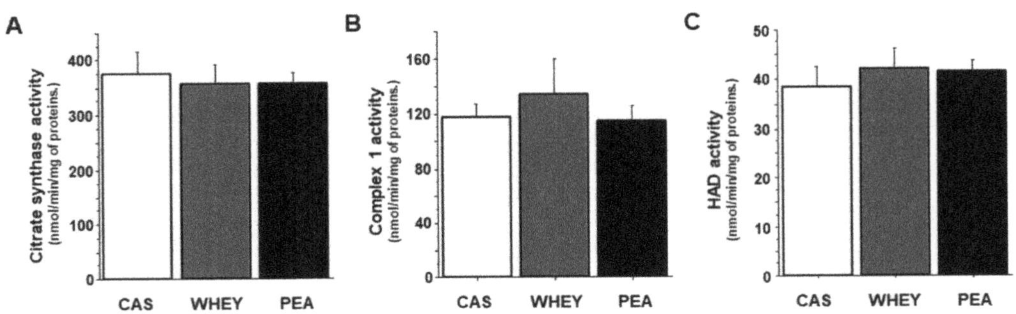

Figure 2. Mitochondrial enzyme activity in skeletal muscles of old rats after 16 weeks of different diets. Mitochondrial function was assessed by measuring citrate synthase (**A**), complex 1 (**B**) and 3-hydroxyacyl-CoA dehydrogenase (**C**) activities in plantaris muscles. Statistical significance was assessed by ANOVA, followed by a Tukey-Kramer test or a Kruskal-Wallis test followed by a Steel–Dwass test, depending on homogeneity of variance and normality. Data are expressed as means ± SEM.

4. Discussion

Protein quality is an important component of protein intake to support growth, development, and maintenance of essential body tissues and functions [19]. The nutritional value of a protein depends on how its AA balance matches to needs, in particular EAA, and on its digestibility, i.e., on the release of AA and small peptides ready for intestinal absorption [20]. Proteins from alternative sources, such as plant proteins, are often described as having less balanced EAA profiles and lower digestibility than animal-sourced proteins [21]. However, there is a lack of data directly comparing the nutritional values of animal and plant proteins under the same experimental conditions, especially in older subjects. Here, we examined the effects of a 16-week pea protein diet on protein digestibility, body weight and composition, tissue weight, metabolic indexes, and muscle protein turnover and metabolism in old rats. Pea protein was compared to two dairy proteins, i.e., whey protein and casein, that are considered to be among the best-quality proteins, especially for maintaining body composition and muscle mass and function during aging [22]. Overall, we clearly showed that in old rats, a 16-week ingestion of milk proteins or pea protein did not influence protein assimilation and nitrogen retention, particularly in skeletal muscle. It should therefore be possible to use such plant-based protein sources for older people, which would make it possible to diversify intake and more easily attain the nutritional recommendations for this population.

4.1. Nitrogen Balance, Digestibility and Rate of Utilization

When studies set out to compare the nutritional quality of several dietary proteins, the first issue to consider is usually how effectively the proteins are assimilated by the body. In particular, it is important to measure nitrogen balance, digestibility and rate of utilization to get a picture of the capacity of the protein to get digested and absorbed and to get assimilated in the tissues. Overall, the data on nitrogen balance, true digestibility and net protein utilization showed that the three proteins tested in this work had a similar effect in old rats. First, the apparent and true digestibilities of pea proteins were in the same range of values of the other proteins. Recent studies have reported that pea protein is highly digestible in rats [18,23]. However, this work represents one of the first studies to show that pea protein is also highly digestible in old rats. It has been suggested that the digestibility of plant proteins is impaired due to the presence of both anti-nutritional factors and indigestible fractions in their sequence [23]. However, the pea protein used here was a protein isolate, and protein isolates are generally well-digested [24]. In addition, protein isolates are particularly low in anti-nutritional factors, due to the manufacturing process used to extract the protein [25]. High protein digestibility induces a high quantity of AA available for intestinal absorption and, thus, improves the nutritional value of the protein source [26]. Hence, net protein utilization was equivalent between old rats fed pea protein, casein or whey protein. Urinary and fecal nitrogen excretion in old rats did not differ between the three groups, leading to an equivalent whole-body nitrogen retention. This observation contrasts with other studies done in pigs that reported increased urinary nitrogen excretion and plasma urea levels in response to soybean protein compared to casein [27]. We previously showed in young rats that protein utilization increased after feeding animals with wheat pasta enriched with fava bean flour as compared to an isoproteic wheat pasta enriched with gluten. However, in this work, protein utilization still remained lower than that measured in rats fed casein [28]. However, when the same study was carried out in old rats, there was no difference between the group fed wheat pasta enriched with fava bean and the group fed casein [18].

Evaluation of the nutritional quality of dietary proteins relies not only on protein digestibility but also on its AA composition, notably its EAA content. The EAA composition of the pea protein used in this study was close to casein and to the needs of rats, according to National Research Council [29]. The AA composition of pea protein is characterized by a limiting content of methionine (Met) [30], but the total sulfur AA content is adequate [29]. Consequently, the net protein utilization and biological value measured in old rats were equivalent regardless of the protein used in the diet. Note that this result could be explained not only by EAA composition, in particular a high leucine content, but also by the high digestibility of the pea protein. To sum up, we showed that the biological value of ingested nitrogen, in particular nitrogen retention, did not differ in old rats, regardless of whether the protein in the diet was casein, whey, or pea protein.

4.2. Body Composition and Skeletal Muscle Mass

In the present study, although we observed an age-related physiological trend towards increased body fat and reduced lean mass between the first and last month of the study, the protein source in the diet did not significantly change body composition in old rats. This result was also confirmed by the tissue weights at the end of the 16-week period. In accordance with the whole-body composition measurements, the weight of tissues constituting the lean mass, i.e., skeletal muscles, liver and heart, and of tissues resulting from the fat mass did not differ between different dietary protein groups. Few studies have focused on comparing the effects of animal versus plant proteins on body composition in old rats. We previously evaluated (also in old rats) the nutritional value of pasta made from a mix of wheat semolina and legume flours, i.e., fava bean, lentil, or pea flour [18]. Two groups were fed diets with casein or whey protein as protein source, and three groups were fed diets made with fava bean pasta, lentil pasta or pea pasta as protein source. The study found that body weight and composition, i.e., fat mass and lean mass were not significantly different between groups at each timepoint, i.e., the beginning, the middle, and the end of the experiment [18]. The effect of dietary protein sources on body composition and tissue weight has been evaluated in other works, but these studies were generally done in young rats. A lower lean mass gain was observed in young rats given soy protein for 28 days than in young rats fed whey protein [31]. At the muscular level, other studies found that, compared to casein, 16 to 20 days of ad libitum consumption of proteins from legumes, i.e., beans or lentils provoked lower muscle weights in young rats [32–34]. In addition, Alonso et al. found that muscle mass and muscle protein content were lower in young rats fed seed peas than in young rats receiving casein. In this latter study, peas were extruded and cooked to reduce the antinutritional factor content [35]. The change in lean mass or skeletal muscle mass after long-term consumption of plant-based meals has not been thoroughly assessed in older people. The rare studies available have shown that the consumption of plant proteins, when provided at sufficient amounts in each meal (i.e., >30 g/meal), should be able to maintain lean and muscle mass, and therefore increase the potential to mitigate sarcopenia in older subjects [5,36,37]. Taken together, the data presented here showed that some plant proteins, e.g., pea proteins, promoted a similar effect on body composition and muscle mass to casein and even whey protein in old rats, and could therefore be tested in the elderly as an intervention to counteract sarcopenia.

4.3. Mechanisms

Several mechanisms may explain the similar action of milk proteins and pea protein on body composition and muscle mass in old rats. First, analysis of the AA content of each protein showed equivalent leucine contents between pea protein and casein. There is clear evidence that during aging, the leucine content of dietary proteins is an important parameter impacting its anabolic effect on lean mass, and specifically skeletal muscle mass [38]. It is now well recognized that leucine acts as an anabolic signal by stimulating protein synthesis and inhibiting protein breakdown at muscle level. For instance, leucine supplementation for 10 days attenuated the decrease in expression of eukaryotic translation initiation factors in young and old rat muscles [39]. In addition, this supplementation decreased the levels of ubiquitinated proteins and inhibited proteasome activity in old rats [40]. The leucine content of pea protein could thus explain its effectiveness on muscle protein turnover and therefore on muscle mass and lean body mass in old rats. Nevertheless, we did not measure the effects of pea protein under postprandial conditions and therefore we cannot draw conclusions on the role of the leucine content on protein anabolism in old rats. Note that a second mechanisms may be involved, as we did not observe any difference between the three dietary proteins in terms of their effect on muscle protein synthesis and degradation, although we measured the rate of muscle protein turnover in postabsorptive condition. The changes observed for plantaris muscle protein synthesis in old rats were relatively in line with the changes that were observed in muscle mass. Although muscle mass tended to be higher in the whey-protein group than the pea protein group, we suggest that pea protein intake could enhance postprandial muscle protein anabolism (although we did not measure it) in old rats, which would translate into muscle protein accumulation and increased skeletal muscle mass. The influence of plant-based proteins and animal-based proteins on muscle protein synthesis has been investigated in several studies. The rate of protein synthesis in gastrocnemius muscle was lower in young rats fed raw fava bean intake than in young rats fed milk protein [41]. In addition, a lower muscle protein synthesis rate was observed in young rats when fed beans and lentils than when fed casein [34]. However, to our knowledge, the long-term effects of plant protein intake on muscle protein synthesis rate in old rats has never before been investigated.

Mitochondrial abnormalities have also been singled out as key factors in muscle changes during aging. Research on the mitochondrial electron transport chain (ETC) in skeletal muscle clearly demonstrated deficient ETC activity in muscles exhibiting the greatest loss of muscle mass with age [42]. Here, citrate synthase activity, complex 1 activity and HAD activity did not differ between dietary protein sources in old rats. Additionally, once more in old rats, we previously demonstrated that maintained mitochondrial function in skeletal muscle was associated with maintained muscle protein synthesis and muscle mass as animals aged [13]. This previous study also demonstrated that one of the mechanisms behind this action was the ability of protein intake to maintain protein turnover at the mitochondrial level [13]. This makes is tempting to postulate that pea protein, like milk proteins, could potentially help to prevent the age-related alteration of mitochondrial functional capacities in skeletal muscle, thus helping to maintain muscle mass.

4.4. Metabolic Parameters

We also measured metabolic parameters related to aging-related changes in muscle mass, in particular plasma pro-inflammatory and anti-inflammatory cytokine levels [43]. The increase in blood pro-inflammatory factors and the decrease in blood anti-inflammatory factors during aging causes inflammatory conditions conducive to muscle protein catabolism [44]. Here too, we showed that pea protein consumption by old rats did not modify some of the markers of the inflammatory system compared to milk proteins. It has been reported that milk protein has anti-inflammatory properties that might be effective in reducing the circulation of pro-inflammatory cytokines, such as interleukin-6 (IL-6) and tumor necrosis factor (TNF-α) [45]. A recent study on pea protein reported that a tripeptide, LRW (Leu-Arg-Trp), characterized from the pea protein legumin, and its previously studied isomer IRW (Ile-Arg-Trp) exerted strong anti-inflammatory effects by modulating the nuclear factor-κB pathway [46]. Hence, the consumption of such proteins could help keep inflammation at a level that prevents muscle protein catabolism in old rats. In addition to inflammation, insulin resistance has been described as another cause of decline in muscle protein anabolism and muscle mass in older people [47]. Here we found no between-group differences in HOMA-IR except a trend towards a reduction in insulin resistance in the PEA group compared to the CAS group. Recent studies have shown that pea glycoproteins and peptides have antidiabetic activities, in particular by reducing insulin resistance [48,49]. Therefore, it may be possible that long-term pea protein consumption could improve age-related insulin resistance in old rats. However, further studies are needed to bridge the gap between age-related inflammation and insulin resistance and pea protein intake.

5. Conclusions

This study, carried out in old rats, showed that, under our experimental conditions, e.g., use of protein isolates, the body uses nitrogen with the same efficiency regardless of whether it is provided by pea protein, casein or whey. This result is partly due to the high digestibility of the pea protein, together with its EAA composition, which is close to that found in milk proteins. The divergence between our results and studies using growing rats or young rats, however, has posed unresolved questions. Here, we found evidence that plant proteins would be more effective in very old animals than in young animals. Further research is warranted to find out whether this is due to an increase in the metabolic efficiency of plant proteins or a decrease in the metabolic efficiency of milk proteins with age. In addition, clinical studies should be set up to assess the quality of plant proteins in humans, in particular the elderly, taking into consideration their pathophysiological situation and their nutritional status.

Author Contributions: Conceptualization: J.S., C.M.-F., Y.B., H.J. and S.W.; Data curation: J.S. and S.W.; Formal analysis: J.S., C.M.-F., H.J. and S.W.; Funding acquisition: H.J. and S.W.; Investigation: J.S., C.G. (Christelle Guillet), O.L.B., C.M.-F., C.G. (Christophe Giraudet), V.P., A.B., P.D., C.P., M.G., Y.B., H.J. and S.W.; Methodology: J.S., C.M.-F., Y.B., H.J. and S.W.; Project administration: J.S., C.M.-F., Y.B., H.J. and S.W.; Resources: H.J. and S.W.; Supervision: H.J. and S.W.; Validation: J.S., C.M.-F., H.J. and S.W.; Visualization: J.S., C.M.-F., H.J. and S.W.; Writing—original draft: J.S. and S.W.; Writing—review & editing: J.S., C.M.-F., V.P. and S.W. All authors have read and agreed to the published version of the manuscript.

Funding: This study was partly supported by a grant from COSUCRA SA, Warcoing, Belgium.

Institutional Review Board Statement: The study was approved by the local institutional animal care and use committee (Comité d'Ethique en Matière d'Expérimentation Animale Auvergne: C2EA-02) (protocol code: APAFIS#5329-2016051115541284 v2 approved on 23 january 2017).

Acknowledgments: The authors thank all the staff at the INRAE–Theix animal facilities for their valuable care and cooperation.

Conflicts of Interest: C.M.F. and H.J. are employees of COSUCRA-Groupe SA (Belgium).

References

1. New Nutrition Business. *10 Key Trends in Food, Nutrition Health 2020*; New Nutrition Business: London, UK, 2020.
2. Berrazaga, I.; Micard, V.; Gueugneau, M.; Walrand, S. The role of the anabolic properties of plant- versus animal-based protein sources in supporting muscle mass maintenance: A critical review. *Nutrients* **2019**, *11*, 1825. [CrossRef]
3. WHO; FAO; UNU. *Protein and Amino Acid Requirements in Human Nutrition*; Report of the Joint FAO/WHO/UNU Expert Consultation; WHO: Geneva, Switzerland, 2007; Volume 935, pp. 1–265.
4. Gorissen, S.H.M.; Witard, O.C. Characterising the muscle anabolic potential of dairy, meat and plant-based protein sources in older adults. *Proc. Nutr. Soc.* **2018**, *77*, 20–31. [CrossRef] [PubMed]
5. Van Vliet, S.; Burd, N.A.; van Loon, L.J. The skeletal muscle anabolic response to plant- versus animal-based protein consumption. *J. Nutr.* **2015**, *145*, 1981–1991. [CrossRef] [PubMed]
6. Banaszek, A.; Townsend, J.R.; Bender, D.; Vantrease, W.C.; Marshall, A.C.; Johnson, K.D. The effects of whey vs. pea protein on physical adaptations following 8-weeks of high-intensity functional training (HIFT): A pilot study. *Sports* **2019**, *7*, 12. [CrossRef]
7. Babault, N.; Paizis, C.; Deley, G.; Guerin-Deremaux, L.; Saniez, M.H.; Lefranc-Millot, C.; Allaert, F.A. Pea proteins oral supplementation promotes muscle thickness gains during resistance training: A double-blind, randomized, placebo-controlled clinical trial vs. whey protein. *J. Int. Soc. Sports Nutr.* **2015**, *12*, 3. [CrossRef]
8. Zhao, Y.; Cholewa, J.; Shang, H.; Yang, Y.; Ding, X.; Wang, Q.; Su, Q.; Zanchi, N.E.; Xia, Z. Advances in the role of leucine-sensing in the regulation of protein synthesis in aging skeletal muscle. *Front. Cell Dev. Biol.* **2021**, *9*, 646482. [CrossRef]
9. Chanet, A.; Verlaan, S.; Salles, J.; Giraudet, C.; Patrac, V.; Pidou, V.; Pouyet, C.; Hafnaoui, N.; Blot, A.; Cano, N.; et al. Supplementing breakfast with a vitamin D and leucine-enriched whey protein medical nutrition drink enhances postprandial muscle protein synthesis and muscle mass in healthy older men. *J. Nutr.* **2017**, *147*, 2262–2271. [CrossRef]
10. Overduin, J.; Guerin-Deremaux, L.; Wils, D.; Lambers, T.T. NUTRALYS((R)) pea protein: Characterization of in vitro gastric digestion and in vivo gastrointestinal peptide responses relevant to satiety. *Food Nutr. Res.* **2015**, *59*, 25622. [CrossRef]
11. Dumas, A. Stickstoffbestimmung Nach Dumas. Die Praxis des Org. Chemikers. In *N-Determination According to Dumas*, 41st ed.; Schrag: Nuremberg, Germany, 1962.

12. Proll, J.; Petzke, K.J.; Ezeagu, I.E.; Metges, C.C. Low nutritional quality of unconventional tropical crop seeds in rats. *J. Nutr.* **1998**, *128*, 2014–2022. [CrossRef]
13. Zangarelli, A.; Chanseaume, E.; Morio, B.; Brugere, C.; Mosoni, L.; Rousset, P.; Giraudet, C.; Patrac, V.; Gachon, P.; Boirie, Y.; et al. Synergistic effects of caloric restriction with maintained protein intake on skeletal muscle performance in 21-month-old rats: A mitochondria-mediated pathway. *FASEB J.* **2006**, *20*, 2439–2450. [CrossRef]
14. Salles, J.; Chanet, A.; Berry, A.; Giraudet, C.; Patrac, V.; Domingues-Faria, C.; Rocher, C.; Guillet, C.; Denis, P.; Pouyet, C.; et al. Fast digestive, leucine-rich, soluble milk proteins improve muscle protein anabolism, and mitochondrial function in undernourished old rats. *Mol. Nutr. Food Res.* **2017**, *61*, 1700287. [CrossRef]
15. Medja, F.; Allouche, S.; Frachon, P.; Jardel, C.; Malgat, M.; Mousson de Camaret, B.; Slama, A.; Lunardi, J.; Mazat, J.P.; Lombes, A. Development and implementation of standardized respiratory chain spectrophotometric assays for clinical diagnosis. *Mitochondrion* **2009**, *9*, 331–339. [CrossRef]
16. Gutierrez Cortes, N.; Pertuiset, C.; Dumon, E.; Borlin, M.; Hebert-Chatelain, E.; Pierron, D.; Feldmann, D.; Jonard, L.; Marlin, S.; Letellier, T.; et al. Novel mitochondrial DNA mutations responsible for maternally inherited nonsyndromic hearing loss. *Hum. Mutat.* **2012**, *33*, 681–689. [CrossRef]
17. Tardif, N.; Salles, J.; Guillet, C.; Tordjman, J.; Reggio, S.; Landrier, J.F.; Giraudet, C.; Patrac, V.; Bertrand-Michel, J.; Migne, C.; et al. Muscle ectopic fat deposition contributes to anabolic resistance in obese sarcopenic old rats through eIF2alpha activation. *Aging Cell* **2014**, *13*, 1001–1011. [CrossRef]
18. Berrazaga, I.; Salles, J.; Laleg, K.; Guillet, C.; Patrac, V.; Giraudet, C.; Le Bacquer, O.; Gueugneau, M.; Denis, P.; Pouyet, C.; et al. Anabolic properties of mixed wheat-legume pasta products in old rats: Impact on whole-body protein retention and skeletal muscle protein synthesis. *Nutrients* **2020**, *12*, 1596. [CrossRef]
19. Gwin, J.A.; Carbone, J.W.; Rodriguez, N.R.; Pasiakos, S.M. Physiological limitations of protein foods ounce equivalents and the underappreciated role of essential amino acid density in healthy dietary patterns. *J. Nutr.* **2021**, *151*, 3276–3283. [CrossRef]
20. Floret, C.; Monnet, A.F.; Micard, V.; Walrand, S.; Michon, C. Replacement of animal proteins in food: How to take advantage of nutritional and gelling properties of alternative protein sources. *Crit. Rev. Food Sci. Nutr.* **2021**. Published online ahead of print. [CrossRef]
21. Young, V.R.; Pellett, P.L. Plant proteins in relation to human protein and amino acid nutrition. *Am. J. Clin. Nutr.* **1994**, *59*, 1203S–1212S. [CrossRef] [PubMed]
22. Berrazaga, I.; Mession, J.L.; Laleg, K.; Salles, J.; Guillet, C.; Patrac, V.; Giraudet, C.; Le Bacquer, O.; Boirie, Y.; Micard, V.; et al. Formulation, process conditions, and biological evaluation of dairy mixed gels containing fava bean and milk proteins: Effect on protein retention in growing young rats. *J. Dairy. Sci.* **2019**, *102*, 1066–1082. [CrossRef] [PubMed]
23. Guillin, F.M.; Gaudichon, C.; Guerin-Deremaux, L.; Lefranc-Millot, C.; Azzout-Marniche, D.; Khodorova, N.; Calvez, J. Multi-criteria assessment of pea protein quality in rats: A comparison between casein, gluten and pea protein alone or supplemented with methionine. *Br. J. Nutr.* **2021**, *125*, 389–397. [CrossRef] [PubMed]
24. Lhoste, E.F.; Mouzon, B.; Andrieux, C.; Gueugneau, A.M.; Fiszlewicz, M.; Corring, T.; Szylit, O. Physiological effects of a pea protein isolate in gnotobiotic rats: Comparison with a soybean isolate and meat. *Ann. Nutr. Metab.* **1998**, *42*, 44–54. [CrossRef] [PubMed]
25. Chamba, M.V.; Hua, Y.; Murekatete, N.; Chen, Y. Effects of synthetic and natural extraction chemicals on yield, composition and protein quality of soy protein isolates extracted from full-fat and defatted flours. *J. Food Sci. Technol.* **2015**, *52*, 1016–1023. [CrossRef] [PubMed]
26. Sun-Watherhouse, D.Z.; Zhao, M.; Waterhouse, G.I.N. Protein modification during ingredient preparation and food processing: Approaches to improve food processability and nutrition. *Food Bioprocess Technol.* **2014**, *7*, 1853–1893. [CrossRef]
27. Lohrke, B.; Saggau, E.; Schadereit, R.; Beyer, M.; Bellmann, O.; Kuhla, S.; Hagemeister, H. Activation of skeletal muscle protein breakdown following consumption of soyabean protein in pigs. *Br. J. Nutr.* **2001**, *85*, 447–457. [CrossRef]
28. Laleg, K.; Salles, J.; Berry, A.; Giraudet, C.; Patrac, V.; Guillet, C.; Denis, P.; Tessier, F.J.; Guilbaud, A.; Howsam, M.; et al. Nutritional evaluation of mixed wheat-faba bean pasta in growing rats: Impact of protein source and drying temperature on protein digestibility and retention. *Br. J. Nutr.* **2019**, *121*, 496–507. [CrossRef] [PubMed]
29. NRC National Research Council. *Nutrient Requirements of Laboratory Animals*, 4th ed.; National Avademic Press: Washington, DC, USA, 1995.
30. Leterme, P.M.; Monmart, T.; Baudart, E. Amino acid composition of pea (*Pisum sativum*) proteins and protein profile of pea flour. *J. Sci. Food Agric.* **1990**, *53*, 107–110. [CrossRef]
31. Wroblewska, B.; Juskiewicz, J.; Kroplewski, B.; Jurgonski, A.; Wasilewska, E.; Zlotkowska, D.; Markiewicz, L. The effects of whey and soy proteins on growth performance, gastrointestinal digestion, and selected physiological responses in rats. *Food Funct.* **2018**, *9*, 1500–1509. [CrossRef]
32. Combe, E.; Pirman, T.; Stekar, J.; Houlier, M.L.; Mirand, P.P. Differential effect of lentil feeding on proteosynthesis rates in the large intestine, liver and muscle of rats. *J. Nutr. Biochem.* **2004**, *15*, 12–17. [CrossRef]
33. Pirman, T.; Combe, E.; Patureau Mirand, P.; Stekar, J.; Oresnik, A. Differential effects of cooked common bean (Phaseolus vulgaris) and lentil (Lens esculenta puyensis) feeding on protein and nucleic acid contents in intestines, liver and muscles in rats. *Ann. Nutr. Metab.* **2004**, *48*, 281–287. [CrossRef]

34. Pirman, T.; Combe, E.; Ribeyre, M.C.; Prugnaud, J.; Stekar, J.; Patureau Mirand, P. Differential effects of cooked beans and cooked lentils on protein metabolism in intestine and muscle in growing rats. *Ann. Nutr. Metab.* **2006**, *50*, 197–205. [CrossRef]
35. Alonso, R.; Grant, G.; Fruhbeck, G.; Marzo, F. Muscle and liver protein metabolism in rats fed raw or heat-treated pea seeds. *J. Nutr. Biochem.* **2002**, *13*, 611–618. [CrossRef]
36. Brown, E.C.; DiSilvestro, R.A.; Babaknia, A.; Devor, S.T. Soy versus whey protein bars: Effects on exercise training impact on lean body mass and antioxidant status. *Nutr. J.* **2004**, *3*, 22. [CrossRef] [PubMed]
37. Joy, J.M.; Lowery, R.P.; Wilson, J.M.; Purpura, M.; De Souza, E.O.; Wilson, S.M.; Kalman, D.S.; Dudeck, J.E.; Jager, R. The effects of 8 weeks of whey or rice protein supplementation on body composition and exercise performance. *Nutr. J.* **2013**, *12*, 86. [CrossRef]
38. Rondanelli, M.; Nichetti, M.; Peroni, G.; Faliva, M.A.; Naso, M.; Gasparri, C.; Perna, S.; Oberto, L.; Di Paolo, E.; Riva, A.; et al. Where to find leucine in food and how to feed elderly with sarcopenia in order to counteract loss of muscle mass: Practical advice. *Front. Nutr.* **2020**, *7*, 622391. [CrossRef] [PubMed]
39. Pereira, M.G.; Silva, M.T.; da Cunha, F.M.; Moriscot, A.S.; Aoki, M.S.; Miyabara, E.H. Leucine supplementation improves regeneration of skeletal muscles from old rats. *Exp. Gerontol.* **2015**, *72*, 269–277. [CrossRef]
40. Combaret, L.; Dardevet, D.; Rieu, I.; Pouch, M.N.; Bechet, D.; Taillandier, D.; Grizard, J.; Attaix, D. A leucine-supplemented diet restores the defective postprandial inhibition of proteasome-dependent proteolysis in aged rat skeletal muscle. *J. Physiol.* **2005**, *569*, 489–499. [CrossRef]
41. Martinez, J.A.; Goena, M.; Santidrian, S.; Larralde, J. Response of muscle, liver and whole-body protein turnover to two different sources of protein in growing rats. *Ann. Nutr. Metab.* **1987**, *31*, 146–153. [CrossRef]
42. Bua, E.A.; McKiernan, S.H.; Wanagat, J.; McKenzie, D.; Aiken, J.M. Mitochondrial abnormalities are more frequent in muscles undergoing sarcopenia. *J. Appl. Physiol.* **2002**, *92*, 2617–2624. [CrossRef]
43. Guillet, C.; Masgrau, A.; Walrand, S.; Boirie, Y. Impaired protein metabolism: Interlinks between obesity, insulin resistance and inflammation. *Obes. Rev.* **2012**, *13* (Suppl. 2), 51–57. [CrossRef]
44. Tardif, N.; Salles, J.; Landrier, J.F.; Mothe-Satney, I.; Guillet, C.; Boue-Vaysse, C.; Combaret, L.; Giraudet, C.; Patrac, V.; Bertrand-Michel, J.; et al. Oleate-enriched diet improves insulin sensitivity and restores muscle protein synthesis in old rats. *Clin. Nutr.* **2011**, *30*, 799–806. [CrossRef]
45. Sugawara, K.; Takahashi, H.; Kashiwagura, T.; Yamada, K.; Yanagida, S.; Homma, M.; Dairiki, K.; Sasaki, H.; Kawagoshi, A.; Satake, M.; et al. Effect of anti-inflammatory supplementation with whey peptide and exercise therapy in patients with COPD. *Respir. Med.* **2012**, *106*, 1526–1534. [CrossRef] [PubMed]
46. Wang, X.; Bhullar, K.S.; Fan, H.; Liao, W.; Qiao, Y.; Su, D.; Wu, J. Regulatory effects of a pea-derived peptide Leu-Arg-Trp (LRW) on dysfunction of rat aortic vascular smooth muscle cells against angiotensin II stimulation. *J. Agric. Food Chem.* **2020**, *68*, 3947–3953. [CrossRef] [PubMed]
47. Guillet, C.; Boirie, Y. Insulin resistance: A contributing factor to age-related muscle mass loss? *Diabetes Metab.* **2005**, *31*, 5S20–5S26. [CrossRef]
48. Liu, J.P.; Qian, Y.F.; Qin, G.Y.; Zhao, L.Y.; Chen, G.T. Antidiabetic activities of glycoprotein from pea (*Pisum sativum* L.) in STZ-induced diabetic mice. *Food Funct.* **2021**, *12*, 5087–5095. [CrossRef] [PubMed]
49. Zhu, Y.; Zhang, H.; Wei, Y.; Cai, M.; Gu, R.; Wang, Y.; Ma, Y.; Chen, L. Pea-derived peptides, VLP, LLP, VA, and LL, improve insulin resistance in HepG2 cells via activating IRS-1/PI3K/AKT and blocking ROS-mediated p38MAPK signaling. *J. Food Biochem.* **2020**, *44*, e13454. [CrossRef]

MDPI
St. Alban-Anlage 66
4052 Basel
Switzerland
Tel. +41 61 683 77 34
Fax +41 61 302 89 18
www.mdpi.com

Nutrients Editorial Office
E-mail: nutrients@mdpi.com
www.mdpi.com/journal/nutrients

www.ingramcontent.com/pod-product-compliance
Lightning Source LLC
Chambersburg PA
CBHW061522210926
51850CB00008B/2009
9783036574509